Nonlinear Dynamics

From Lasers to Butterflies

WORLD SCIENTIFIC LECTURE NOTES IN COMPLEX SYSTEMS

Editor-in-Chief: A.S. Mikhailov, *Fritz Haber Institute, Berlin, Germany*

H. Cerdeira, *ICTP, Triest, Italy*
B. Huberman, *Hewlett-Packard, Palo Alto, USA*
K. Kaneko, *University of Tokyo, Japan*
Ph. Maini, *Oxford University, UK*

AIMS AND SCOPE

The aim of this new interdisciplinary series is to promote the exchange of information between scientists working in different fields, who are involved in the study of complex systems, and to foster education and training of young scientists entering this rapidly developing research area.

The scope of the series is broad and will include: Statistical physics of large nonequilibrium systems; problems of nonlinear pattern formation in chemistry; complex organization of intracellular processes and biochemical networks of a living cell; various aspects of cell-to-cell communication; behaviour of bacterial colonies; neural networks; functioning and organization of animal populations and large ecological systems; modeling complex social phenomena; applications of statistical mechanics to studies of economics and financial markets; multi-agent robotics and collective intelligence; the emergence and evolution of large-scale communication networks; general mathematical studies of complex cooperative behaviour in large systems.

World Scientific Lecture Notes
in Complex Systems– Vol.1

Editor

Rowena Ball
Australian National University

Assistant Editor

Nail Akhmediev
Australian National University

Selected Lectures from the
15th Canberra International Physics Summer School

Australian National University   21 January – 1 February 2002

Nonlinear Dynamics

From Lasers to Butterflies

World Scientific
New Jersey • London • Singapore • Hong Kong

Published by

World Scientific Publishing Co. Pte. Ltd.

5 Toh Tuck Link, Singapore 596224

USA office: Suite 202, 1060 Main Street, River Edge, NJ 07661

UK office: 57 Shelton Street, Covent Garden, London WC2H 9HE

British Library Cataloguing-in-Publication Data
A catalogue record for this book is available from the British Library.

NONLINEAR DYNAMICS: FROM LASERS TO BUTTERFLIES

ISBN 981-238-320-4

Printed in Singapore by Mainland Press

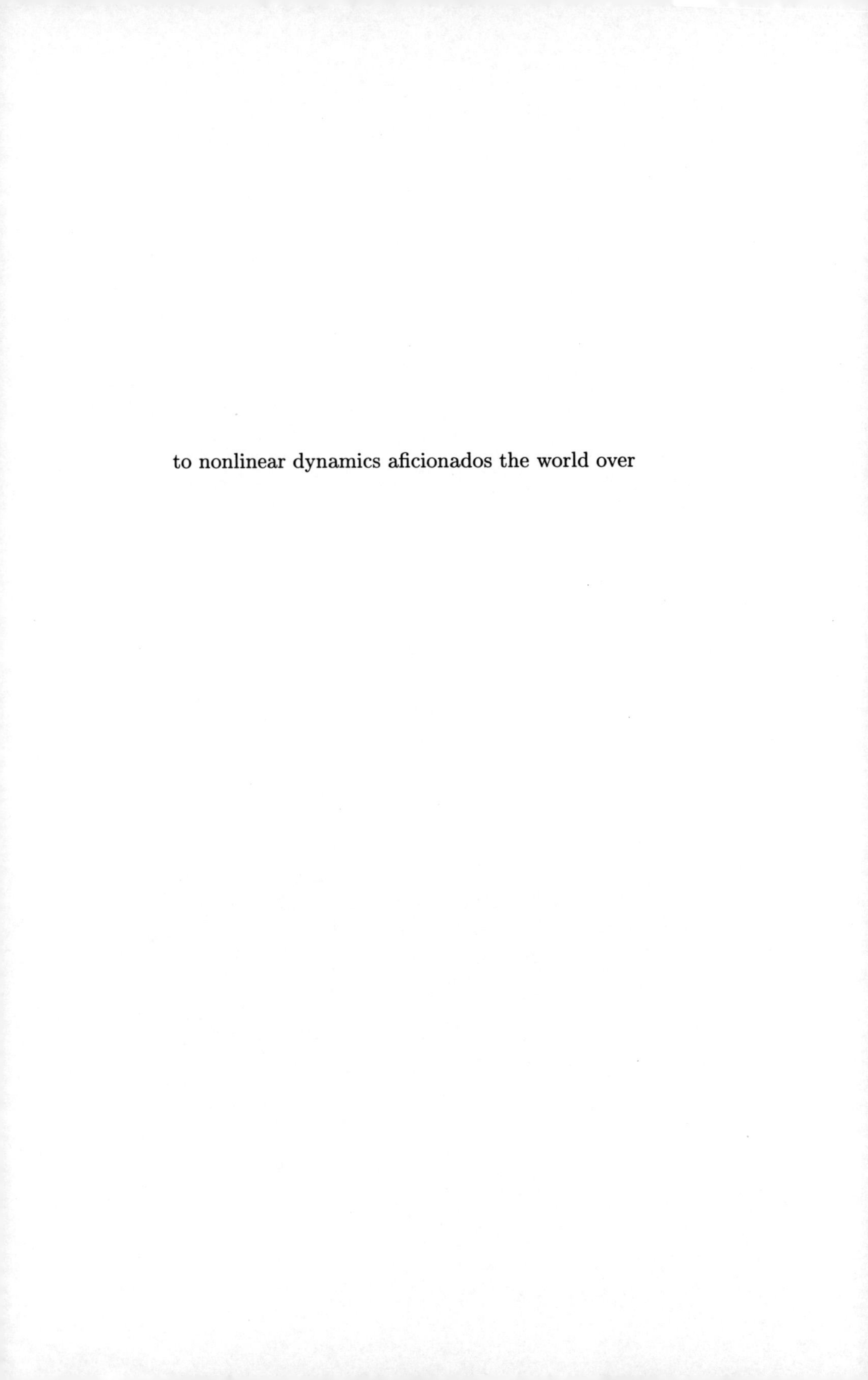

to nonlinear dynamics aficionados the world over

Preface

Do you sometimes hear or read grand pronouncements such as "The world is essentially nonlinear"? Or maybe you have heard that famous smart remark, to the effect that nonlinear dynamics is "like defining the bulk of zoology as the study of non-elephants", attributed to the mathematician Stanislaw M. Ulam[1] (who, however, failed to provide convincing proof that elephants are linear) — the point being that linearity is the exceptional case.

Over the last three or four decades interest in nonlinear phenomena has certainly grown nonlinearly, as you can see in the graphs. The data

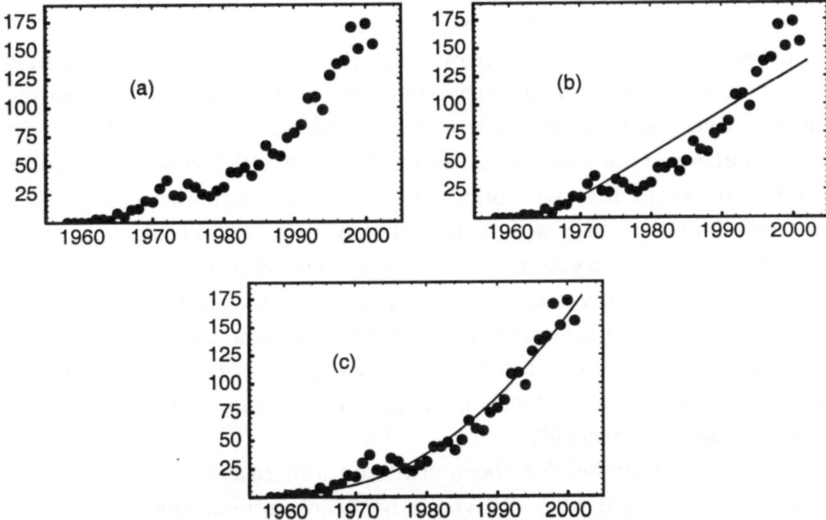

Frequency of the word "nonlinear" (vertical axis) versus year (horizontal axis)

represent the number of times the word "nonlinear" occurs in the titles and abstracts of articles in *Physical Review Letters*, a leading broad-spectrum physics journal, for each of the years between 1958 and 2001. The data do not look very linear (a). In fact a linear fit is really not very good (b), and

<hr>

[1] David K. Campbell: Nonlinear Science from Paradigms to Practicalities. In: From Cardinals to Chaos: Reflections on the Life and Legacy of Stanislaw Ulam, ed. Necia Grant Cooper, Cambridge University Press, 1989, p. 218.

a nonlinear fit is obviously much better (c). (Of course a least-squares fit does not model fluctuations in the data.)

Another inference that can be made from these data is that nonlinear science is increasingly multi-field and multi-disciplinary. A selection of key-words from these articles from 1999–2001 includes terms such as solitons, earthquakes, neural nets, photorefractive materials, laser cavity resonators, lattice dynamics, plasma transport, optical communications, chaos, information encryption and decryption, patterns, Bose–Einstein condensates, superconductivity, gravitational waves, black holes, quasicrystals, thin liquid films, the cochlea, quantum computing, and carbon nanotubes.

Many of the leading-edge research fields associated with these keywords have grown up synergistically with advances in the theory and mathematics of nonlinear dynamics. In disciplines such as the bio-medical sciences and earth sciences many fields have been given new impetus and direction from lateral applications of the paradigms of nonlinear dynamics that are described by the authors of the chapters in this volume.

In a sense this exponential growth of interest in nonlinear phenomena is unique in the history of physics. It is not driven by defence and cold war imperatives, as for example was nuclear physics in the 1950s and 60s, or industrialization, as was thermodynamics in the 19th century. What we are seeing now is the reverse: nonlinear theory and mathematics is actually driving developments in a wide range of very diverse fields, from lasers to butterflies, optics to ecosystems, neuroscience to climatology to engineering.

Such marvellous universality of nonlinear dynamics is all very well, but if, like most graduate students and researchers, you have emerged from the undergrad subspace of simple pendulums and first-order reactions, you might ask: how, exactly, does one begin to describe the dynamics and physics of a nonlinear world?

This book is intended for those students and researchers. It is a collection of lectures that were delivered by nine eminent scientists at DynamicSummer: the 15th Canberra International Physics Summer School, in January 2002. The topics covered in the lectures collectively provide an inspirational introduction to modern research directions and scholarship in nonlinear dynamics, at the same time as guiding the graduate student through the fundamental concepts and methodologies. With each chapter having a comprehensive bibliography, the book will also be a valuable reference for researchers already delving into matters nonlinear. In the overall content a balance is achieved between theory, modelling, mathematics, and experiments. Applications and examples chosen by the authors are drawn from mathematical and physical sciences disciplines.

The association between nonlinear dynamics and zoology noted by Ulam is reflected in several aspects of the book. The butterflies mentioned in the title are of two species. The first kind is introduced in Chapter 1 by Brian Davies. This is the famous Lorenz butterfly, a metaphor for the amazing sensitivity of deterministic nonlinear evolutions to initial conditions. In Chapter 2 this butterfly is also mentioned by Nalini Joshi, who goes on to show that it is but one species of a vast family of unstable and mischievous butterflies, many of whom hide deep within the asymptotic solutions of innocent-looking, ordered, integrable systems that are important in physics, such as Painlevé equations.

A classic problem of Hamiltonian dynamics, the Fermi problem of a particle bouncing between a fixed and an oscillating wall, is described by Mike Lieberman in Chapter 3. With the aid of Poincaré return maps, a useful tool of nonlinear dynamics that was introduced in Chapter 1, it is shown how the Fermi acceleration model describes collisionless electron heating in industrial plasmas. Hamiltonian dynamics remain the theme of Chapter 4, in which Cathy Holmes compares model and experimental dynamics for two systems, cold atoms in a periodic potential coupled Bose–Einstein condensates of alkali gases.

The next few chapters elucidate various aspects of that perennial problem of nonlinear science: turbulence. In Chapter 5 a new analysis of wave turbulence is presented by Biven, Connaughton, and Newell. Renormalized closure theory is elucidated in Chapter 6 by Jorgen Frederiksen, and its applications for subgrid-scale parameterizations of two-dimensional turbulence simulations are developed. Tony Roberts in Chapter 7 describes a flexible, low-dimensional approach to modelling of flows using centre manifold theory.

In the final chapters there is a more explicit theme-shift to the spatial structures that grow out of nonlinear processes. Weiss and co-workers in Chapter 8 discuss the beautiful and intriguing vortices and solitons found in non-linear optical resonators. Mathematical treatments of solitons are developed in Chapter 9 by Ablowitz, Hirooka, and Musslimani, with applications to an optical beam in a waveguide array and optical fiber communications. In Chapter 10 Ercolani, Indik, Newell, and Passot show that pattern formation can be described in terms of the Cross–Newell phase diffusion equation, and incidentally proving that elephants, which are qualitatively hinted at in figure 9 of Chapter 10, are actually nonlinear.

This book could not have been prepared without the valued input and assistance of a great many people. As Convenor of DynamicSummer I would like to thank them all a thousand times.

To the lecturers, all internationally recognised champions in their research fields, who gave their time and expertise so freely, communicated their lectures with such clarity and inspirational quality, and prepared the generously written and richly illustrated manuscripts that form this book — thank you all.

To the participants of DynamicSummer, the students and researchers whose interest and enthusiasm drove them to come from many parts of the world to learn, special thanks are owed. The tutelage and nurturing of new talent in nonlinear dynamics was the major motivation for this event and for this book, and the liveliness and enthusiasm of the participants ensured that DynamicSummer developed a memorable dynamics of its own.

Sincerest thanks go to the other three members of the DynamicSummer Production Team: Prof. Nail Akhmediev — Associate Editor and Casting Director, Dr Vanessa Robins — Technical Director, Site Coordinator, Schedule Planner, and Continuity Manager, and Prof. Bob Dewar, Casting Director, Expert Adviser, and Finance and Budget Officer (for those students requesting financial assistance), and Complaints Officer (for those students requesting financial assistance). I am also deeply grateful to Nyssa Gyorgi-Faul and Heli Jackson for their administrative expertise, and Dr Mihajlo Mudrinic, who managed our web site.

Finally I thank, in advance, the readers of this volume. Some of you will have already heard, as DynamicSummer participants, the lectures on which the chapters are based. I have no doubt that these readers will enjoy dipping into the volume, revisiting topics that they so often engaged the lecturers in lively discussions and arguments. That is good — learning should be enjoyable and argumentative. For other readers this book will be a guided but exciting adventure into the nonlinear realm. I am confident that all readers will see that opportunities for students who are trained in methods of tackling nonlinear problems are rich and diverse, that non-elephants are on the increase, and that many more butterflies await discovery.

Rowena Ball
Convenor, DynamicSummer 2002.

Contents

Chapter 4   Large Resonances in Hamiltonian Systems, with Applications

165

C. Holmes

Chapter 5   Structure Functions, Cumulants and Breakdown Criteria for Wave Turbulence

197

L. J. Biven, C. Connaughton, and A. C. Newell

Chapter 6   Renormalized Closure Theory and Subgrid-scale Parameterizations for Two-Dimensional Turbulence

225

J. S. Frederiksen

Chapter 7 Low-Dimensional Modelling of Dynamical Systems Applied to Some Dissipative Fluid Mechanics 257
A. J. Roberts

Chapter 8 Vortices and Spatial Solitons in Optical Resonators, and the Relations to Other Fields of Physics 315
C. O. Weiss, K. Staliunas, M. Vaupel, V. B. Taranenko,
G. Slekys, and Ye. Larionova

Chapter 9 Nonlinear Waves and (Interesting) Applications 369
M. J. Ablowitz, T. Hirooka, and Z. H. Musslimani

Chapter 10 Global Description of Patterns Far from Onset: A Case Study 411

N. Ercolani, R. Indik, A. C. Newell, and T. Passot

Chapter 1

Nonlinearity and Complexity: An Introduction

Brian Davies[1]

Mathematical Sciences Institute, Australian National University
Canberra ACT 0200 Australia

Abstract.   These lectures provide an elementary introduction to some of the key ideas of complex and chaotic behaviour in nonlinear dynamical systems. They are mainly concerned with an exposition of the behaviour of discrete-time systems in one and two dimensions, although some simple nonlinear differential equations are considered. Only elementary mathematics is used. The lectures are illustrated using computer software which is freely available for readers wishing to make their own numerical experiments; the same software was used to generate the printed illustrations.

1   Poincaré, Lorenz, Butterflies

To the Greeks, *chaos* signified the infinite formless space which existed before the universe was created. Since the latter part of the 20th century, however, the word has been adopted by science to denote a situation described in a paper of Phillip Holmes [15]:

> We thus see that deterministic dynamical systems can give
> rise to motions which are essentially random.

In this lecture I shall discuss some simple deterministic dynamical systems and demonstrate that they may exhibit just such behaviour.

---

[1] `http://wwwmaths.anu.edu.au/~briand`

1.1   *Dynamical models*

In the physical sciences we can predict the motion of a space craft so as to enable it to make a journey of several years. This comes about through the use of a *dynamical model*, in which the state of the system is described by state variables. A *deterministic* dynamical model is one whose future states are uniquely determined from its present state by prescribed laws of evolution. For example, a model of the solar system may consider the sun and planets as point masses moving in otherwise empty space, under the sole effect of mutual gravitational forces. Despite omitting an untold wealth of detail, this model has been one of the most fruitful in the history of the physical sciences.

Population models

In population dynamics, it is desirable to predict trends in populations due to external influences. Let's begin by considering the simplest population model, for a single species of a seasonally breeding organism whose generations do not overlap. We seek to understand how the size x_{k+1} of a population in generation $k+1$ is related to the size x_k of the population in the preceding generation, assuming a scalar relationship of the form

$$x_{k+1} = r(x_k) \cdot x_k.$$

Here $r(x)$ is the reproductive rate, and is a function of the population x and of an adjustable parameter r. I assume that it decreases from an initial value $r(0) = r$ to $r(x) = 0$ at a limiting population number K. Using x to measure the population as a fraction of K, then $r(x) = 0$ at $x = 1$.

A simple example is the *logistic model*, which employs a linear decrease of $r(x)$ with increasing x:

$$r(x) = r(1 - x).$$

Starting from an initial population x_0, this gives rise to the sequence of populations, at successive generations k,

$$x_{k+1} = rx_k(1 - x_k).$$

To ensure that $x_k \leq 1$ always, I impose the restriction $r \leq 4$. Clearly, if $r < 1$, the population gradually dies out, since the reproductive rate is insufficient for any positive x.

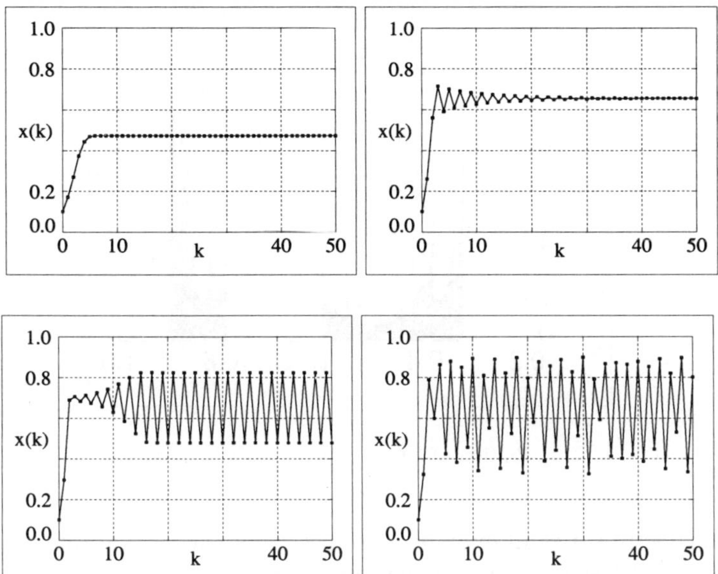

Fig. 1   Iterations of the logistic map, with parameters (from top left) (i) $r = 1.9$, (ii) $r = 2.9$, (iii) $r = 3.3$, (iv) $r = 3.6$.

Examples of behaviour

Using scChaos for Java[2], we can examine the solution in a number of ways. Figure 1 shows the first 50 generations, commencing from an initial population $x_0 = 1/10$, with four different values of the parameter r. Observe the following properties:

(i) For $r = 1.9$ the population rises rapidly to a steady value of about 0.47 (47% of the carrying capacity), a figure determined by crowding.
(ii) For $r = 2.9$ the population again stabilises, this time through a sequence of small boom and bust cycles which die out.
(iii) Increasing r to 3.3 changes the behaviour fundamentally. Now the system stabilises on a permanent boom and bust cycle which alternates between good and poor seasons.
(iv) At $r = 3.6$ the behaviour has become extremely complex, with no apparent pattern or simple repetition. It is in fact chaotic.

---

[2]Most of the illustrations were produced with this cross-platform software, which is free for download at **http://sunsite.anu.edu.au/education/chaos**.

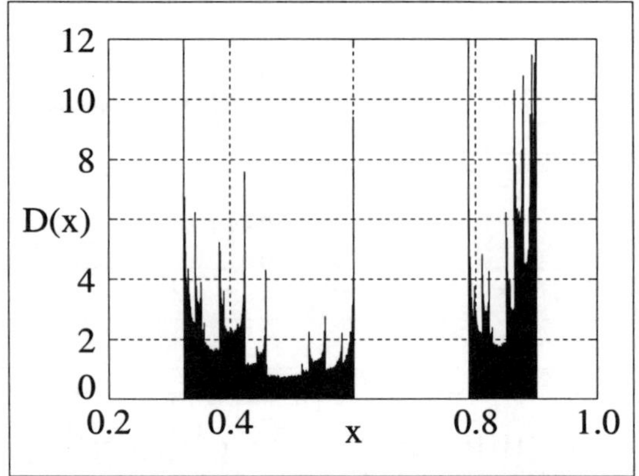

Fig. 2   Distribution of 10^6 iterations, logistic map, $r = 3.6$.

In a later lecture, I shall explain more precisely what "chaotic" means; for now simply observe that the iterations appear to be randomly distributed as shown in Fig. 2.

Imagine the implications for population control policies if such a simple model can generate such disparate outcomes, depending only on the policy settings! Holton and May[3] discuss some problems with drawing conclusions from such a model, but conclude:

> Despite this possibility, there is a rationale for constructing overly simplified models: to capture the essence of observed patterns and processes without being enmeshed in the details.

The point is that, although simple models cannot replace careful simulation of real-life phenomena, they may have much to tell about real-life behaviour. If a simplified model turns out to be essentially unpredictable, and to display surprisingly complex behaviour, that alone serves to prevent the belief that the situation will be remedied by adding layers of complication. Moreover, understanding the origins and structure of the complexity provides an essential backdrop for understanding the workings and behaviour of real-life systems.

[3]David Holton and Robert M. May, "Chaos and one-dimensional maps", in [22], p101.

1.2   *Poincaré and the birth of chaos*

One of the earliest dynamical systems to be studied is the solar system —
for a delightful account see the book by Ivars Peterson [26]. Here I shall be
brief.

Kepler and Newton

Johannes Kepler reduced a mass of observations to three laws:

(i) Planetary orbits are plane ellipses with the sun at one focus.

(ii) A line joining a planet with the sun sweeps out area at a rate constant in time.

(iii) The squares of the orbital periods are proportional to the cubes of
the mean radii.

One sees that the very statement of the laws takes us a long way toward
reducing the data to a dynamical model. They define the important state
variables as the positions and velocities of the solar bodies, and they state
some empirical relationships.

Isaac Newton's theory explains these laws as the consequence of a simple
dynamical model for which he gives the *equations of evolution*. Newton's
second law of motion states that the rate of change of momentum of a body
is equal to the sum of the forces acting on it; his gravitational theory states
that the force acting between any pair is proportional to the product of
their masses, inversely proportional to the square of the distance between
them, and directed along the line joining them at any instant of time. The
constant of proportionality, G, is a *universal constant* of nature.

Laplace

Newton was acutely aware of various deficiencies of his theory. He was
unhappy about his inability to give a proper account for the observed motion of the moon. There were other discrepancies too, particularly in the
motion of the two largest planets, Jupiter and Saturn. In some brilliant
work, Pierre Simon de Laplace accounted for this latter as a mutual near
resonant interaction resulting in periodic changes which take approximately
900 years for each cycle. So confident was he of the validity of the under-

lying methods of dynamics that he wrote [4]

> Assume an intelligence that at a given moment knows all
> the forces that animate nature as well as the momentary
> position of all things of which the universe consists, and
> further that it is sufficiently powerful to perform a calcu-
> lation based on these data. It would then include in the
> same formulation the motions of the largest bodies in the
> universe and those of the smallest atoms. To it, nothing
> would be uncertain. Both future and past would be precise
> before its eyes.

This is an extreme statement of the view that the solar system — even the
entire universe — is a predictable *clockwork* system.

The birth of chaos

In November 1890 Henri Poincaré's memoir on the three-body problem was
published as the winning entry in an international competition to honour
the 60th birthday of King Oscar II of Sweden and Norway. There were
four questions from which the contestants might choose; Poincaré's choice
was the one whose solution would, it was hoped, lead to a resolution of the
question of the stability of the solar system. In part, the question read:

> A system being given of a number whatever of particles
> attracting one another mutually according to Newton's law,
> it is proposed ... to expand the coordinates of each particle
> in a series ... according to some known functions of time ...

Poincaré's winning entry was on "the problem of three bodies and the
equations of dynamics". In fact, his investigations are concerned with the
"restricted circular three body problem". This version has two of the bodies in known circular orbits about their centre of mass. It seeks only to
explain the motion of a third body whose mass is too small to influence
the two primaries. This simplification of the original question — concerned
with an arbitrary number of bodies moving in three dimensions — to three
bodies moving in a plane, two of them in fixed circular motion, illustrate

[4]Taken from Peterson [26], p229.

the importance of simple models to making progress in fundamental understanding. It underscores the comments of Holton and May, made on population models, and quoted above.

An exposition of Poincaré's work is the subject of the monograph by June Barrow-Green [5]. Here I just mention a few salient points.

(i) Poincaré gave prominence to the geometric properties of the orbits as smooth curves in space, defined by the evolution of the state variables.
(ii) He showed that individual orbits may be investigated via the set of points at which they pierce a two-dimensional *transverse surface.*
(iii) The dynamics is now encoded as the *map* which relates the successive piercings of this surface, determined solely by the equations of motion.
(iv) Periodic orbits show up as isolated points in this map.
(v) Entire families of orbits may now be represented as curves in a surface of section, each point on the curve representing an orbit.
(vi) Certain families of orbits lie on curves which intersect themselves infinitely often in the neighbourhood of a single point.

Thus Poincaré instigated a new way to study dynamical systems which emphasised qualitative and geometric features, not just analytical formulae. His method (ii) is widely used today and is known as the method of *Poincaré sections.*

The study of maps, instituted in (iii), is used in the theory of dynamical systems and chaos. The *homoclinic tangles* identified in (vi) play an important part in advanced studies of chaotic systems. Although I shall not be concerned with the theory of such orbits herein, it is important to understand that it was in the process of exploring the *infinite complexity* of such bizarre objects that Poincaré arrived at the doorway to chaos.

Understandably, Poincaré's work, particularly the memoir of 1890, has drawn unceasing admiration for more than a century. One hundred years later, one reviewer [16] wrote that the memoir was

... the first textbook in the qualitative theory of dynamical systems ...

The driven pendulum

Because of the sophistication of even the restricted circular three body problem, I will consider a favourite of introductory texts, the *driven plane*

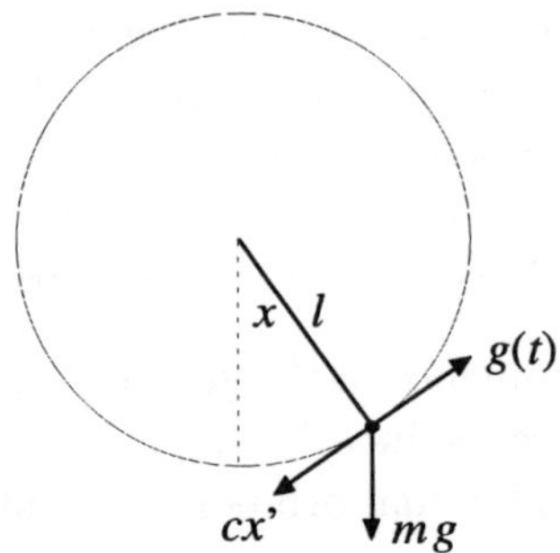

Fig. 3   Driven plane pendulum.

pendulum. As a dynamical model it represents an idealised system consisting of a mass m constrained to move in a vertical plane at a constant distance l from a fixed point (figure 3). Only one coordinate is needed to specify its state, the angular position $\theta(t)$, measured from the bottom (rest) position. Three forces combine to produce the motion:

(i) Gravity, acting vertically downward with magnitude mg; the tangential component $mg \sin\theta$ acts as a restoring force.
(ii) A damping force $c\theta'$, proportional to the angular velocity.
(iii) An applied periodic driving force, $k \cos\Omega t$, acting tangentially.

Then the equation of motion is the second order nonlinear differential equation[5]

$$ml\theta'' + c\theta' + mg \sin\theta = k \cos\Omega t.$$

There are several constants here. They do not play an equal rôle, however. If we change the unit of time by the substitution $t \to t\tau$, $\tau = \sqrt{l/g}$, and divide by mg, we obtain the equation (the constants are rescaled)

$$\theta'' + c\theta' + \sin\theta = k \cos\Omega t. \tag{1.1}$$

This leaves only three parameters to be chosen, the damping factor c, the driving force amplitude k and frequency Ω. With these choices the frequency Ω is measured in units of the natural frequency $\sqrt{g/l}$.

---

[5]Since $\sin\theta$ is not proportional to θ, it provides the nonlinearity.

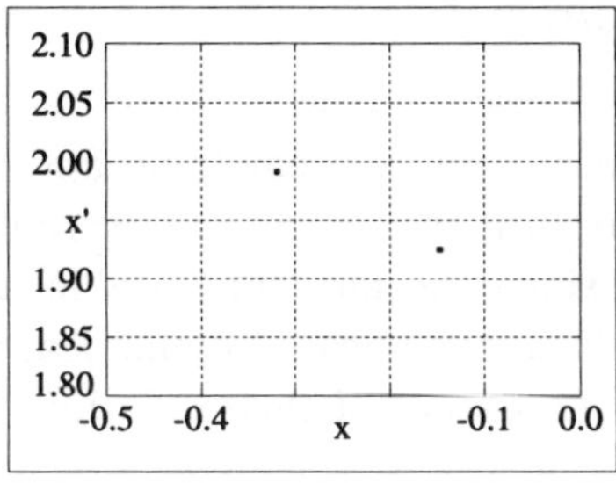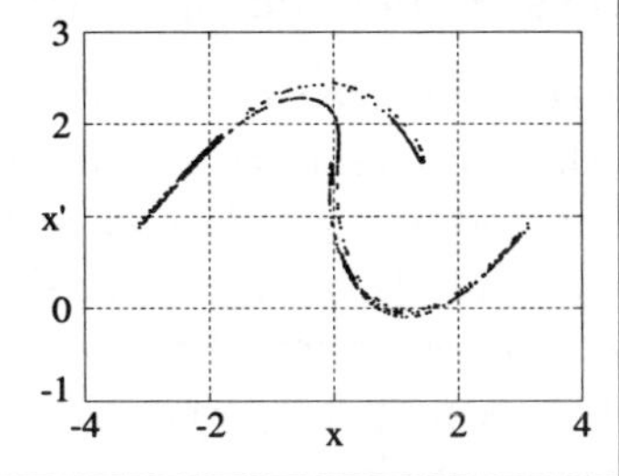

Fig. 4   Poincaré sections for the driven pendulum. $k = 1.07$, period 2 (left), $k = 1.5$ strange attractor (right).

Poincaré sections

Solutions of these equations are curves in space, described by coordinate functions $\theta(t)$ and $\theta'(t)$. It is an important property of these solutions that they remain solutions under the time translation

$$t \rightarrow t + 2\pi/\Omega.$$

I shall show in the last lecture that this means that the points of intersection of the solution with the transverse surfaces specified by the time sequence

$$t_0, t_0 + 2\pi/\Omega, t_0 + 4\pi/\Omega, \cdots$$

completely characterise the behaviour of the system. This is an example of a Poincaré section. Two such sections are shown in Fig. 4.[6] Both have the choice $c = 1/2$, $\Omega = 2/3$, in both of them t_0 (the time of the first section) has been set to a large value so as to allow transient behaviour to die out. The left-hand picture is for the amplitude $k = 1.07$, at which value the steay-state solution is of period 2. This shows on the section as a simple pair of points, which are visited alternately.[7] For $k = 1.5$, the solution is chaotic, and the section has an obviously complex structure. The orbit is not periodic, as evidenced by the fact that the same point is never revisited.

---

[6] The Roman letter x is used instead of the Greek θ in scChaos for Java.

[7] Since the *exact* orbit cannot be computed, each of the two "points" is actually an infinite number of points converging toward the exact solution.

1.3   *Lorenz: the end of weather prediction?*

Despite the importance of Poincaré's work, and other work in the first half of the 20th century, the implications for unpredictable and chaotic behaviour were not widely appreciated until the advent of electronic computation. This is hardly surprising, since the fact that usable analytic formulae cannot be found for relatively uncomplicated dynamical models means that a proper appreciation of the nature of their solutions had to await such a development.

Thus it was not until Edward Lorenz' 1963 paper [17] that a new era opened in nonlinear dynamics and chaos. Lorenz considered the relatively harmless looking differential equations[8]

$$\frac{dx}{dt} = \sigma(y - x),$$
$$\frac{dy}{dt} = rx - y - xz, \qquad (1.2)$$
$$\frac{dz}{dt} = xy - bz.$$

Here x, y and z are the state variables, σ, b and r are parameters which control the types of behaviour. Were it not for the two nonlinear terms (xz in the second equation and xy in the third), the complete set of solutions would be expressible using only the exponential, sine, and cosine functions, and a few constants easily computed from the coefficients b, σ and r, together with the initial values of x, y and z. That is to say, not only would it be a deterministic dynamical system, but more importantly, all possible behaviour patterns would be simple to understand. One facet I want to emphasise here is that, because of the nature of the equations, were they linear then at most one natural frequency would be required for the description of the motion.

---

[8]In the model from which the Lorenz equations are distilled, interest is on convective fluid motion driven by heating from below, such as might occur locally over warm terrain. Lorenz took a set of seven coupled differential equations (derived by a colleague), ignored four apparently insignificant variables, and investigated solutions of the remaining three coupled equations. This gave him his first real glimpse of infinitely complex behaviour in a simple deterministic system.

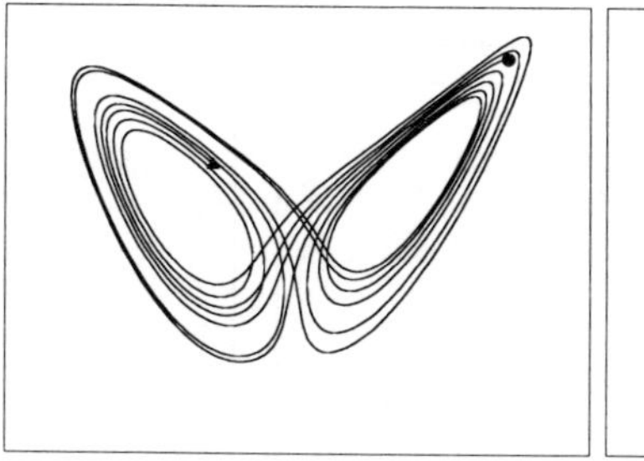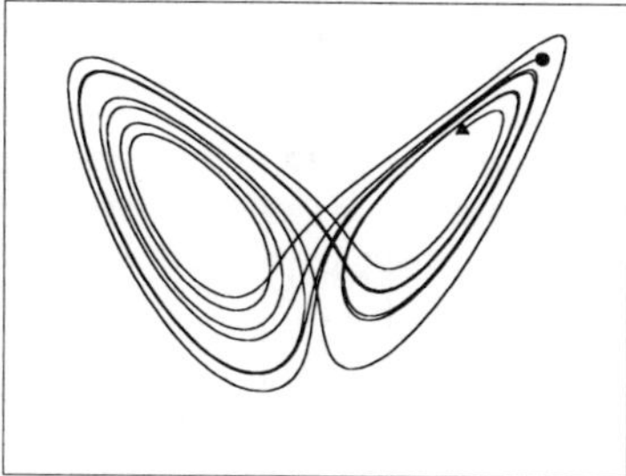

Fig. 5   Orbits of the Lorenz equations, both with $\sigma = 10$, $b = 8/3$, and $r = 28$. The initial positions (circle) differ in the fourth significant place; the final positions (triangle) are qualitatively different after a few circuits.

Strange attractors

What Lorenz found are solutions which are *nonperiodic*, that is, they cannot be represented using any finite number of frequencies. These solutions are also *sensitively dependent* on initial conditions, which means that for all practical purposes, prediction of the state of the system is limited to relatively short times. Furthermore, in the regime where chaotic solutions exist, then regardless of the initial conditions, they are all attracted to some region of state space whose dimension is not an integer! It resembles a surface with two wings, but it is more like a "thick" surface, with an infinite number of sheets.[9] Such objects are generally called *strange attractors*, and again we are confronted with the infinite when examining the behaviour of a simple dynamical system.

Two typical solutions are shown in Fig 5, numerically generated by scChaos for Java. Using Lorenz' choice for the parameters, namely $\sigma = 10$, $b = 8/3$ and $r = 28$, one finds orbits which have become one of the icons of chaos. Each orbit commences from the point displayed as a small circle and ends at the point displayed as a small triangle. The initial position of the two differs only in the fourth significant place of the x-coordinate; it is clear that the final point is on a different wing. In fact, the generating point makes one or more circuits around one of the wings before switching to the other: this process of making circuits then switching continues indefinitely.

[9]It is, in fact, a fractal object.

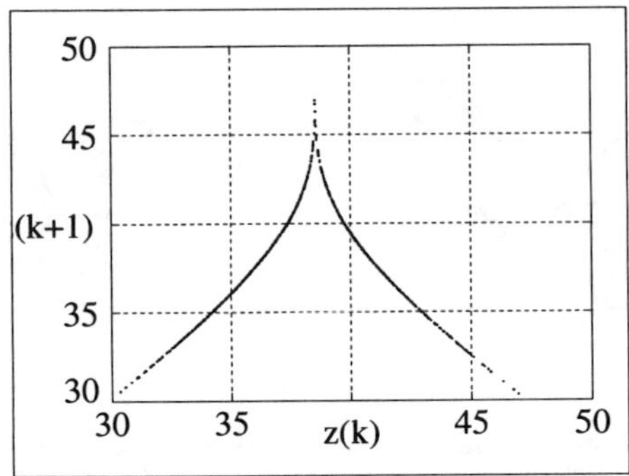

Fig. 6   Maximum in z return map, Lorenz equations, $r = 28$.

More Poincaré sections

In Fig 5, orbits of the Lorenz equations circle indefinitely about two organising centres, switching erratically from one to the other. On each circuit they attain a maximum value for the variable z. In fact, it follows from the last of the equations 1.2 that these maxima must occur when $z' = xy - bz = 0$, that is, they mark places at which the orbit pierces the surface $z = xy/b$, which is therefore a simple example of a Poincaré transverse surface of section.

Now, because the Lorenz attractor for $r = 28$ is like a thick surface, its intersection with this transverse surface is a "thick" line, and the information about the points of intersection is almost contained in the maximum z values alone, which I call z_k.[10] For a one-dimensional discrete system $z_{k+1} = f(z_k)$, and if one plots pairs (z_k, z_{k+1}) in a plane, they will fall on the graph of f. Therefore, even though the sequence of maximum z values is not exactly one-dimensional, it is instructive to examine such a plot. In fact this was done by Lorenz in his original paper. Figure 6 shows the plot of 1000 pairs (z_k, z_{k+1}). It gives good motivation for investigating the properties of some simple one-dimensional maps.

---

[10]The corresponding value of the product $x_k y_k$ is given by bz_k. If the points fell on a simple smooth curve in the surface of section this would determine x_k and y_k separately.

1.4 *Butterflies*

Lorenz noticed that when he attempted to recompute a given orbit, using the same program on the same computer, he got a different result from the original. This was because his recorded values of x, y and z were less accurate than the internal representation used by the computer, so he was comparing two solutions which differed in their initial state by a small amount. The surprising effect is that, after a while, the two solutions don't seem to have much correlation with each other at all.

For example, the two orbits shown in Fig 5 differ only by a change of initial values of x in the fourth significant place. Even in the short time span (20 units) of the displayed orbits, one sees that they no longer agree except in the most qualitative feature that they are both organised by the same strange attractor. It is not just the growth of error that is involved. What we are facing in equations such as Lorenz' is the fact that the relative error quickly becomes as large as the quantities themselves, and that different solutions only have similar qualitative behaviour over relatively short time intervals. That being said, a strange attractor does supply a recognisable structure for the solutions.

This effect, *sensitive dependence* of the evolution of a system to the most infinitesimal changes of initial state, is known as the *butterfly effect*, after the title of a talk by Lorenz:[11]

> Predictability: Does the flap of a butterfly's wings in Brazil
> set off a tornado in Texas?

It encapsulates the question: if Lorenz' equations do not allow long time prediction, why should more complicated dynamical models of the atmosphere do any better? Debate over such questions continues, as does research into weather and climate prediction. An entire chapter of Lorenz' book [18] is devoted to an informed but non-technical discussion of the weather and the implications of chaos for forecasting.

A simple map $\cdots$

To help understand why the Lorenz equations are unpredictable, even though they are completely deterministic, I follow the spirit of Lorenz'

---

[11]The text of the talk is reprinted in Lorenz [18]. He points out that one might equally well ask whether the butterfly can prevent a tornado in Texas.

original paper and investigate a map similar to, but simpler than, the return map shown in figure 6. Explicitly, my toy map is the "fully chaotic tent map", whose dynamics are given by the simple formula

$$x_{k+1} = \begin{cases} 2x_k, & 0 \le x_k \le \tfrac{1}{2}, \\ (1 - x_k), & \tfrac{1}{2} \le x \le 1. \end{cases} \tag{1.3}$$

If we work with binary arithmetic, there is a very simple way to express the dynamics of this map. Every number $0 \le x \le 1$ has a representation

$$x = 0.d_1 d_2 d_3 \ldots \tag{1.4}$$

where the d_j are either 0 or 1. The meaning is that

$$0.d_1 d_2 d_3 \ldots = d_1/2 + d_1/2^2 + d_1/2^3 + \cdots$$

This representation is not unique; sequences ending with the recurring number 1 are equivalent to a terminating sequence, that is,

$$0.d_1 d_2 d_3 \ldots d_k \dot{1} \equiv 0.d_1 d_2 d_3 \ldots (d_k + 1)\, 0.$$

If $d_k = 1$, then $(d_k + 1) = 0$, carry 1; this process continues until the unit can be added without a carry. The extreme example is that $0.\dot{1} = 1.0$.

Now it is easy to see that, for the map 1.3, a number with the base 2 representation $0.d_1 d_2 d_3 \ldots$ is mapped to

$$f(0.d_1 d_2 d_3 \ldots) = \begin{cases} 0.d_2 d_3 \ldots & (x \le \tfrac{1}{2}) \\ 0.\bar{d}_2 \bar{d}_3 \ldots & (x \ge \tfrac{1}{2}) \end{cases} \tag{1.5}$$

where $\bar{d}_j = 1 - d_j$ is the 2's complement. At first glance, it seems that I forgot about the first digit d_1; a little reflection shows that if $x \le \tfrac{1}{2}$ we may take $d_1 = 0$, and if $x \ge \tfrac{1}{2}$ we may take $d_1 = 1$ so that $\bar{d}_1 = 0$. So, the dynamics is a simple shift of one to the left, together with a complement operation if $d_1 = 1$.

$\cdots$ *with complex behaviour*

From this analysis we can draw come surprising conclusions:

(i) *Sensitive dependence* — suppose we know the initial value x_0 to N binary places. Consider the (uncountable) collection of random numbers which agree with x_0 to this accuracy. Then, after N iterations, the orbit is essentially random, in no way related to the initial condition.

(ii) *Mixing* — consider the range of initial values which differ first at the Nth binary place. At the Nth iteration, they are spread over the whole interval $[0, 1]$. Thereafter the whole interval is stretched to twice its length at each iteration and then folded back on itself.

(iii) *Dense periodic orbits* — the binary representation of a rational number ends with a recurring group, and so generates an orbit which eventually repeats periodically. Irrational numbers never recur. Therefore, orbits with periodic behaviour are *dense* in the set of chaotic orbits.

Deterministic dynamical systems may exhibit regular behaviour for some values of their control parameters and irregular behaviour for others. One speaks of *regular* and *chaotic* behaviour in such a system. In the present context, chaos in a dynamical system is a situation where one sees:

(i) Sensitive dependence on the initial conditions, making long-term prediction impossible — *the butterfly effect.*

(ii) Interweaving of the states of the system on ever finer scales so that the trajectories which it may follow become inextricably tangled — *mixing.*

(iii) A pathological structure which is just as bad as that of the rational numbers in the reals — *dense unstable periodic orbits.*

2   One-dimensional systems

This lecture I discuss discrete one-dimensional dynamical systems, particularly properties of stability and periodicity of orbits. The systems all take the form

$$x_{k+1} = f(x_k; \mu); \tag{2.1}$$

the coefficient μ, which is not affected by the iteration, is a *parameter.*

2.1   *Fixed points and stability*

As a function of x, f must have the property that the domain (input) space is *mapped* to itself, so as to allow for iteration; for this reason I shall refer to *maps of an interval* rather than to functions. Furthermore, I shall call a particular sequence of values x_k, $k = 0, 1, \ldots$, generated by the system 2.1, an orbit *generated* by the *initial value* x_0. Two important maps are:

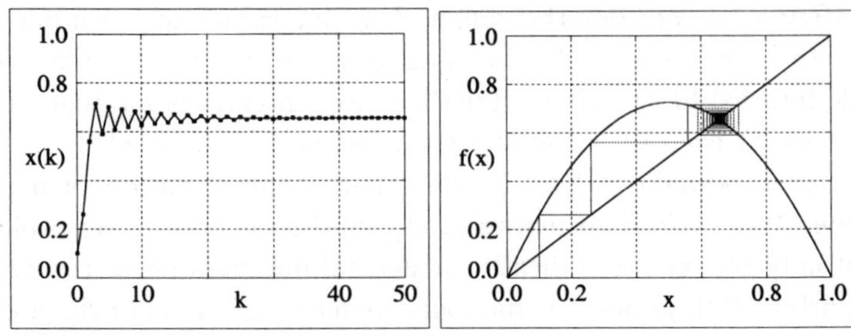

Fig. 7   Iterations of the logistic map with $r = 2.9$, $x_0 = 0.1$, viewed as a plot of x_k versus k (left) and as a cobweb (right). Lines joining successive points serve only as a guide to the eye.

(i) The logistic map,

$$f(x) = rx(1 - x), \qquad (0 \le r \le 4). \qquad (2.2)$$

(ii) The tent map,

$$f(x) = \begin{cases} 2tx, & (x \le 1/2) \\ 2t(1 - x), & (x \ge 1/2) \end{cases} \qquad (0 \le t \le 1). \qquad (2.3)$$

Both maps are *unimodal,* meaning that they have a single maximum in the interior of the interval.[12] Both were encountered in the first lecture.

Let's look at what happens when we iterate the logistic map; Fig 7 shows two views of the first 50 iterations. One is a plot of x_k as a function of k, in which the actual values are joined by straight lines simply as a guide to the eye. The other is a *cobweb plot,* in which each vertical line guides the eye from x_k to $f(x_k)$, each horizontal line from $f(x_k)$ to x_{k+1}. The cobweb plot shows most clearly what is going on: the system approaches the limiting point at which the graph of $y = f(x)$ intersects the line $y = x$, that is, solutions of $f(x) = x$.

Fixed points

Any value x^* for which $f(x^*) = x^*$ is called a *fixed point* of f. It is *stable* if all nearby orbits converge to x^* as k increases toward infinity; it is *unstable* if all nearby orbits are repelled in this limit.[13] A stable fixed point will

---

[12]Of course, they have extrema at the end-points.

[13]The definition I have given is generally called asymptotic stability.

also be called an *attractor*, an unstable fixed point a *repeller*. The set of all initial states whose orbits converge to a given attractor is called the *basin of attraction.*

Rather than consider the fixed point equation directly, define a function

$$\phi(x) = x - f(x); \tag{2.4}$$

then the fixed points are precisely its zeros. For the logistic map, this reduces to the quadratic equation

$$\phi(x) = x - rx(1 - x) = 0,$$

whose two solutions are $x_0^* = 0$ and $x_1^* = (1 - 1/r)$. For $0 < r < 1$, x_1^* is not in the interval $[0, 1]$ so there is only one fixed point,[14] x_0^*. Therefore, the formulae for the fixed points, in the interval $[0, 1]$, are

$$\begin{aligned} x_0^* &= 0, & (0 \leq r \leq 4), \\ x_1^* &= 1 - 1/r, & (1 \leq r \leq 4). \end{aligned} \tag{2.5}$$

Stability analysis

The stability of a fixed point depends on the first derivative of the map. To see this, let $\delta_k = x_k - x^*$; it is interesting to investigate how it compares with the next difference $\delta_{k+1} = x_{k+1} - x^*$. Using linear approximation,

$$\begin{aligned} \delta_{k+1} &= f(x_k) - x^* = f(x^* + \delta_k) - x^* \\ &\approx f(x^*) + \delta_k f'(x^*) - x^* = \delta_k f'(x^*). \end{aligned} \tag{2.6}$$

An important feature is that δ_{k+1} will have the same sign as δ_k if $f'(x^*)$ is positive, the opposite sign if $f'(x^*)$ is negative. The distance changes by the "magnification factor" $|f'(x^*)|$.

This is only a linear approximation. However, it is easy to produce an exact argument which shows that x^* is stable if $|f'(x^*)| < 1$ and unstable if $|f'(x^*)| > 1$. For the logistic map, application of the test is simple. Since $f'(0) = r$, x_0^* is stable in the range $0 \leq r < 1$ and unstable if $1 < r \leq 4$. For the other fixed point x_1^*,

$$f'(x_1^*) = 2 - r.$$

---

[14]Of course, the other solution is relevant if we remove the restriction $0 \leq x \leq 1$.

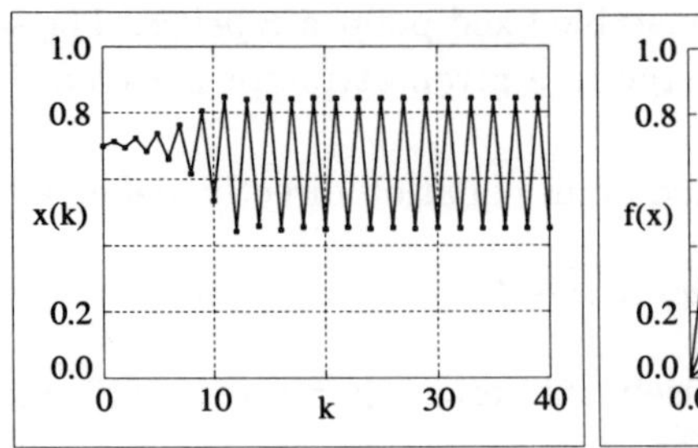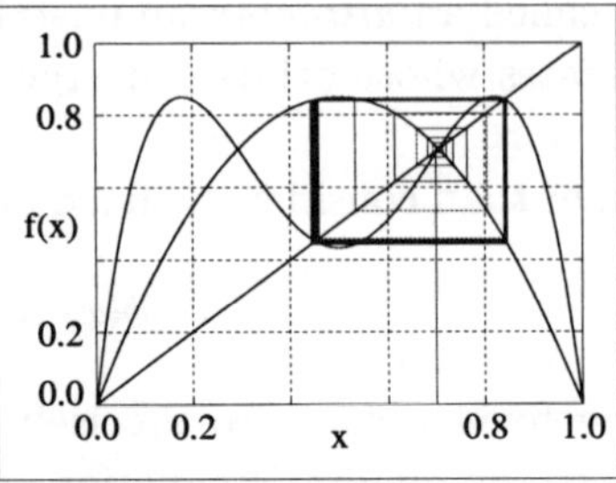

Fig. 8   Iteration of the logistic map (left), cobweb plot with second composition superposed (right), $r = 3.4$, $x_0 = 0.7$.

So x_1^* is a stable fixed point in the range $1 < r < 3$, but unstable in the range $3 < r \le 4$. Instability occurs because $f'(x_1^*)$ passes through -1.

2.2   *Periodic orbits*

What happens to the logistic map when $r > 3$? Both of its fixed points are unstable, and a new period 2 orbit comes into being. That is, there are a *pair* of values, $x_\pm^*$, with the property that

$$f(x_+^*) = x_-^*, \qquad f(x_-^*) = x_+^*. \tag{2.7}$$

Since the two points are distinct, neither is a fixed point of f. However, equation 2.7 shows that they are fixed points of the *second composition map*, defined as

$$f_2(x) = f(f(x)).$$

In fact, we can find period 2 orbits by finding fixed points of f_2 and checking that they are *not* fixed under f.

Figure 8 shows an example of iterations of the logistic map. Although x_0 is quite close to the (now unstable) fixed point $x_1^* \approx 0.7059$, the iterations move quickly away, and converge to a rectangular path on the cobweb plot — a period 2 orbit at the new fixed points of f_2, also shown in the figure.

Formula for the orbit

Now let's find the fixed points of f_2, for the logistic map. Substituting $f(x) = rx(1 - x)$ for x into $f(x)$ itself, gives[15]

$$f_2(x) = r^2 x(1 - x)(1 - rx + rx^2).$$

Extending the notation introduced in 2.4, the fixed points are the solutions of the quartic equation

$$\phi_2(x) = x - f_2(x) = 0. \tag{2.8}$$

Fixed points of f must be fixed points of f_2, so two factors of $\phi_2(x)$ are $(x - x_0^*)$ and $(x - x_1^*)$. Using this knowledge,

$$\phi_2(x) = x(1 - r + rx)(1 + r - rx - r^2 x + r^2 x^2). \tag{2.9}$$

The new fixed points are the roots of the quadratic factor. Calling them $x_\pm^*$,

$$x_\pm^* = \frac{1 + r \pm \sqrt{(r + 1)(r - 3)}}{2r}. \tag{2.10}$$

For real solutions the quantity under the square root sign cannot be negative, which gives the condition $r \geq 3$ for the existence of the new fixed points $x_\pm^*$.

Stability

It is easy to see that stability of period 2 orbits is equivalent to stability of the corresponding fixed points of f_2, to which we must apply the derivative test. First, recall the chain rule for a function of a function, $f(g(x))$:

$$\frac{df}{dx} = \frac{df}{dg}\frac{dg}{dx} = f'(g(x)) \cdot g'(x).$$

In the present case f and g are the same, which gives

$$f_2'(x) = f'(f(x)) \cdot f'(x). \tag{2.11}$$

---

[15] This simple substitution is possible only because the map is defined by a single formula over the entire interval.

Now we must find the value of f_2' at each of the two fixed points. The important fact is that $f(x_\pm^*) = x_\mp^*$, so that

$$f_2'(x_\pm^*) = f'(x_+^*) \cdot f'(x_-^*). \qquad (2.12)$$

This shows that the derivative used to test stability assumes the *same* value at either point, since the product may be taken in either order.

For the logistic map, a simple calculation gives

$$f_2'(x_\pm^*) = 4 + 2r - r^2. \qquad (2.13)$$

This is a decreasing function of r for $r > 1$.[16] At $r = 3$, $f_2'(x_\pm) = 1$, and when $r = 1 + \sqrt{6} \approx 3.4494897$, $f_2'(x_\pm) = -1$. The conclusion is that the period 2 orbit is stable in the range $3 < r < 1 + \sqrt{6} \approx 3.4494897$ and unstable beyond that point. We shall see in the next lecture that at the critical value, a further period doubling takes place.

Higher periods

A *periodic orbit* of period n is a sequence $\{x_0^*, \ldots, x_{n-1}^*\}$ for which

$$x_1^* = f(x_0^*), \qquad \ldots \qquad x_{n-1}^* = f(x_{n-2}^*), \qquad x_0^* = f(x_{n-1}^*),$$

with the property also that all the points are distinct from each other. It is *stable* if each point on it attracts nearby orbits; a stable periodic orbit will also be called a *periodic attractor*. Moreover, the set of initial values from which iterations converge to a periodic attractor are called its *basin of attraction*.

Following the precedent set for period 2, the *n-fold composition* f_n of the map f is defined inductively by $f_1(x) = f(x)$ and

$$f_n(x) = f_{n-1}(f(x)), \qquad n = 2, 3, 4, \ldots.$$

Obviously each member x_k^* of a period n orbit is a fixed point of the n-fold composition f_n. Because the points are distinct, they cannot be fixed points of any composition f_m for which $m < n$. This gives an alternative method of defining a periodic orbit; any fixed point of f_n, which is not a fixed point of f_m for all $m < n$, generates a period n orbit. It is possible to catalogue the number of periodic orbits simply by careful counting of

---

[16] Its value has no interest for $r < 3$ since $x_\pm^*$ are complex in this case.

n	fixed points	period $m < n$	new points	period n
1	2	–	2	2
2	4	2	2	1
3	2	2	–	–
4	8	4	4	1
5	12	2	10	2
6	16	4	12	2
7	30	3	28	4
8	32	8	24	3

Table 1.1   Table of periodic orbits, logistic map, $r = 3.8$.

fixed points of $f, f_2, \cdots$, as in table 1.1, which is for the logistic map with $r = 3.8$.

Derivatives of compositions

Stability of fixed points of f_n is equivalent to stability of the corresponding period n orbit. All we need is a general formula for the derivative of an n-fold composition f_n. It is quite simple: if $x_0, x_1, x_2, \ldots$ are successive iterates of the map, starting with x_0, then

$$f_n'(x_0) = f'(x_{n-1}) \cdots f'(x_1) \cdot f'(x_0). \tag{2.14}$$

This has already been demonstrated for $n = 2$; for arbitrary n, the chain rule gives

$$
\begin{aligned}
f_n'(x_0) &= f_{n-1}'(f(x_0)) \cdot f'(x_0) \\
&= f_{n-1}'(x_1) \cdot f'(x_0) \\
&= f'(x_{n-1}) \cdots f'(x_1) \cdot f'(x_0).
\end{aligned}
$$

2.3   *Lyapunov exponents*

In general, if x_0 is the initial point of an orbit, and if the system is started from a nearby point whose distance from x_0 is δ_0, then after one iteration the distance between the two is approximated by

$$\delta_1 \approx |f'(x_0)|\delta_0 = M_0 \cdot \delta_0.$$

M_0 is the magnification factor for this step. At the next step

$$\delta_2 \approx |f'(x_1)|\delta_1 = M_1 \cdot \delta_1 \approx M_1 M_0 \cdot \delta_0,$$

where $M_1 = |f'(x_1)|$ is the magnification factor for the second iteration. Lyapunov exponents extend this to arbitrary orbits. After n iterations, the total magnification factor is the product

$$M_{n-1} \cdots M_1 M_0.$$

Since this is an accumulation of magnification factors, it makes sense to consider an average. For this purpose, note:

(i) Because it is a product, we should use the geometric average:

$$M = (M_{n-1} \cdots M_1 M_0)^{1/n}.$$

(ii) Taking logarithms leads to the more common arithmetic average:

$$L = \ln M = \frac{1}{n} \left(\ln M_{n-1} + \cdots + \ln M_0 \right).$$

(iii) The stability test for a periodic orbit is

$$L < 0 \quad \text{(stable)}, \qquad L > 0 \quad \text{(unstable)}.$$

The infinite limit

The crucial step is to take a "long-term" average. For a given initial point x_0, the *Lyapunov exponent* $L(x_0)$ of a map f is given by

$$L(x_0) = \lim_{k \to \infty} \frac{1}{k} \sum_{i=0}^{k-1} \ln |f'(x_i)|,$$

provided the limit exists. In the case that any of the derivatives in the calculation is zero, it is conventional to write $L(x_0) = -\infty$.

For an arbitrary orbit, the existence of the limit is an open question in general. If the orbit is periodic, then the average involves only the n points

x_i^* on it, giving,

$$L(x_0^*) = \frac{1}{n} \sum_{i=0}^{n-1} \ln |f'(x_i^*)| = \frac{1}{n} \ln |f_n'(x_0^*)|.$$

Another simple example is the tent map, for which $L(x_0) = \ln 2t$ for almost all x_0. The exceptions are initial values which lead to the point $1/2$ being on the orbit.

Basins of attraction

Suppose that x_0 be within the basin of attraction of a periodic orbit. Then the Lyapunov exponent will be exactly the same as for the orbit itself. To see this, note that we may choose an integer m and split the defining formula as

$$L(x_0) = \lim_{k \to \infty} \frac{1}{k} \left(\sum_{i=0}^{m-1} \ln |f'(x_i)| + \sum_{i=m}^{k-1} \ln |f'(x_i)| \right). \tag{2.15}$$

Now the orbit is approaching the periodic attractor as k increases, so the differences between $\ln |f'(x_i)|$ and $\ln |f'(x_i^*)|$ are converging to zero. Choose a number m large enough so that the differences are less than some chosen ϵ for all $i > m$. Then, in the infinite limit, $L(x_0)$ is within ϵ of $L(x_0^*)$, because the first m terms are divided by k and don't contribute. The only effect of choosing a smaller ϵ is to require a larger m, but this is of no consequence when an infinite limit is taken.

As an example, for the logistic map the fixed point $x_1^* = (1 - 1/r)$ is stable in the parameter range $1 < r < 3$, and the interval $B = (0, 1)$ is its basin of attraction. It follows that

$$L(x_0) = \ln |2 - r|, \quad (0 < x_0 < 1), \quad (1 < r < 3). \tag{2.16}$$

Notice that $L \to 0$ at $r \to 1$ and $r \to 3$, $L \to -\infty$ as $r \to 2$. As for the fixed points themselves, $L(x_0^*) = \ln |f'(x_0^*)| = \ln r$ for $0 \leq r \leq 4$, $L(x_1^*) = \ln |f'(x_1^*)| = \ln |2 - r|$ for $1 < r \leq 4$.

Numerical estimation

Numerical computations are made by iterating the map to try to achieve convergence to any attracting set of states, after which the average value of $|\ln f'(x_i)|$ is computed over some sufficiently large sample. Numerical

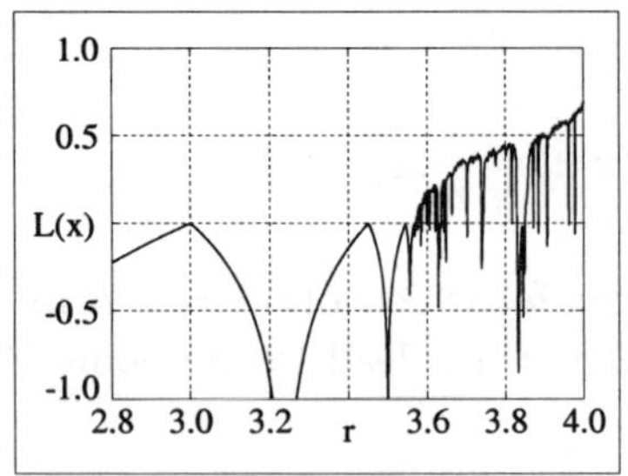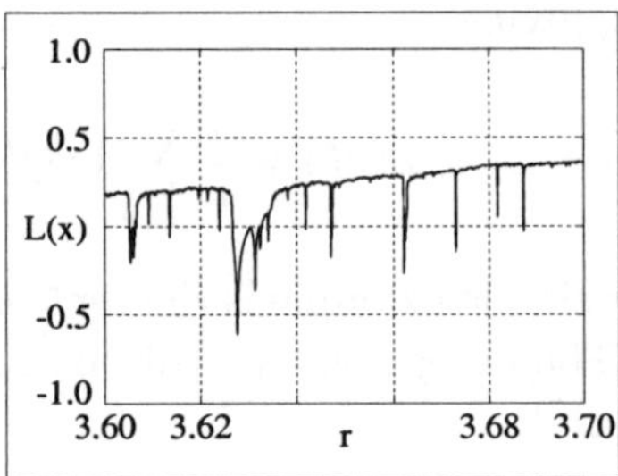

Fig. 9   Lyapunov exponents for the logistic map, $x_0 = 0.499999$. The right hand figure is a close up of part of the left hand one.

results must be treated with caution. The use of a finite sample means that the exponent is only an estimate.

Windows of periodic behaviour are readily seen from graphs displayed in Fig 9. However, there is also fractal structure, since there are an infinity of such windows. Hints of this are seen in the pictures. The more complex the structure, and the finer the scale, the more calculation is required to see it.

2.4   *Fourier analysis*

Fourier analysis applies both to continuous and discrete functions. If a signal is periodic then the spectrum contains the basic frequency plus integer multiples thereof — the *harmonics*. Otherwise, it contains a continuum of frequency components. This is a familiar idea in many areas of science. The discrete case is appropriate for discrete dynamical systems. It is easy to understand the meaning, and what it says about the *frequency domain*. Since actual calculations are always done using a computer package, that is all we shall require.

General representation

The general hypothesis is that, given an orbit x_k, $k = 0, \ldots, N - 1$, there is a trigonometric representation

$$x_k = \sum_m \left[a_{m/N} \cos\left(2\pi \cdot {}^m\!/_N \cdot k\right) + b_{m/N} \sin\left(2\pi \cdot {}^m\!/_N \cdot k\right) \right]. \qquad (2.17)$$

There are general principles which serve to make such a representation useable:

(i) An orbit of length N requires about $N/2$ cosine functions and $N/2$ sine functions. Therefore it is natural that m should be restricted by $m \leq N/2$.

(ii) The original data set x_k is represented as a linear combination of the coefficients $a_{m/N}$ and $b_{m/N}$. Provided there are exactly N of these coefficients, it should be possible to solve the linear equations to obtain formulae for them.

(iii) The observable frequency components should be in the range 0 to $1/2$, since it is not reasonable to attempt to extract periodic components whose variation is faster than that, from a discrete process. This is consistent with the restriction on m.

(iv) The trigonometric functions $\cos\left(2\pi \cdot {}^{m}\!/_{N} \cdot k\right)$ and $\cos\left(2\pi \cdot {}^{(N-m)}\!/_{N} \cdot k\right)$ take the same values for each integer k.[17] This is known as *aliasing*.[18] Similarly with the sine function, with the additional fact that it is zero if $m = 0$ or $m = N/2$.

(v) This last fact shows that the restriction $0 \leq m \leq N/2$ should be applied to the cosine terms, and $0 < m < N/2$ to the sine terms. Simple counting shows that this guarantees that there are exactly N terms in the sum.

From now on, I assume that N is an even number in using the representation 2.17, so that the maximum value of m is $N/2$, an integer. This is no real restriction, since we shall make N large in typical computations, and there are other reasons to make N divisible by small numbers, particularly powers of 2. The frequencies which may be detected in a finite-length sample (even N) are

$$ {}^{m}\!/_{N} = {}^{0}\!/_{N}, {}^{1}\!/_{N}, {}^{2}\!/_{N}, \ldots, {}^{1}\!/_{2}, $$

that is, a discrete set of frequencies from 0 to $1/2$ in steps of $1/N$.

---

[17]Of course, they do differ if k is a continuous variable. The vital point is that, for a discrete dynamical system, the concept of intermediate values of the state variables has been deliberately abandoned, even if the underlying system is continuous in time.

[18]It is the reason why CDs use a sampling rate of 44kHz in order to attain a (theoretical) maximum frequency content of 22kHz.

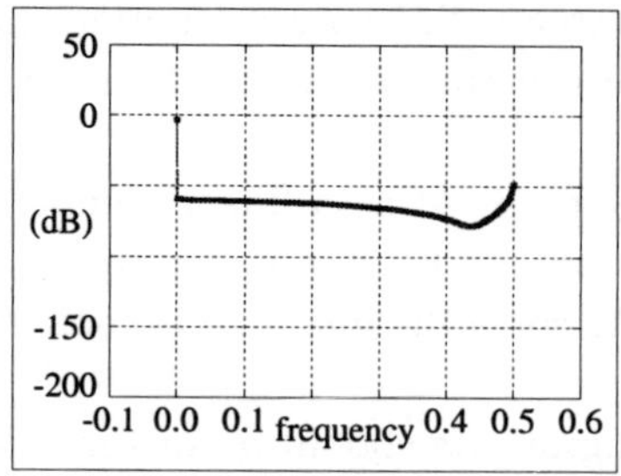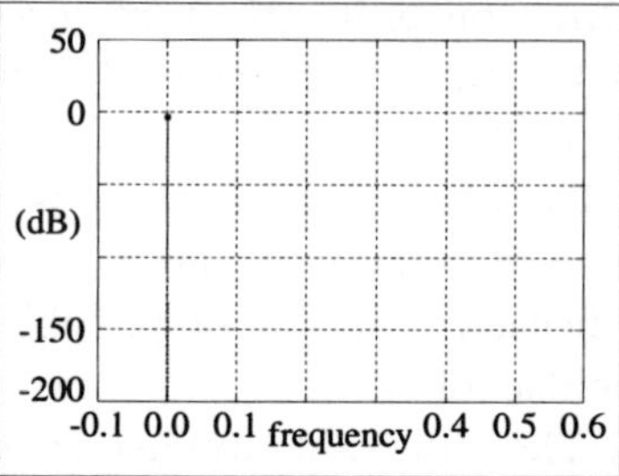

Fig. 10   Fourier amplitudes of 1000 iterations of the logistic map, $r = 2.99$. Initial transient (left), after convergence (right).

Amplitude and phase

Observe that terms with frequency m/N come in sine and cosine pairs. Using a trigonometric identity, such pairs may always be combined to single cosine terms, to give a simpler general form than equation 2.17, namely

$$x_k = \sum_m A_{m/N} \cos \left(2\pi \cdot m/N \cdot k - \phi_{m/N}\right).\qquad(2.18)$$

We shall be interested solely in the amplitudes when analysing orbits of dynamical systems. Moreover, because of the large range of values, a logarithmic amplitude scale is normally used — the *decibel*. For an amplitude A, the decibel (dB) value is calculated as $20 \log A$, where the logarithm is taken to the base 10. Since $\log 2 \approx 0.3$, each doubling or halving of an amplitude shows up as an increase or decrease of about 6dB on this scale.

Some examples

Figure 10 (left) shows the spectrum[19] generated by 1000 iterations of the logistic map with $r = 2.99$, starting from $x_0 = 0.1$. Because r is close to the critical value $r = 3$, convergence is quite slow. The zero-frequency component is dominant, because the system is converging to a fixed point. There is also a pronounced period 2 peak, and in fact all frequency components are significant, due to the fact that the system is transient. To get the long-term behaviour, I have discarded the initial 10^5 iterations (right). Clearly only the zero-frequency component survives this limit, indicating

---

[19]For visual clarity, the frequency scale is chosen from -0.1 to 0.6. When interpreting such figures remember that possible frequencies are restricted to the range 0.0 to 0.5.

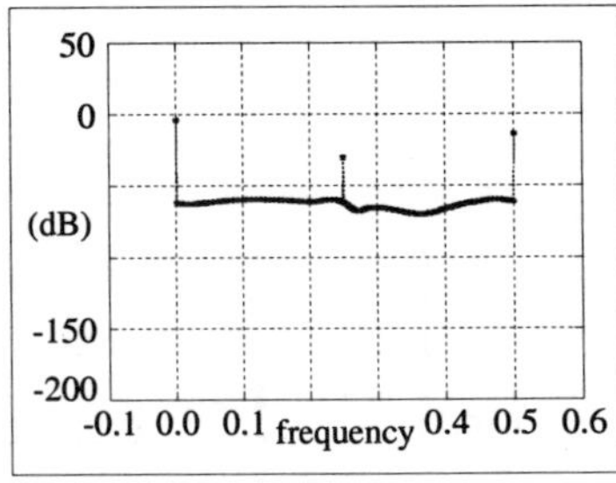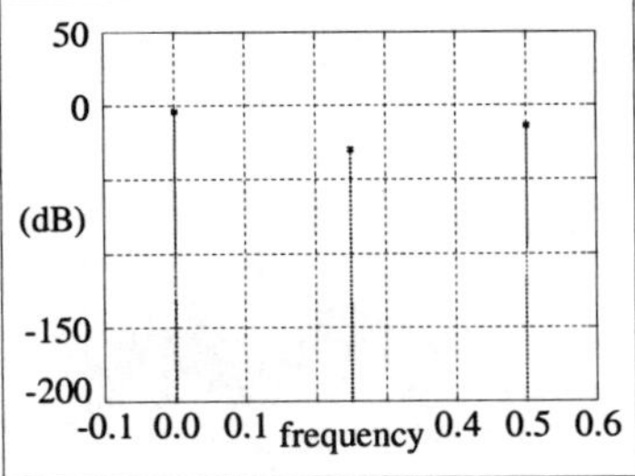

Fig. 11 Fourier amplitudes of 1000 iterations of the logistic map, $r = 3.5$. Initial transient (left), after convergence (right).

the existence of the stable fixed point quite clearly.

A similar pair of results is shown in Fig 11, for which $r = 3.5$. The long-term behaviour shows frequency components only at 0, 1/4 and 1/2, which indicates convergence to a stable period 4 orbit. It is often a cause of confusion that such a spectrum does not seem to include all of the expected frequencies: why does the frequency 3/4 not occur? This cannot be: 1/2 is the maximum observable frequency in a discrete system. In fact, 3/4 > 1/2 is aliased with 1/4.

2.5 *Chaos*

A *chaotic orbit* of a bounded system is one which is not periodic or eventually periodic, and which has positive Lyapunov exponent. A dynamical system is said to be *chaotic* when it is in a regime with chaotic orbits.[20] Note that the property of having a positive Lyapunov exponent implies that the orbit never falls within the basin of attraction of any periodic orbit. In case the orbit is periodic, or eventually periodic, it must be unstable.

Numerical evidence for chaos

Actually showing whether an orbit is periodic, or its Lyapunov exponent positive, may be a difficult problem for any particular system. Often numerical calculation is used as the main evidence. Fourier analysis can provide strong evidence of periodic or nonperiodic behaviour. Given this, together with a positive numerically estimated Lyapunov exponent, one can

---

[20]Some definitions of chaos include a requirement that there be a dense set of unstable periodic orbits.

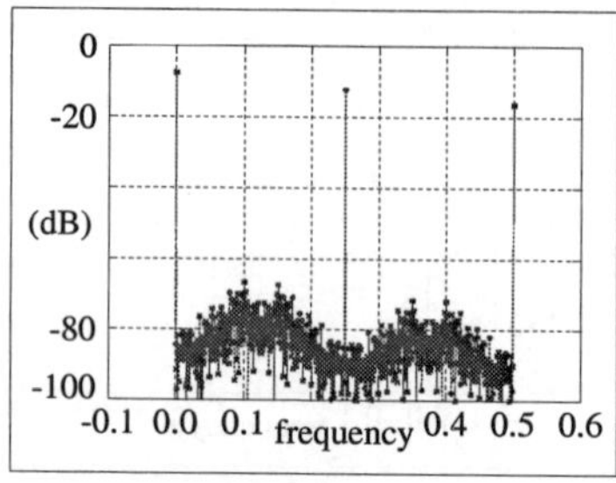
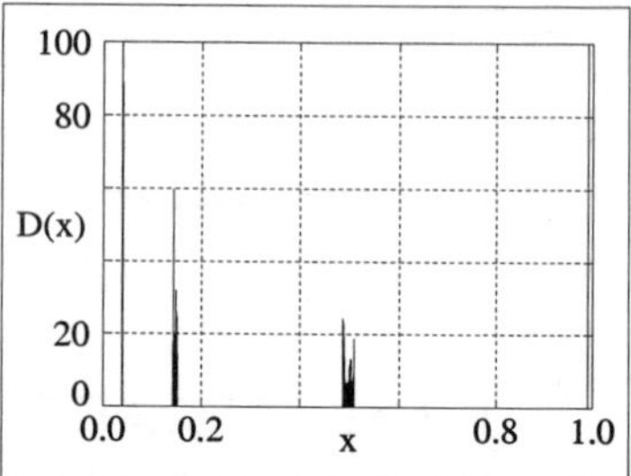

Fig. 12   Fourier amplitudes for the logistic map, $r = 3.9615$ (left). There is a strong period 4 component, but the orbit is not periodic. Relative density of same 10^5 iterations (right).

be rather certain that a particular system is exhibiting chaos, even though a theoretical demonstration is not feasible.

As an example which is not completely trivial, consider the Fourier spectrum shown in Fig 12 (left), for the logistic map with $r = 3.9615$. The spectrum shows clear peaks at frequencies of $1/4$ and $1/2$, suggesting the possibility of a period 4 orbit. But the sample size, and the time allowed for the system to reach a stable situation, are both large, despite which there is a significant amount of what looks like noise. It is easy to estimate the Lyapunov exponent of this orbit using scChaos for Java. Using the same initial value and sample size, the result is $L \approx 0.1261$, so the system is chaotic. I leave it to you to investigate precisely what is going on here; to assist, I have shown in the right hand figure the orbital density of 10^5 iterations. They indicate that the orbit visits four *intervals*, rather than four points.

Orbital density

A way to sample a single orbit over a long time is to calculate the statistics of where the iterations fall. That is, one divides the x-axis into N equal small small intervals, or *bins*, and records what fraction of the iterates of this single orbit falls into each bin. Next, we follow a single orbit of length m, and count the number of times m_k the orbit visits each I_k. Finally, define the fraction of the times the orbit visits each subinterval,

$$\mu_k = m_k/m.$$

If the process were random, we would interpret the function μ_k as an orbital probability distribution or orbital density, which is why it is normalised so that $\sum_{k=1}^{N} \mu_k = 1$. Some examples have already been shown in figures 2 and 12. The main defining feature of the *probability measure*, which is approximated by such computations, is that it is *invariant* under iteration of the map. However, it turns out that in most cases it has a dense set of singularities, making it a particularly difficult function to investigate.

The above assumes that there are *ergodic orbits*, orbits with the property that every point in the set of accessible states is approached arbitrarily closely by some iteration x_k. Unstable periodic orbits are not ergodic; fortunately straightforward numerical iteration cannot reproduce such an orbit. But we expect chaotic orbits to be ergodic.

Ergodic hypothesis

The use of probability densities relates two distinct types of statistics:

(i) Assign a probability density to each subinterval, based on the dynamics, and then compute average values with respect to this density. This kind of average is known in statistical mechanics as an *ensemble average*.

(ii) Use a single representative orbit over a long time period — the *time average*. This allows for simple numerical experimentation. It assumes that the orbit is *ergodic*, meaning that every point in the set of accessible states is approached arbitrarily closely by some iteration x_k.

It is a much used assumption in equilibrium statistical mechanics that the two methods give equivalent results, an assumption known as the *ergodic hypothesis*. It is an intestesting subject for advanced research.

3 Bifurcations

In this lecture I discuss some important aspects of bifurcations in one-dimensional systems. In its most general form, bifurcation theory is concerned with equilibrium solutions of nonlinear systems. In the present context, the equilibrium solutions of interest are fixed points and periodic orbits, stable or unstable.

In the previous lecture we noted that period doubling of the logistic map first occurs $r = 3$. This is an example of a *bifurcation*, which is a change in the structure of the set of periodic orbits of the map at some critical value of a parameter. The restriction to periodic orbits may appear to be rather limiting; but it turns out that they play a vital rôle in understanding chaos.

3.1   *Bifurcation diagrams*

Period doubling is not the first bifurcation to occur for the logistic map. Earlier, two fixed points were found, one of which (x_1^*) is negative for $r < 1$. If we regard f as a map of the real numbers, rather than just the interval $[0, 1]$, then x_1^* is unstable for $0 \leq r < 1$ and stable for $1 < r < 3$, the exact reverse of x_0^* in the vicinity of $r = 1$.

This behaviour is best appreciated using a diagram which shows a plot of the orbits, as functions of r, indicating also their stability. Figure 13 shows two such diagrams. The left-hand one is the case under consideration. Stability is indicated by showing stable orbits as dark lines and unstable ones as light lines.[21] We see that, as the parameter is increased, the two fixed points collide, whereupon they exchange their stability. This type of bifurcation is known as *transcritical*. The period doubling is shown in the right-hand figure. The fixed point lies on a smooth curve, but changes its stability. The period 2 orbit comes into existence discontinuously as this happens.

The critical condition

Bifurcation diagrams are produced by finding solutions of the equation $x = f_n(x; r)$, as functions of a parameter, which I denote here by r. Proceeding as in equation 2.8, this is better written as

$$\phi_n(x; r) = x - f_n(x; r) = 0. \tag{3.1}$$

Generally, the solution curves are smooth, since

$$\frac{\partial \phi_n}{\partial r} + \frac{\partial \phi_n}{\partial x} \frac{dx}{dr} = 0,$$

---

[21]Blue and red are the default in scChaos for Java.

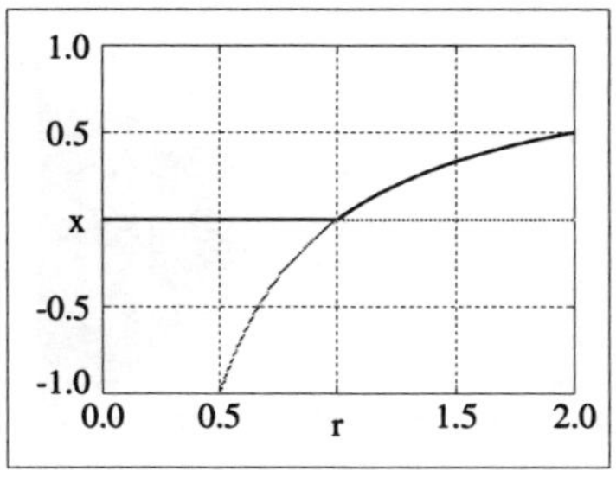 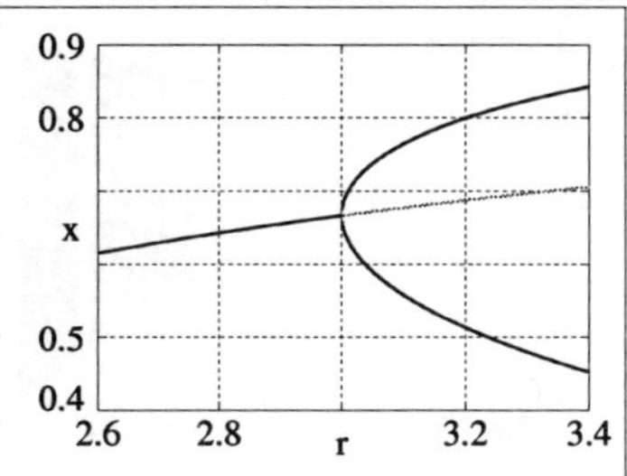

Fig. 13 Two bifurcations of the logistic map. Transcritical at $r = 1$ (left), period doubling at $r = 3$ (right).

which has a smooth solution for dx/dr provided that $\phi'_n(x) \neq 0$.[22] This is equivalent to

$$f'_n(x^*) \neq +1. \tag{3.2}$$

What about the other condition for loss of stability, namely $f'_n(x^*) = -1$, for which $\phi'_n(x^*) = 2$? Here it is the fixed points of f_{2n} which change their geometric structure, since

$$f'_{2n}(x^*) = f'_n(x^*)^2 = +1. \tag{3.3}$$

This is the origin of period doubling.

3.2 *Final state diagrams*

The bifurcation diagrams of Fig 13 were produced by finding all solutions of equations 3.1, as functions of a parameter, by a process of interval searching. The computational limit imposed by this indicates that one must complement such calculations by simple observation, using numerical iteration.

Forward limit sets

Ideally, we would like to observe the limiting set of points to which the system is attracted. More formally, the *forward limit set* is the set of points which the orbit approaches arbitrarily closely infinitely often. This definition is quite subtle:

[22]I am assuming that the partial derivatives are themselves smooth.

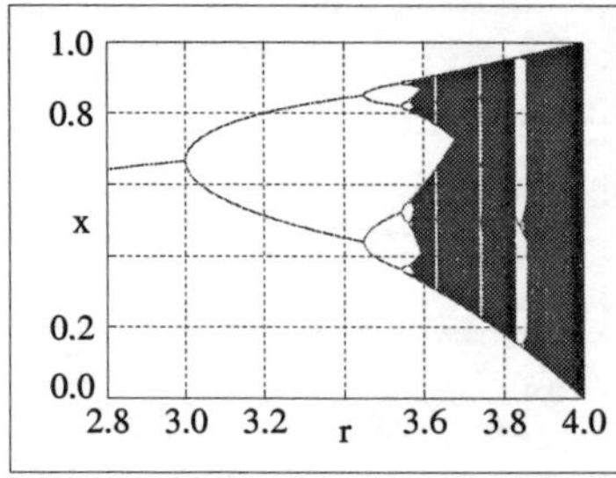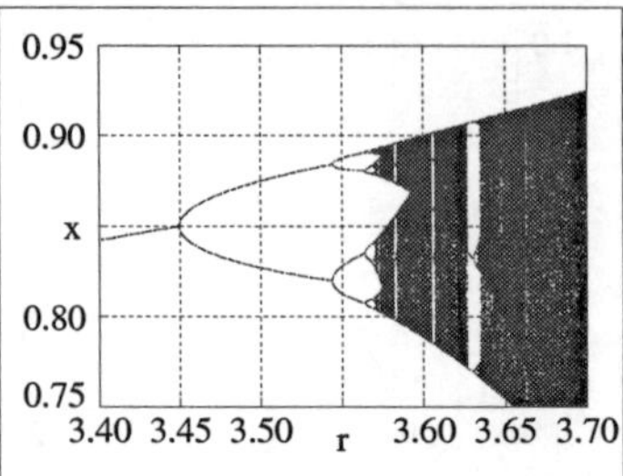

Fig. 14   Final state diagrams for the logistic map.

(i) To return arbitrarily closely means that for any distance δ, no matter how small, and for any point ξ in the forward limit set, there are points on the orbit which satisfy $|x_k - \xi| < \delta$.

(ii) Infinitely often means that there are always recurrences of this event: for any integer N, no matter how large, there remain values of $k > N$ for which this inequality is again satisfied.

Numerical approximation

In practice one first iterates the map a few hundred or thousand times, then takes a sample of a similar number of iterations. The sample itself is put into *bins,* which are intervals whose size is the x (vertical) resolution represented by one pixel on the graphical device used to present the data. Obviously the resolution comes to the rescue when finding the limit sets of reasonably stable orbits, since it does not require too many iterations to converge within one pixel. Near to points of bifurcation, however, the rate of convergence is too slow for this. In fact, simple iteration can never replace more exact methods at such points, as you can easily discover by experimenting with scChaos for Java.

The parameter range is traversed in increments whose size is the (horizontal) pixel resolution, and the process is repeated for all values of the parameter which are represented at this resolution. This gives a global picture of the final limit sets of the system. I follow the book by Peitgen, Jürgens and Saupe [25] in referring to these as *final state diagrams.*

As an example, two final state diagrams for the logistic map, shown in figure 14, exhibit quite clearly the birth and death of the period 2 orbit already investigated. It also shows that the period doubling repeats itself,

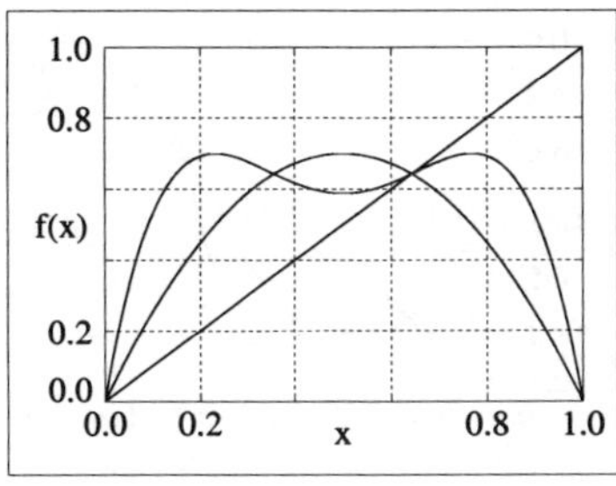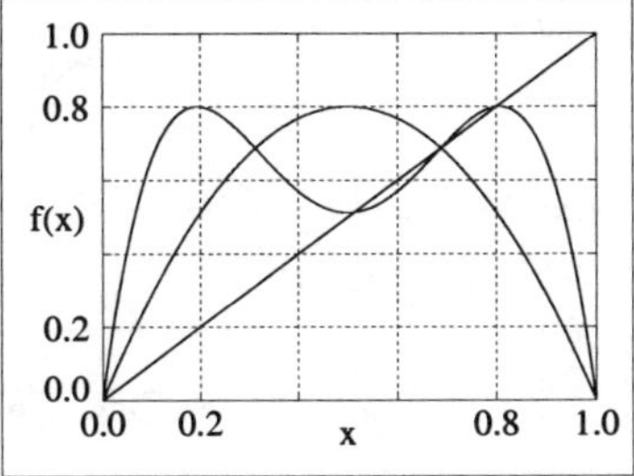

Fig. 15   Fixed points of the logistic map f and second composition f_2, parameter value $r = 2.8$ (left), $r = 3.2$ (right).

on an ever finer scale, up to the resolution of the picture. What it does not show is that periodic orbits do not really die, they simply become unstable. Eventually there are an infinite number of such unstable orbits in the background, creating chaos.

3.3   *Period doubling mechanism*

Looking at bifurcation and final state diagrams in some detail, it is apparent that each stable period doubled orbit gives birth to a new stable period doubled orbit as it loses its own stability, and that there is an infinite cascade of orbits with periods $1 \to 2 \to 2^2 \to 2^3 \to \cdots$, ending in chaos. It is known as the *period doubling route to chaos*, and I want to throw some light on the mechanism, despite the complexity of the functions f_{2^n}.

Properties of graphs

Here I shall simply examine the mechanisms from a graphical point of view. This will suggest what calculus is required to complete the analysis. Figure 15 shows superposed graphs of f and f_2 for values of r a little below, and a little above, the critical value s = 3. It is clear what happens: as r increases through s = 3, the derivative $f_2'(x^*)$ increases through the value $+1$. So long as $f_2'(x^*) < 1$, the graph only intersects the line $y = x$ at one point in the vicinity of x^*. But when $f_2'(x^*) > 1$, it intersects at three nearby points. By simple graphical reasoning we expect that $f_2'(x_\pm^*) < 1$ at the two new fixed points, at least while r is close to s. This produces a stable period 2 orbit, since the new fixed points of f_2 are not fixed points of f.

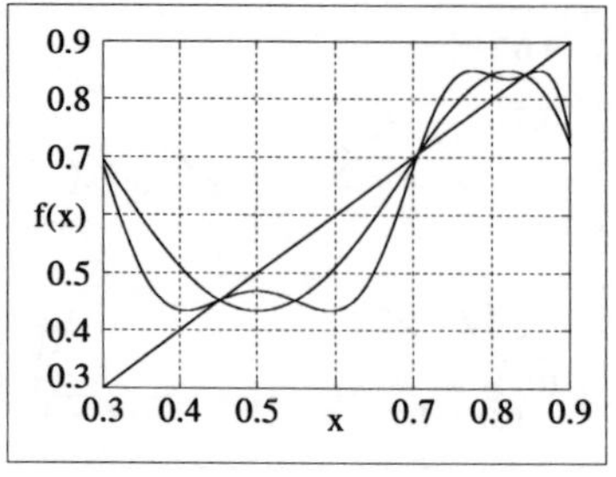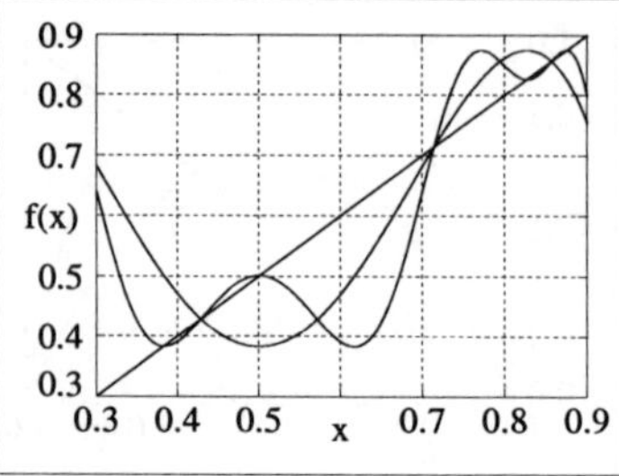

Fig. 16   Fixed points of the compositions f_2 and f_4 of the logistic map; parameter value $r = 3.4$ (left), $r = 3.5$ (right).

Recall that the period 2 orbit loses stability at $r = 1 + \sqrt{6} \approx 3.45$. It period doubles to a new stable period 4 orbit. The mechanism is seen in the graphs of f_2 and f_4 for $r = 3.4$ and $r = 3.5$ (Fig. 16). Evidently it is a duplication of what happened for f and f_2 at $r = 3$ except that now it is $f_2'(x_{\pm}^*)$ which passes through the critical value -1, while $f_4'(x_{\pm}^*)$ passes through $+1$, since

$$f_4'(x_{\pm}^*) = f_2'(x_{\pm}^*)^2.$$

Furthermore, period doubling seems to be moving much faster this time, considered as a function of r.

Properties of derivatives

The fact that $\phi_2(x)$ is approximated by a cubic near to x^* is equivalent to the coincidence that, when $r = s$,

$$\phi_2(x^*) = 0, \qquad \phi_2'(x^*) = 0, \qquad \phi_2''(x^*) = 0. \tag{3.4}$$

The first derivative condition follows directly from the fact that $f_2'(x^*) = +1$. The second derivative condition, however, seems a little more mysterious. Let's calculate this derivative, both as a general formula and evaluated at the fixed point itself. First, the general formula

$$\begin{aligned} f_2''(x) &= \frac{d}{dx}\left[f'(f(x)) \cdot f'(x)\right] \\ &= f''(f(x)) \cdot f'(x)^2 + f'(f(x)) \cdot f''(x), \end{aligned} \tag{3.5}$$

which gives

$$\phi_2''(x^*) = -f_2''(x^*) = -f''(x^*)[f'(x^*)^2 + f'(x^*)] = 0.$$

So the coincidence 3.4 depends only on the fact that loss of stability arises from the passage of $f'(x^*)$ through the value -1.

Period doubling cascades

I have been careful in the foregoing not to assume too much about the function f beyond a reasonable amount of differentiability. In fact we may replace f by any composition f_n, and f_2 by the corresponding doubled composition f_{2n}, at each step of the argument.

Let's apply this first to the cascade which emanates from the fixed point x_1^*. It becomes unstable and period doubles at some critical parameter value. At the birth of this new orbit, $f_2'(x_\pm^*) = 1$; immediately afterward it has decreased below this value. It is reasonable to expect that further increase of the parameter will result in continued decrease in $f_2'(x_\pm^*)$, leading to loss of stability when it passes through the critical value -1.

At this point it is appropriate to pick ourselves up by the bootstraps. Setting f_2 into the place previously occupied by f, its second composition f_4 will take the place previously occupied by f_2. Since $f_2'(x_\pm^*) = -1$ for either point on the period 2 orbit, $f_4'(x_\pm^*) = +1$ and $f_4''(x_\pm^*) = 0$. Therefore,[23] each of the two fixed points of f_2 undergoes a period doubling bifurcation, to produce four new stable fixed points $(x_0^*, \ldots, x_3^*)$ of f_4. They are not fixed points of either f or f_2, hence they form a stable period 4 orbit of f. The bootstrap process continues. At the critical point where $f_4'(x_i^*)$ reaches the critical value -1, f_8 acquires eight new stable fixed points (as four pairs), which together form a stable period 8 orbit. And so on, in an infinite cascade.

And — there is more. Suppose that we commence with a stable period 3 orbit, that is, three connected stable fixed points of f_3 which are not fixed points of f. Their stability is determined by $f_3'(x_i^*)$. As the parameter value is increased, this may also pass through the critical value -1; if it does so, the composition map f_3 period doubles to give a stable period 6 orbit. Once started, this cascade will generally continue to chaos. In fact,

---

[23] Actually, we need a condition on the third derivative, which can be checked using the theory of the "Schwarzian derivative".

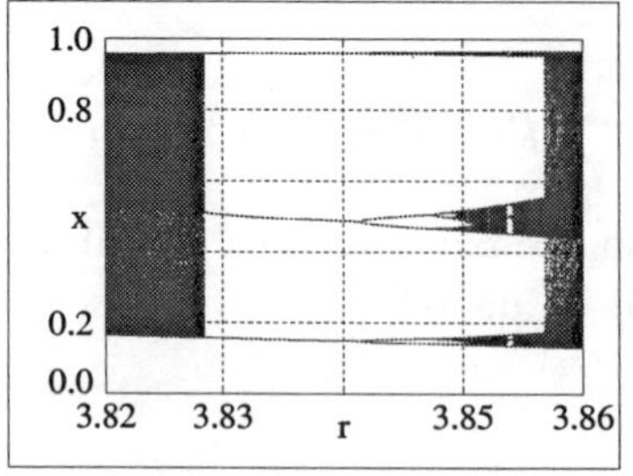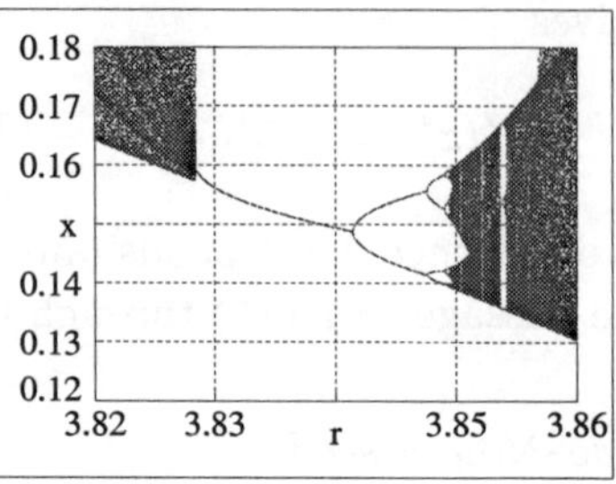

Fig. 17   Period doubling in the period 3 window of the logistic map (left) and detail of the bottom branch (right).

it is the most common mechanism for the destruction of stable periodic orbits of any period. The period 3 cascade is shown in Fig. 17.

3.4   *Feigenbaum's universal constants*

Define r_n as the value of r at the point where the nth period doubling occurs. Then the sequence of values converges geometrically to a limiting value r_∞ according to the *scaling law*,[24]

$$r_n - r_\infty \sim K\delta^{-n}, \qquad n \to \infty. \tag{3.6}$$

This remarkable fact was discovered by Mitchell Feigenbaum [10]. It provides the first opportunity, in these brief introductory lectures, to observe a remarkable fact about nonlinearity and chaos: there is a surprising degree of *universality* in the modes of behaviour of apparently different systems.[25]

Given accurate numerical values, it is not difficult to estimate δ and r_∞. Taking the difference of two adjacent equations,

$$(r_n - r_\infty) - (r_{n+1} - r_\infty) = r_n - r_{n+1} \sim K\delta^{-n}(1 - \delta^{-1}).$$

The ratio of two such relations gives

$$\delta \sim \frac{r_n - r_{n+1}}{r_{n+1} - r_{n+2}}.$$

---

[24]The exact meaning of the symbol $\sim$ is given in equation 3.7.

[25]In the interest of brevity, I only consider the logistic map in this lecture. However, it is easy to check that other maps provided in scChaos for Java exhibit the same bifurcations and scaling, although there are some additional features which are not mentioned here.

n	bifurcation	r_n	δ	r_∞
1	$1 \to 2$	3.0000000000		
2	$2 \to 4$	3.4494897428		
3	$4 \to 8$	3.5440903596	4.751446	3.56930747
4	$8 \to 16$	3.5644072661	4.656251	3.56996403
5	$16 \to 32$	3.5687594195	4.668242	3.56994586
6	$32 \to 64$	3.5696916098	4.668739	3.56994570
7	$64 \to 128$	3.5698912594	4.669132	3.56994567
8	$128 \to 256$	3.5699340184	4.669183	3.56994567

Table 1.2 Estimates of the Feigenbaum constant δ, and r_∞, for the logistic map. Data is from the main period doubling cascade.

Further simple algebra gives an equation for r_∞,

$$r_\infty \sim \frac{r_n r_{n+2} - r_{n+1}^2}{r_n - 2r_{n+1} + r_{n+2}}.$$

Notice that these formulae involve differences which decrease rapidly as n increases, with consequent loss of numerical accuracy.

For the main period doubling sequence of the logistic map, the first few r_n values are given in table 1.2. The second column indicates the actual bifurcation, the third column the corresponding value of r_n. They were obtained using scChaos for Java to locate, to ten decimal places, the values of r at which the derivative $f_n'(x_i^*)$ attains the critical value -1. The obvious convergence of these numbers is strong evidence to support the scaling hypothesis. As always in a good theory, that does not settle the question: rather it presents a challenge for understanding.

Scaling relations

Equation 3.6 is an example of a relation which gives important and general information about how a dynamical system behaves in an infinite or infinitesimal limit. Its precise meaning is that

$$\lim_{n \to \infty} \frac{r_n - r_\infty}{\delta^{-n}} = K, \tag{3.7}$$

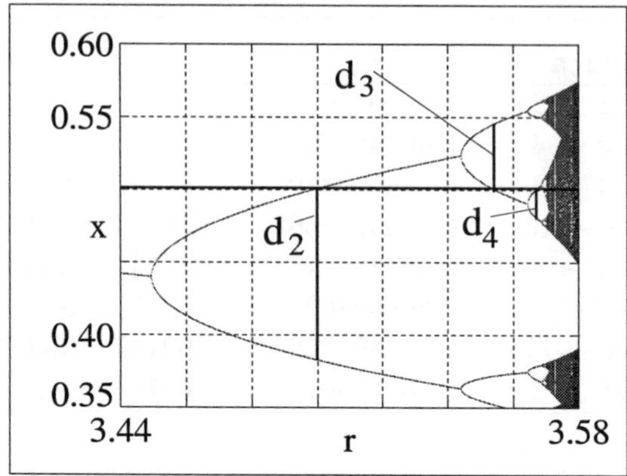

Fig. 18   Final state diagram showing the definition of d_n as the difference between $x_{\max}$ and the $f_{2^{n-1}}(x_{\max})$.

which shows the different rôles of δ and K. Relations of this kind are an important feature of the theory, giving laws which separate *universal constants* (here δ), which depend only on the general character of the system, from *amplitudes* (here K), which depend on the details. Specifically, δ is the same for all period doubling cascades of a smooth unimodal map,[26] whereas K depends on the particular cascade, as does r_∞. Scaling relations are used to extend the concept of dimension to strange attractors and other fractal objects, and in many other places.

As a further example from period doubling, one can examine how distances between points on period 2^n orbits scale with increasing n. A convenient way to choose a set of quantities d_n for this purpose will be described shortly, and is shown in Fig. 18; it will be seen that they also scale geometrically, this time to zero:

$$|d_n| \sim L\alpha^{-n}, \qquad n \to \infty.$$

Theory and numerical experiment both give the value

$$\alpha \approx 2.502908\ldots \tag{3.8}$$

The constants K, L depend on the actual map, but α and δ are *universal constants*, known as Feigenbaum's constants. They are the same for all period doubling cascades of a smooth unimodal map.[26] This remarkable fact

[26]More precisely, α and δ are the same for all smooth maps with a quadratic maximum,

n	$\bar{r}_n$	δ	r_∞	$|d_n|$	α
0	2.0000000000				
1	3.2360679775			0.30901699	
2	3.4985616993	4.708943	3.56933489	0.11640177	2.65475
3	3.5546408628	4.680771	3.56987657	0.04597521	2.53184
4	3.5666673799	4.662960	3.56995066	0.01832618	2.50872
5	3.5692435316	4.668404	3.56994579	0.00731843	2.50411
6	3.5697952937	4.668954	3.56994568	0.00292368	2.50316
7	3.5699134654	4.669156	3.56994567	0.00116809	2.50296

Table 1.3   Estimates of Feigenbaum constants α and δ, and r_∞, for the logistic map. Data is from the superstable orbits of the main period doubling cascade.

was discovered by Mitchell Feigenbaum [10], and originated as a conjecture based on numerical experiments performed with a hand calculator.[27]

Superstable orbits

Stability of a periodic orbit is determined by the derivative test; the smaller the magnitude $|f'_n(x_i^*)|$, the faster the rate of convergence. A period n orbit is said to be *superstable* if the derivative f'_n is zero.

Recall the life of a period doubled orbit. It is born with derivative $f'_n(x^*) = +1$, and it dies (becomes unstable) when the derivative passes through the critical value -1. In between there must be a point at which it is superstable. Let's denote the values of r at which the period 2^n orbit is superstable by $\bar{r}_n$. The Lyapunov exponent has the value $-\infty$ wherever a derivative used in its computation becomes zero, so one can spot the superstable orbits by looking for sharp dips in numerical computations of the exponent, although this is not a method for accurate determination.

Numerical values for the first few $\bar{r}_n$ are shown in table 1.3. Clearly the parameter values $\bar{r}_n$ should scale with the same universal Feigenbaum constant δ. The data up to $n = 7$ is shown, together with the estimates of

the usual case. In the general case they depend only on the order of the maximum.

[27]The story is recounted in the Glieck's popular book [11], p177ff.

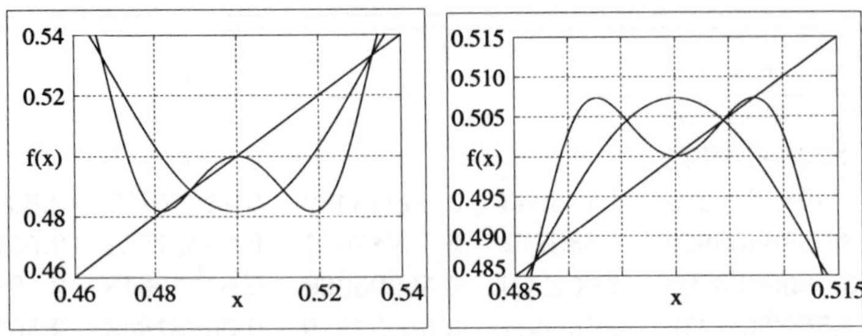

Fig. 19   Logistic map, showing self-similarity of compositions near $x = x_{\max}$. f_8, f_{16} for $r = \bar{r}_4$ (left), f_{16}, f_{32} for $r = \bar{r}_5$ (right). The two appear to be related by the scaling factor $\alpha \approx 2.5$.

δ, r_∞ and α.

Scaling and self-similarity

Using the formula for the derivative in the condition for superstability,

$$f_n'(x_i) = \prod_{j=0}^{n-1} f'(x_{i+j}) = 0,$$

we see that an orbit is superstable if and only if $f'(x_k^*) = 0$ for some point x_k^* on it. That is, the maximum point $x_{\max}$ of f must belong to the orbit.

The period 2^n orbit comes into existence at a critical value r_n at which the period 2^{n-1} orbit period doubles; in fact each of the 2^{n-1} fixed points of $f_{2^{n-1}}$ which are on that orbit simultaneously period double, to produce 2^n new fixed points of f_{2^n}. These new fixed points occur in pairs, each with a common parent, now unstable. As r is increased, the new orbit becomes superstable at a critical value $\bar{r}_n$, and we have just observed that one of the corresponding fixed points of f_{2^n} must be at $x = x_{\max}$. Suppose we fix our attention on the pair to which this belongs; when $r = \bar{r}_n$ they must be at $x = x_{\max}$ and $x = f_{2^{n-1}}(x_{\max})$. Denote by d_n the difference between them, that is,

$$f_{2^{n-1}}(x_{\max}) = x_{\max} + d_n, \qquad f_{2^{n-1}}(x_{\max} + d_n) = x_{\max}. \qquad (3.9)$$

Figure 18 shows the meaning of d_2, d_3 and d_4, on a portion of a bifurcation diagram for the logistic map; the thick horizontal line is at $x = x_{\max}$. Again, Fig. 19 shows how the values of d_n may be obtained from the graph of f_{2^n} with $r = \bar{r}_n$.

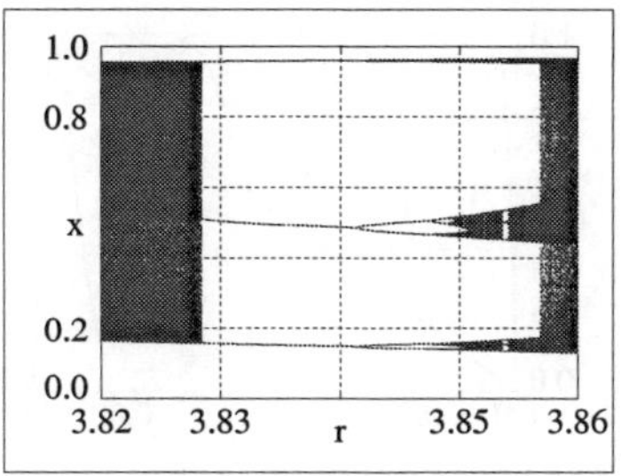 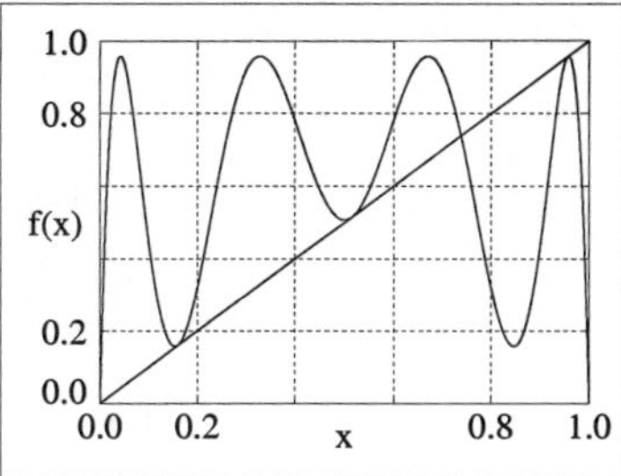

Fig. 20   Tangent bifurcation of the logistic map. Final state diagram (left). Graph of f_3, $r = 3.8284$ (right).

Feigenbaum scaling comes from *self similarity* of the functions f_{2^n}. The graphs become extremely complicated with increasing n, when viewed in the large. However, one can observe self-similarity in the functions f_{2^n} near to $x = x_{\max}$, by expanding the horizontal and vertical scale by the factor α for each increment of n. This is illustrated in Fig. 19. The point to observe is that the similarity involves scaling the graph and turning it upside down, which is why the signs alternate in equation 3.9.

3.5   *Tangent bifurcations*

Periodic windows are a common and obvious feature of final state diagrams, for example the period 3 window of the logistic map shown in figure 20 (left). The chaotic behaviour suddenly switches, at $r = 1 + \sqrt{8} \approx 3.8284$, to a stable period 3 orbit, which takes the period doubling route back to chaos.

The mechanism which gives birth to this period 3 orbit is seen in Figs. 20 and 21. On the right of the former is a graph of f_3, with $r = 3.8284$. There are three points where the graph seems to be touching the line $y = x$. The other two graphs (Figs 21) show the situation for $r = 3.8$ and $r = 3.85$. It's quite clear what is happening. For $r = 3.8$ the function has two minima and one maximum which do not quite meet the line, and there are only two fixed points of f_3, both unstable. With a slight increase in r, the three peaks push through the line, creating a further six fixed points. Initially three are stable, three unstable, so the collision gives birth to a pair of period 3 orbits, one stable and one unstable. At the critical value s for which the collision takes place, the line $y = x$ is tangent to the graph. For this reason, this kind of bifurcation is known as a *tangent bifurcation*.

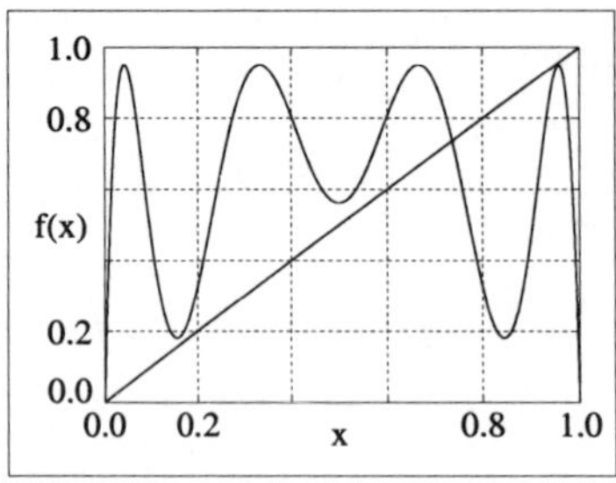
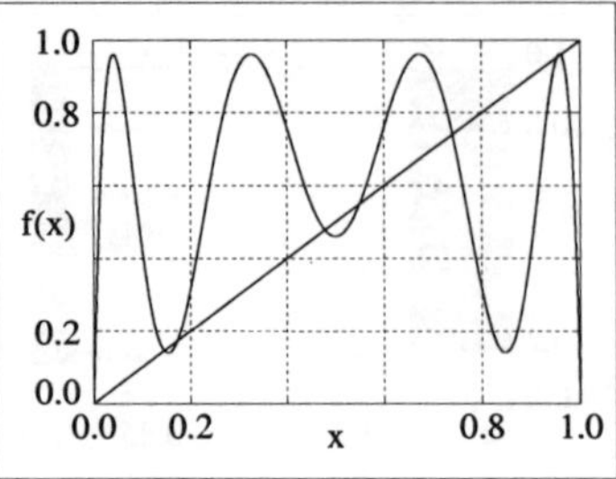

Fig. 21   Tangent bifurcation of the logistic map. Graph of f_3, $r = 3.8$ (left), $r = 3.85$ (right).

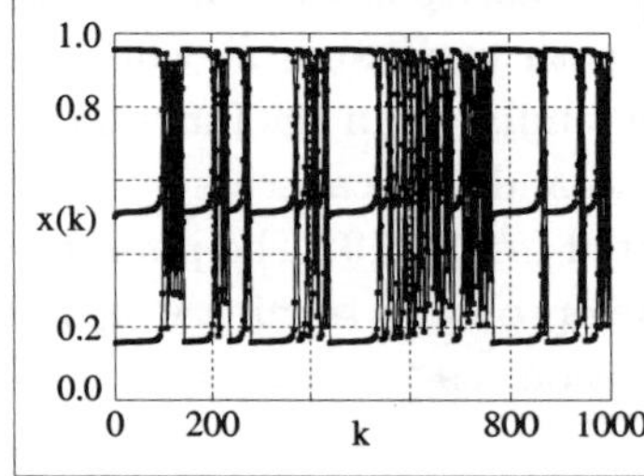
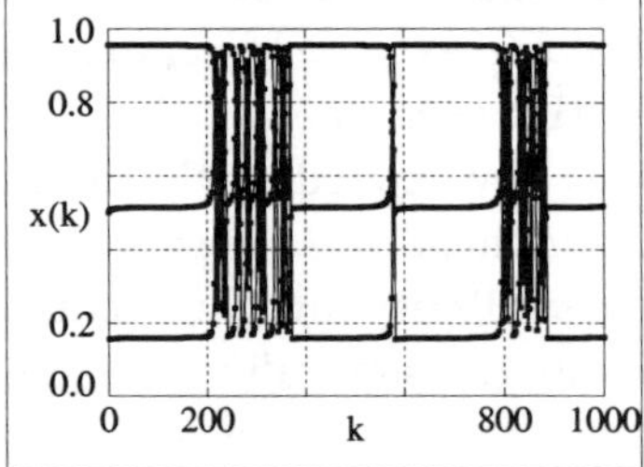

Fig. 22   Intermittent behaviour of logistic map just before the tangent bifurcation. Parameter values, $r = 3.8283$ (left), $r = 3.8284$ (right), Every third point joined to show laminar regions.

Figure 25 shows a detail of the period 3 tangent bifurcation of the logistic map. Periodic orbits up to period 3 have been added to the final state diagram; as in previous figures the unstable branch is a light line. It is clear that the stable orbit period doubles back to chaos, it also clear that the unstable orbit plays an interesting rôle, which I shall discuss shortly.

Intermittent behaviour

There is an interesting phenomenon associated with tangent bifurcations, which manifests itself as *intermittent* behaviour of the system. It is quite easily observed for the logistic map just before the period 3 tangent bifurcation.

Figure 22 shows two pictures, in which the iterations are joined so as to show the action of the third composition map. That is, $x_0, x_3, x_6, \ldots$ are joined by straight lines, similarly for $x_1, x_4, x_7, \ldots$ and $x_2, x_5, x_8, \ldots$.

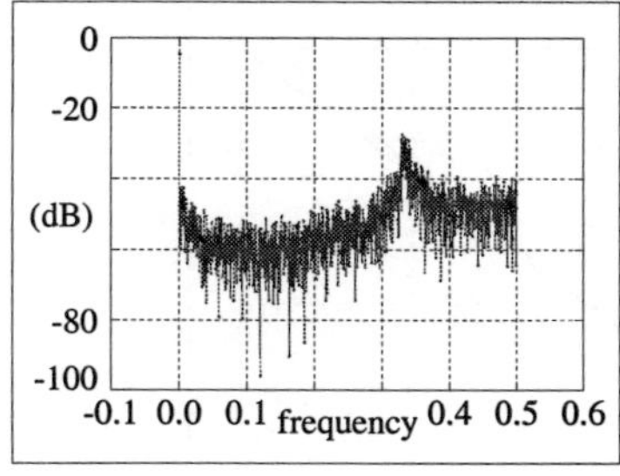
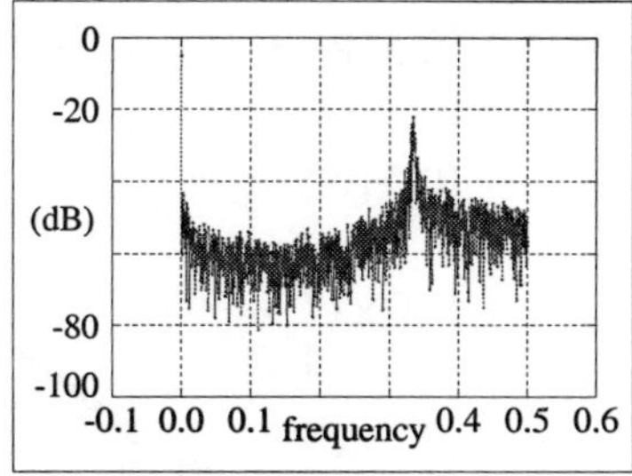

Fig. 23   Fourier spectra for the iterations of the previous figure. A peak is developing at frequency $1/3$. Initial value $x_0 = 0.5$, sample size 3×10^3 points, initial 10^4 points discarded.

Looking at 1000 iterations you can see that there are long stretches, called *laminar regions*, when it appears that the orbit is almost stable period 3, interspersed by *chaotic bursts*. Observe also that the length of the laminar regions is bigger for $r = 3.8284$, which is closer to $1 + \sqrt{8}$, than for $r = 3.8283$. The influence of these laminar regions, which prefigure the period 3 orbit, is seen quite clearly in the accompanying Fourier spectra shown in figure 23. A sharp peak is developing at frequency $1/3$.

A scaling relation

We can see what is going on in the laminar region by considering the cobweb plot. Evidently we must examine the behaviour of the third composition map f_3 just below the critical value of r. It is clear that the progress made in a single iteration is $\Delta x_k = (x_{k+1} - x_k) = \phi_3(x_k)$. Since the number of iterations required to get through the channel is large, one can think of the rate of progress as if the iteration count k is a continuous quantity, just as with large numbers of insects in a population model. Approximating ϕ_3 near to the extrema by a quadratic, whose extreme value is proportional to $\Delta r = (s - r)$, this gives

$$\frac{dx}{dk} \approx A\Delta r + K(x - a)^2. \tag{3.10}$$

Taking the reciprocal,

$$\frac{dk}{dx} \approx \frac{1}{A\Delta r + K(x - a)^2},$$

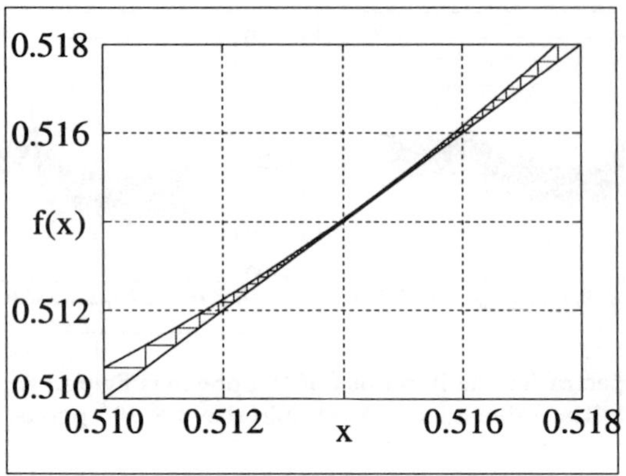

Fig. 24   Cobweb plot near to tangent bifurcation of the logistic map. Approximately 180 iterations of f are required to make the passage. ($r = 3.8284$, $x_0 = 0.51$.)

from which we can obtain an estimate of the number of iterations to progress from $x = a - \Delta x$ to $x = a + \Delta x$, as

$$k(a + \Delta x) - k(a - \Delta x) \approx \int_{a-\Delta x}^{a+\Delta x} \frac{dx}{A\Delta r + K(x-a)^2}.$$

When Δr is small compared with Δx, which is the present case, we might as well replace $\pm\Delta x$ by $\pm\infty$, after which our estimate of the total number L of iterations required to navigate the bottleneck becomes

$$L \approx \int_{-\infty}^{\infty} \frac{dx}{A\Delta r + K(x-a)^2} = \frac{\pi}{\sqrt{AK\Delta r}}. \tag{3.11}$$

3.6   *Unstable orbits and crises*

One of the common features of final state diagrams is the fact that a chaotic attractor may change its size discontinuously, or even appear or disappear suddenly, at a critical value of a parameter. Such an occurrence is an example of a *crisis*. A tangent bifurcation may precipitate a crisis, as for example the sudden disappearance of chaos with the emergence of a periodic window, but this is not the only form a tangent bifurcation can take.

In terms of the earlier strict definition, crises do not qualify as bifurcations, since there does not need to be a change in the structure of the periodic orbits. However, crises are intimately connected with unstable periodic orbits, and it is appropriate to deal with the phenomenon here.

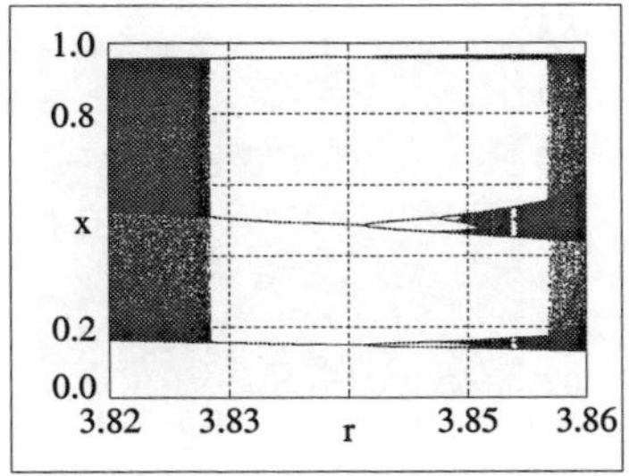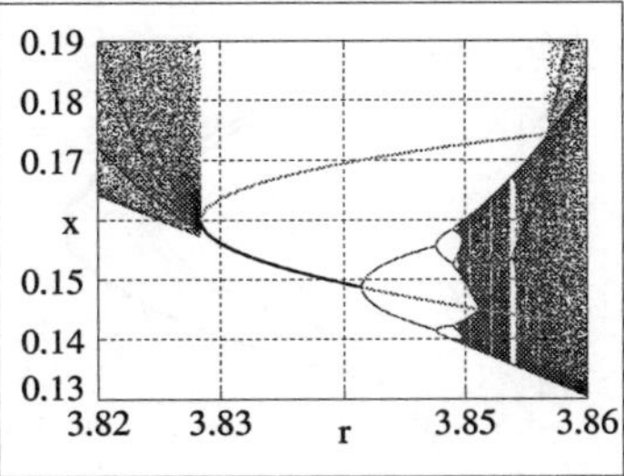

Fig. 25   An interior crisis (left). Detail showing period 3 orbits, stable and unstable; the latter collides with the chaotic attractor.

An interior crisis

An example of an interior crisis of the logistic map is shown in figure 25. At the value $r \approx 3.8568$, the system switches from chaos with a strong period 3 component, in which the three bands are visited in regular order, to chaos in a much wider band. The periodic order within this three-band attractor has its origin in the fact that it is at the end of the period 3 window, and is evident from the Fourier spectra, which I have not shown. There is a strong peak at frequency $1/3$.

These pictures show that the final limit set suddenly jumps from three disjoint intervals to a larger single interval which contains them. The crisis is precipitated by the collision of a chaotic attractor with an unstable orbit. However, it is an *interior* crisis because the former attractor is a subset of the latter, so the effect is a sudden jump in size. The mechanism is seen quite clearly if we use scChaos for Java to superpose periodic orbits (unstable as well as stable) on the final state diagram and then zoom in on one of the bands.

4   Hénon's map

In this lecture I discuss a two-dimensional system of the form

$$x_{k+1} = f(x_k, y_k; \mu),$$
$$y_{k+1} = g(x_k, y_k; \mu),$$
(4.1)

In the one-dimensional case it is usually convenient to restrict attention to maps of an interval; for typical two-dimensional systems this is rarely so, and I simply assume that x_k, y_k are real numbers and regard Eq. 4.1 as a

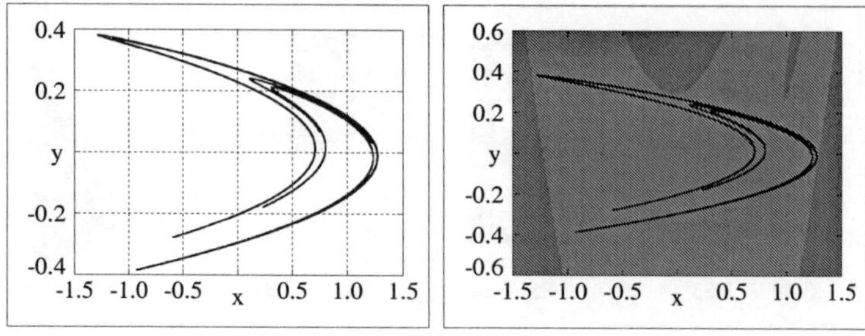

Fig. 26   The Hénon attractor, $a = 1.4$, $b = 0.3$ (left). Same attractor, showing basin of attraction (right).

map from the two-dimensional (real) plane to itself — the *state space*.

4.1   *The map: an overview of behaviour*

The precise form of the two-dimensional Hénon map is

$$f(x,y) = 1 - ax^2 + y,$$
$$g(x,y) = bx, \tag{4.2}$$

in which a and b are parameters. The map was introduced by M. Hénon [13], as a model which exhibits much of the interesting dynamical behaviour of more complicated systems, but using simple algebraic functions. It can be regarded as an extension of the logistic map to two dimensions, produced by applying a linear feedback loop.[28]

The Attractor and its basin

Hénon concentrated much attention on the parameter values $a = 1.4$, $b = 0.3$. It is interesting to examine the set of points in the x-y plane which are visited by a single long orbit of the map — see Fig. 26. A forward limit set with the property that all orbits which start sufficiently close to it, converge to points in it, will be called an *attractor*. The strange[29] attractor of the Hénon map contains an uncountably infinite number of points, so what is meant by convergence for it is a little more complicated.

The set of initial conditions whose orbits converge to a given attractor

---

[28] See Davies [6], p140ff.

[29] Attractors with fractal properties were called strange by D. Ruelle and F. Takens [28].

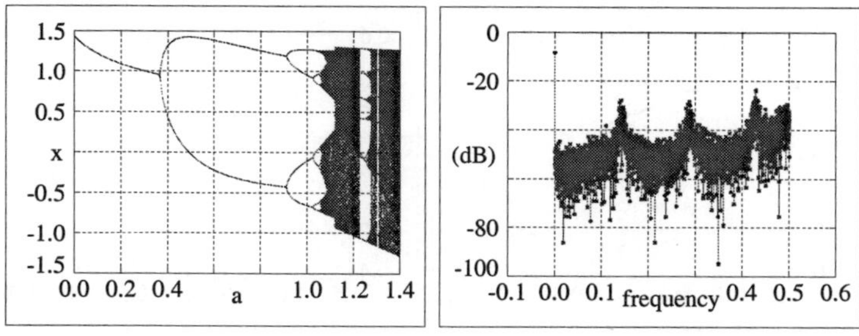

Fig. 27   Final state diagram (left). Fourier spectrum, $a = 1.226$, $b = 0.3$ (right).

constitute its *basin of attraction*. Once a numerical orbit has been computed, it is not difficult to compute a numerical approximation to its basin of attraction, although it is a computationally intensive task. The reason the basin is not the entire plane is that most initial points which are at a large distance from the origin increase without bound; the *point at infinity* is also an attractor. A typical picture is shown in Fig. 26 (right); the *basin of infinity* is the dark grey area, while the bounded attractor is shown sitting in a light area which is its basin.

Bifurcations

The second equation of the Hénon system shows that successive y values are a constant multiple of the previous x values. So the dynamical information is completely contained in the sequence x_k alone, even though it is far richer than for a one-dimensional map. This implies that it is useful to investigate the sequences x_k using tools already developed for the one-dimensional case.

Figure 27 (left) shows a final state diagram which displays quite clearly period doubling cascades and tangent bifurcations. The most prominent periodic window has period 7, unlike the logistic map where period 3 has this distinction. But the general features are exactly as in one dimension, including the fact that periodic windows normally end in a period doubling cascade to chaos.

As further proof that Fig. 27 (left) does indeed show a tangent bifurcation, prefigured by intermittent behaviour, a Fourier spectrum is shown on the right, taken just below the critical value of a. The peaks which prefigure the stable orbit are clearly visible.

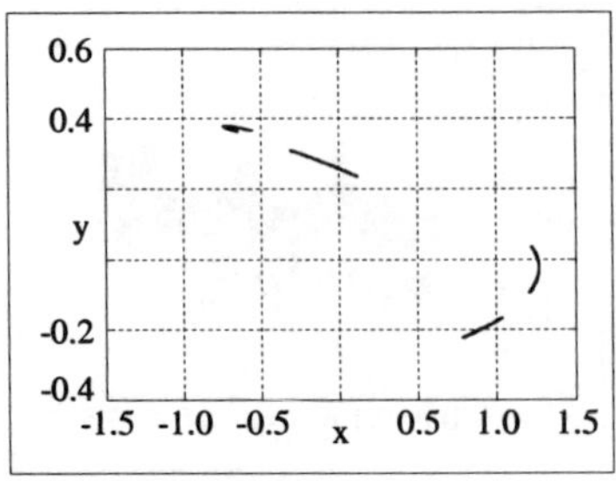 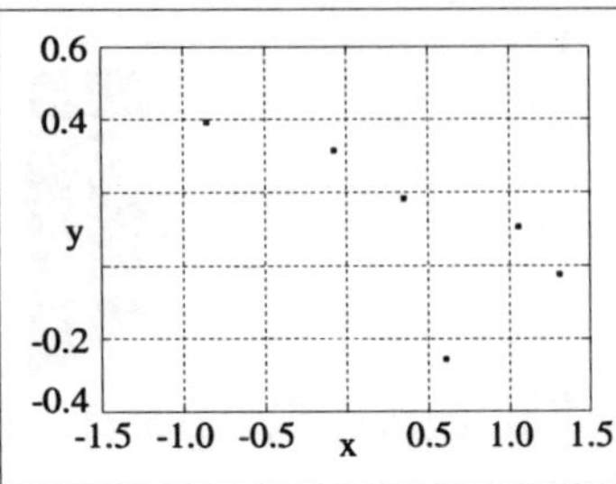

Fig. 28   Coexisting attractors, $a = 1.07$, $b = 0.3$. A strange attractor is produced from the initial state $(0.5, 0.5)$ (left), a period 6 orbit from the initial state $(-0.8, -0.4)$ (right).

Coexisting Attractors

Setting $a = 1.07$ and $b = 0.3$, it is easy to discover that there are two bounded attractors, one periodic, the other a strange attractor. They are shown separately in Fig. 28. The periodic attractor has period 6, hence it appears as just six distinct points (right). The other attractor (left), has four pieces, which the iterations visit in a periodic order, nevertheless the orbit is not period 4. Similar behaviour, for the logistic map, was noted in lecture 2. The Fourier spectrum of the strange attractor is shown in Fig. 29 (left), providing clear evidence of a nonperiodic orbit with a strong period 4 component. In fact, it is a chaotic orbit, since it its *largest* Lyapunov exponent is positive.[30] Moreover it is fractal, which is why the attractor is called strange.

Each orbit has its basin of attraction, and infinity is a third attractor. It is interesting to examine these basins, indicated in Fig. 29 (right). It is evident that the basins of the two bounded attractors are intertwined in an extremely complex, indeed fractal, way.

4.2   *Area contraction and Lyapunov exponents*

In the remainder of this lecture, I consider a *generalised Hénon map*, in which the function f of Eq. 4.2 is generalised to

$$f(x, y) = h(x) + y.$$

---

[30]The appropriate definitions will be given shortly.

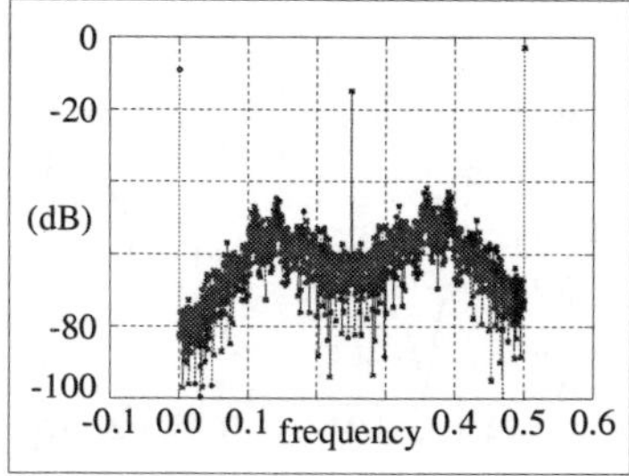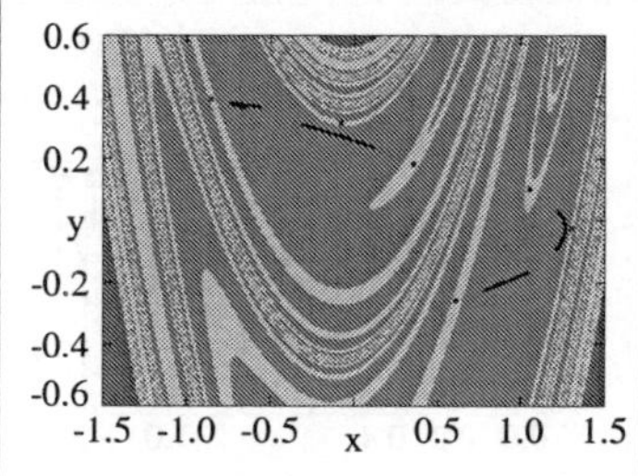

Fig. 29   Fourier spectrum of the previous strange attractor (left). Intertwined basins of attraction have complex structure (right).

For the Hénon map,

$$h(x) = 1 - ax^2,$$

another example will be used in the last lecture. Consider the transformation of a small square region of side $\delta > 0$, with corners at

$$\begin{array}{ll} (x, y + \delta) & (x + \delta, y + \delta) \\ (x, y) & (x + \delta, y) \end{array}$$

which I have laid out so as to indicate their relative positions in the plane, and let x', y', be the image of the bottom left corner under one iteration of the map. Assume for the time being that $b > 0$; in this case we shall see that it is also the bottom left corner of the image. An elementary calculation, using linear approximation, shows that the four corners map to the following four points (in the original order):

$$\begin{array}{ll} (x' + \delta, y') & (x' + h'(x)\delta + \delta, y' + b\delta) \\ (x', y') & (x' + h'(x)\delta, y' + b\delta) \end{array} \tag{4.3}$$

Examination of the formulae shows that the image of the square is a parallelogram, with bottom of length δ. The top left and bottom right corners of the image have been reversed compared to the original, the perpendicular distance between bottom and top is $b\delta$, so the new area is $b\delta^2$. The case $b < 0$ involves exactly the same equations, however there is no reversal of orientation in this case, while the area is transformed to $|b|\delta^2$.

The conclusion is that, for the generalised Hénon map, the area of any simple small region of the x-y plane is multiplied, at each iteration, by the factor $|b|$. Since we are only concerned with the case that $|b| < 1$, this

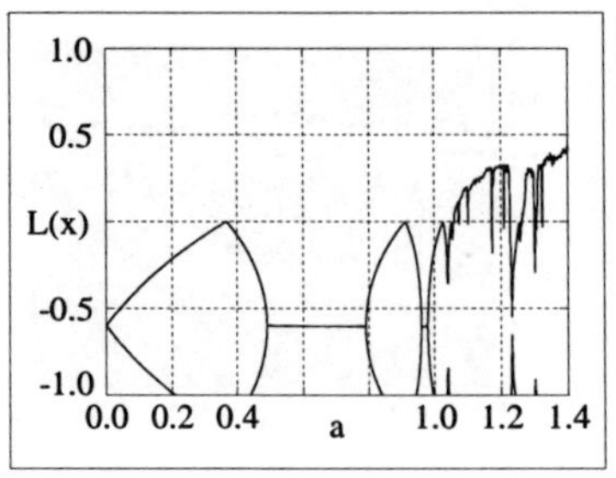
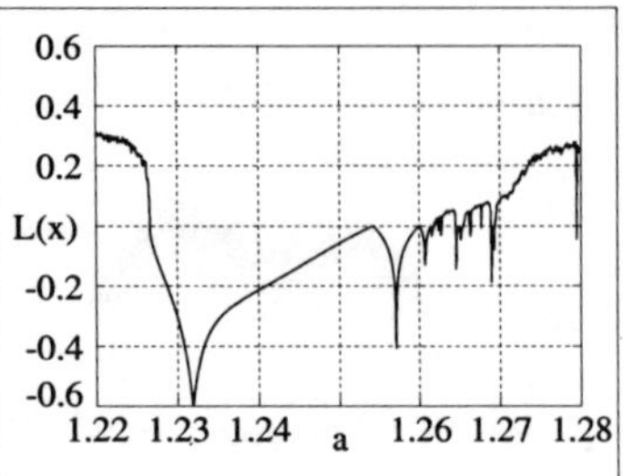

Fig. 30   Lyapunov exponents, $b = 0.3$.

is an area contraction. Two-dimensional maps with the area contracting property are said to be *dissipative*. Other examples will appear in the final lecture.

Lyapunov exponents

For one-dimensional maps the Lyapunov exponent is defined by tracking the image of an interval of negligible length. For two-dimensional maps we see from Eq. 4.3 that expansion and contraction is non-uniform; for this reason it is necessary to track a small ellipse. It is easy to show that, to linear approximation, the image of a small ellipse is another, although the lengths of the axes and their orientation are changed. Suppose that the initial semi-major axis is $\delta_0^{\max}$, the initial semi-minor axis $\delta_0^{\min}$. After k iterations, these are changed to $\delta_k^{\max}$, $\delta_k^{\min}$, and they satisfy the area reduction property

$$\delta_k^{\max}\delta_k^{\min} = |b|^k \delta_0^{\max}\delta_0^{\min} \tag{4.4}$$

The *Lyapunov exponents* of a two-dimensional map are defined as:

$$L_1(x_0, y_0) = \lim_{k\to\infty} \frac{1}{k}\left(\lim_{\delta_0\to 0} \ln|\delta_k^{\max}/\delta_0^{\max}|\right),$$

$$L_2(x_0, y_0) = \lim_{k\to\infty} \frac{1}{k}\left(\lim_{\delta_0\to 0} \ln|\delta_k^{\min}/\delta_0^{\min}|\right),$$

provided the infinite-k limit exists. The limit $\delta_0 \to 0$ is taken first to ensure that we only deal with a small ellipse at every stage of the computation. Taking logarithms of Eq. 4.4 gives the relation

$$L_1(x_0, y_0) + L_2(x_0, y_0) = \ln|b|. \tag{4.5}$$

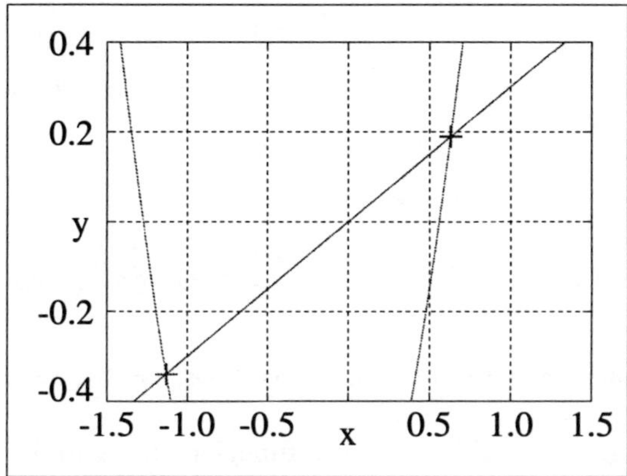

Fig. 31   Zero curves and fixed points, $a = 1.4$, $b = 0.3$.

This implies that Lyapunov exponents of these two-dimensional maps can never be more negative than $\frac{1}{2}\ln|b|$.

Computed values for the Hénon map are shown in Fig 30. The right-hand figure explores the parameter range where the period 7 tangent bifurcation, and subsequent period doubling cascade, takes place.

4.3   *Fixed points and periodic orbits*

Any pair (x^*, y^*) for which

$$
\begin{aligned}
f(x^*, y^*) &= x^*, \\
g(x^*, y^*) &= y^*,
\end{aligned}
\tag{4.6}
$$

is called a *fixed point* of the two-dimensional dynamical system. Following the method of earlier lectures, I define functions $\phi(x, y)$, $\psi(x, y)$, as

$$
\begin{aligned}
\phi(x, y) &= x - f(x, y), \\
\psi(x, y) &= y - g(x, y).
\end{aligned}
$$

Assuming sufficiently smooth functions, the zeros of these functions implicitly define curves in the x-y plane. The fixed points may therefore be regarded as points of intersection of curves in the plane.

The zero-curves of ϕ and ψ, for the Hénon map with $a = 1.4$, $b = 0.3$,

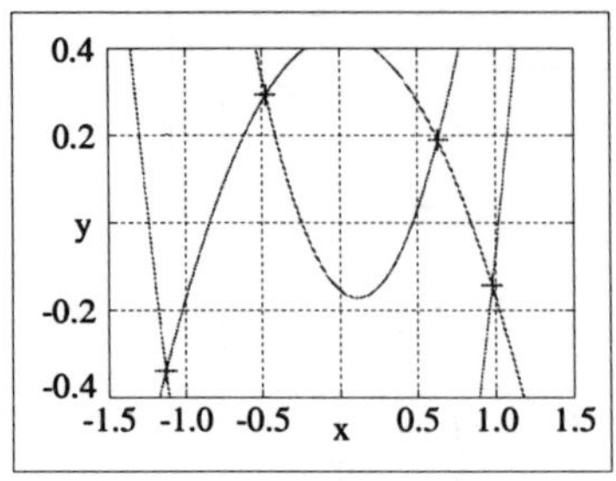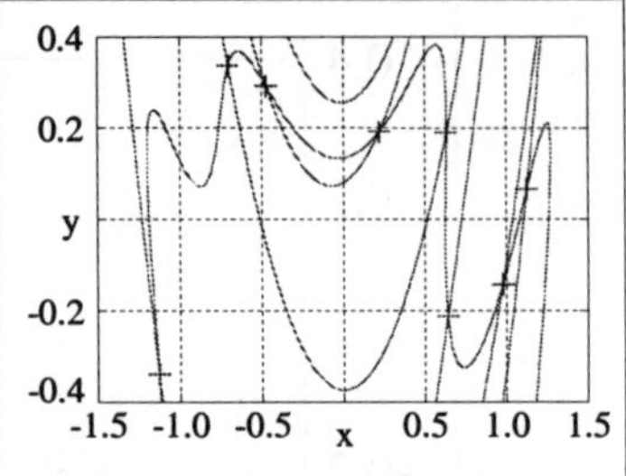

Fig. 32   Second and fourth compositions, $a = 1.4$, $b = 0.3$.

are shown in Fig. 31. In this case, the functions are simply

$$\phi(x,y) = ax^2 + x - 1 - y,$$
$$\psi(x,y) = y - bx,$$

(4.7)

so the corresponding zero-curves are defined explicitly as a parabola (if $a \neq 0$) and a straight line, respectively. One sees that there are two fixed points, provided that the two curves itersect. If $a = 0$, there is only one fixed point.

Compositions and periodic orbits

Compositions are defined by formulae which are easy to define, but difficult to deal with:

$$f_2(x,y) = f(f(x,y), g(x,y)), \qquad g_2(x,y) = g(f(x,y), g(x,y)),$$

and, in general

$$f_n(x,y) = f_{n-1}(f(x,y), g(x,y)), \qquad g_n(x,y) = g_{n-1}(f(x,y), g(x,y)).$$

As previously, a *periodic orbit* is a set of n fixed points of the n-fold composition, each of which is not a fixed point of an m-fold composition for any $m < n$, which are visited in sequence under iteration of the map. If we define functions $\phi_n(x,y)$, $\psi_n(x,y)$, by

$$\phi_n(x,y) = x - f_n(x,y),$$
$$\psi_n(x,y) = y - g_n(x,y),$$

(4.8)

then their zeros are curves in the x-y plane, whose intersections therefore determine the structure and stability of the period n orbits of the map.

n	fixed points	period $m < n$	new points	period n
1	2	–	3	2
2	4	2	2	1
3	2	2	–	–
4	8	4	4	1
5	2	2	–	–
6	16	4	12	2
7	30	2	28	4
8	64	8	56	7

Table 1.4 Table of periodic orbits, Hénon map, $a = 1.4$, $b = 0.3$.

Figure 32 displays the zero-curves of ϕ_n and ψ_n, for the second and fourth compositions. Fixed points are marked by a cross. There are four fixed points with $n = 2$, two of which are fixed points for $n = 1$, indicating that there is a single period 2 orbit. Again, there are eight fixed points with $n = 4$, of which two are fixed points of the map, and two more belong to the period 2 orbit. This leaves four fixed points which constitute a single period 4 orbit. A catalogue of periodic orbits may be constructed, exactly as in lecture 2 — see table 1.4.

Stability analysis

Suppose that the point (x, y) maps to (x', y'), not necessarily close to it. Let's compute the motion of a nearby point, $(x+\xi, y+\eta) \to (x'+\xi', y'+\eta')$, as in Eq. 4.3. In matrix form it may be written as

$$\begin{pmatrix} \xi' \\ \eta' \end{pmatrix} = \begin{pmatrix} h'(x) & 1 \\ b & 0 \end{pmatrix} \begin{pmatrix} \xi \\ \eta \end{pmatrix} \tag{4.9}$$

In general, the matrix operation acts to change both length and direction; there may be, however, directions with the property that vectors in those directions only change their length. Finding them is an eigenvalue problem. If the direction is not changed, then the vector (ξ', η') must be a multiple of (ξ, η). Denoting the multiplying factor by λ, everything will be determined from the solutions of the *eigenvalue equation*

$$\begin{pmatrix} h'(x) & 1 \\ b & 0 \end{pmatrix} \begin{pmatrix} \xi \\ \eta \end{pmatrix} = \lambda \begin{pmatrix} \xi \\ \eta \end{pmatrix} \qquad (4.10)$$

Eliminating ξ, η gives the *characteristic equation* which must be satisfied by λ,

$$\lambda^2 - h'(x)\lambda - b = 0. \qquad (4.11)$$

This will determine, in general, two distinct *eigenvalues*, $\lambda_\pm$. Once the eigenvalues are determined, the corresponding *eigenvectors* may be obtained.

Properties of the eigenvalues and eigenvectors

From Eq. 4.11 is immediate that

$$\lambda_+ + \lambda_- = h'(x), \qquad \lambda_+ \lambda_- = -b. \qquad (4.12)$$

The second relation implies the area reduction, but there is another consequence of great importance: in the dissipative case only one of the eigenvalues can have magnitude greater than unity.[31] There are three distinct possibilities:

(i) The eigenvalues are real but unequal. In this case the eigenvectors may be used to define coordinate axes. Because iteration of the map has such a simple effect on the eigenvectors, this shows that the fixed point is stable if both eigenvalues satisfy $|\lambda| < 1$, otherwise the fixed point is unstable.

(ii) The eigenvalues are complex. Then they are complex conjugate, and satisfy

$$\overline{\lambda}_+ = \lambda_-, \qquad |\lambda_\pm| = \sqrt{|b|}, \qquad (4.13)$$

It follows that the the fixed point is stable, since $|\lambda_\pm^k| < 1$ if $|b| < 1$.

(iii) The eigenvalues are real and equal. This occurs as a transitional case between the previous two. Since $|\lambda_\pm| = |b|^{1/2} < 1$, the fixed point is stable.

---

[31] Obviously, Lyapunov exponents are related to an eigenvalue problem, although this is not straighforward; the condition $\lambda_+ \lambda_- = -b$ is related to Eq. 4.5.

Loss of stability for the Hénon map

For $a > 0$, $b > 0$, we have (using $h'(x) = -2ax$)

$$\lambda_\pm(a) = -ax_+^* \pm \sqrt{(ax_+^*)^2 + b}. \tag{4.14}$$

Both eigenvalues are real since $b > 0$, and it is easy to show that, as functions of a,

$$0 < \lambda_+(a) < b^{1/2}, \qquad \lambda_-(a) < 0.$$

This means that the fixed point becomes unstable because λ_- passes through the critical value -1. Moreover, iterates oscillate from side to side, producing a period doubling bifurcation.

Periodic orbits

Consider two orbits, the first period n, starting from (x_0^*, y_0^*), the other a nearby orbit starting from $(x_0^* + \xi_0, y_0^* + \eta_0)$. After one iteration

$$\begin{pmatrix} \xi_1 \\ \eta_1 \end{pmatrix} = \begin{pmatrix} h'(x_0^*) & 1 \\ b & 0 \end{pmatrix} \begin{pmatrix} \xi_0 \\ \eta_0 \end{pmatrix}, \tag{4.15}$$

and this process may be iterated. For n steps, which is also a single step of the n-fold composition, this gives

$$\begin{pmatrix} \xi_n \\ \eta_n \end{pmatrix} = M(x_0^*) \begin{pmatrix} \xi_0 \\ \eta_0 \end{pmatrix}, \tag{4.16}$$

where

$$M(x_0^*) = \begin{pmatrix} h'(x_{n-1}^*) & 1 \\ b & 0 \end{pmatrix} \cdots \begin{pmatrix} h'(x_1^*) & 1 \\ b & 0 \end{pmatrix} \begin{pmatrix} h'(x_0^*) & 1 \\ b & 0 \end{pmatrix}. \tag{4.17}$$

This should be compared with Eq. 2.14, it is in fact an example of the *chain rule* for partial differentiation.

Equation 4.16 allows us to extend the eigenvalue method. First, we replace Eq. 4.10 by the general form

$$\begin{pmatrix} M_{11} & M_{12} \\ M_{21} & M_{22} \end{pmatrix} \begin{pmatrix} \xi \\ \eta \end{pmatrix} = \lambda \begin{pmatrix} \xi \\ \eta \end{pmatrix}. \tag{4.18}$$

The characteristic Eq. 4.11 for the eigenvalues λ becomes

$$\lambda^2 - T(x_0^*)\lambda + D = 0, \tag{4.19}$$

n	bifurcation	a_n	δ	a_∞
1	$1 \to 2$	0.3675000000		
2	$2 \to 4$	0.9125000000		
3	$4 \to 8$	1.0258554050	4.807887	1.05562399
4	$8 \to 16$	1.0511256620	4.485724	1.05837531
5	$16 \to 32$	1.0565637582	4.646894	1.05805492
6	$32 \to 64$	1.0577308396	4.659569	1.05804975
7	$64 \to 128$	1.0579808932	4.667325	1.05804908
8	$128 \to 256$	1.0580344522	4.668750	1.05804905

Table 1.5   Estimates of the Feigenbaum constant δ, and a_∞, Hénon map, $b = 0.3$. Data is from the main period doubling cascade.

where $T(x_0^*)$ and D are the *trace* and *determinant* of $M(x_0^*)$, respectively:

$$T(x_0^*) = M_{11} + M_{22}, \qquad D = M_{11}M_{22} - M_{12}M_{21}.$$

Since each of the matrices in the product 4.17 has determinant $(-b)$, the latter is simply $D = (-b)^n$.

Earlier arguments about the eigenvalues being either real, or a complex conjugate pair, still hold; again the orbit is stable provided that both of the eigenvalues are smaller than unity in magnitude, while only one eigenvalue can attain the value ± 1, in which case both eigenvalues are real. Finally, we need to know the following fact:

> The eigenvalues of a product of square matrices are invariant under cyclic permutation of the product order.

This ensures that the test is completely independent of the fixed point chosen for its application, even though matrix multiplication is not commutative.

4.4   *Period doubling*

At the first period doubling, the fixed point (x_+^*, y_+^*) of the Hénon map is also a fixed point of the second composition map. Therefore, the matrix $M(x_+^*)$ is the square of a matrix whose eigenvalues arde $\lambda_- = -1$ and

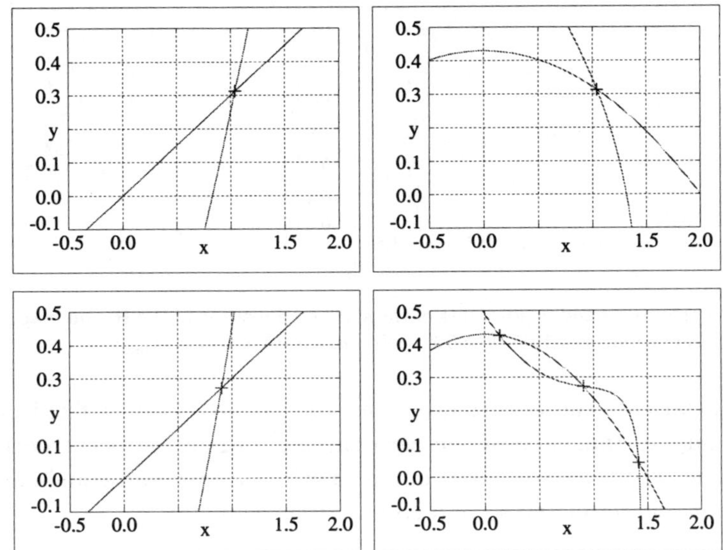

Fig. 33   First period doubling, $b = 0.3$. Zero curves and fixed points are shown as follows (from top left) (i) $a = 0.25$, $n = 1$, (ii) $a = 0.25$, $n = 2$, (iii) $a = 0.45$, $n = 1$, (iv) $a = 0.45$, $n = 2$.

$\lambda_+ = b$, so its eigenvalues are[32]

$$\lambda_+^{(2)} = \lambda_-^2 = +1, \qquad \lambda_-^{(2)} = \lambda_+^2 = b^2. \tag{4.20}$$

This is exactly parallel to the result for one-dimensional maps undergoing period doubling, for which $f'(x^*) = -1$ implies $f_2'(x^*) = f'(x^*)^2 = 1$ at $r = s$.

It is easy to visualise the mechanism graphically, exactly as in figure 15. In Fig. 33, the zero curves of the pairs (ϕ, ψ) and (ϕ_2, ψ_2) are plotted for $a = 0.25 < a_1$ (top pair) and $a = 0.45 > a_1$ (bottom pair). For the second composition map, neither of these curves is a straight line; apart from this difference, the similarity with the one-dimensional case is quite apparent. It can be shown that this is because of the same coincidences. In the one-dimensional case the fact that $f'(x^*) = -1$ implies both $f_2'(x^*) = +1$ and $f_2''(x^*) = 0$. The former is a condition of tangency of $y = f_2(x)$ with $y = x$ at $x = x^*$; the latter is a condition of equal curvature. In the two-

---

[32]The notation $\lambda^{(2)}$ is a reminder that these eigenvalues are associated with the second composition map.

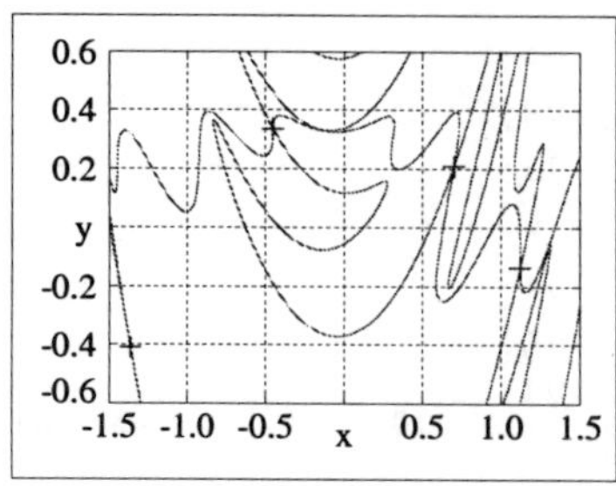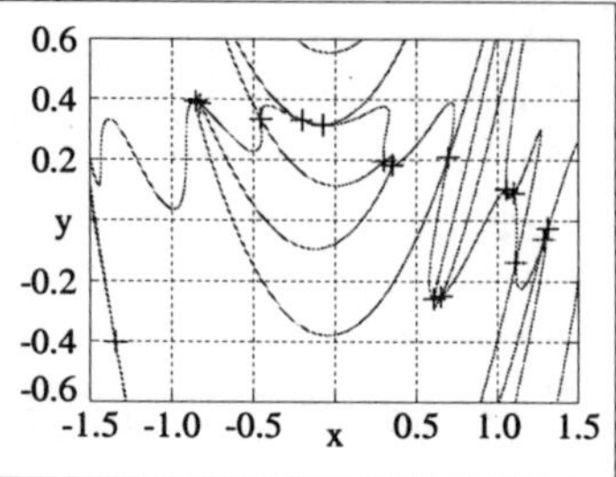

Fig. 34   Zero curves and fixed points of the sixth composition of the Hénon map, showing tangent bifurcation. $a = 1.05$, $b = 0.3$, (left); $a = 1.07$, $b = 0.3$, (right).

dimensional case the occurrence of an eigenvalue $+1$ for a fixed point of a map implies tangency of the zero curves,[33] while the occurrence of an eigenvalue -1 for a fixed point implies both tangency and equal curvature of the zero curves of the second composition map [7].

The period doubling cascade

The new period doubled orbit is a fixed point of the second composition map which is not a fixed point of the map; initially (when $a = a_1$), $\lambda_+^{(2)} = 1$, $\lambda_-^{(2)} = b^2$, however as a is increased, we expect a further period doubling to occur, caused by $\lambda_-^{(2)}$ passing through the value -1. Furthermore, we expect this process to continue in a period doubling cascade to chaos.

Critical values a_n, for the nth period doubling, may be obtained by careful numerical experimentation. This involves searching for the parameter value which gives $\lambda_- = -1$ for one of the fixed points of the 2^n-fold composition map belonging to the period 2^n orbit. Numerical data, obtained using scChaos for Java, is given in table 1.5, for $b = 0.3$. The values are seen to converge geometrically, according to the general scheme for period doubling, with Feigenbaum constant $\delta \approx 4.669$ and $a_\infty \approx 1.05805$.

4.5   Tangent bifurcation

We have already seen evidence of a tangent bifurcation of the Hénon map in the bifurcation diagram of Fig. 27; in figures 28 and 29, we also observed

---

[33]If there is an eigenvalue $+1$, the determinant of the matrix $[M(x_0^*) - I]$ is zero; from Eq. 4.8 this implies that the vectors $(\partial\phi_n/\partial y, -\partial\phi_n/\partial x)$ and $(\partial\psi_n/\partial y, -\partial\psi_n/\partial x)$, which define the tangents to the respective curves, are linearly dependent.

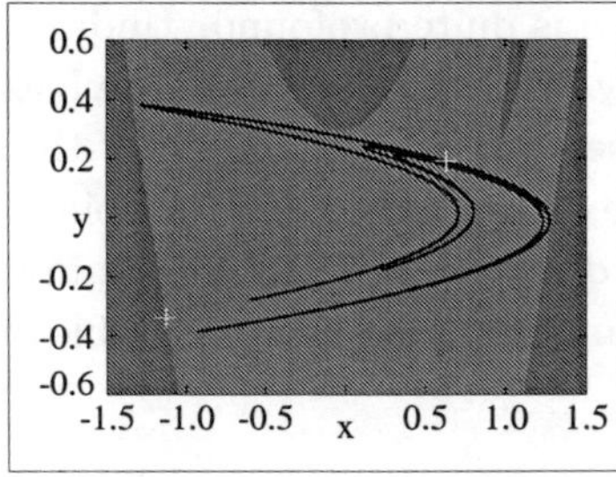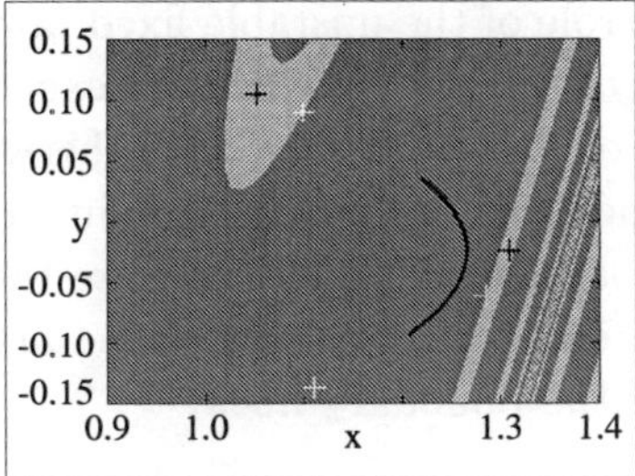

Fig. 35   Two basins for the Hénon map, showing also fixed points. $a = 1.4$, $b = 0.3$ (left), $a = 1.07$, $b = 0.3$ (right). Stable fixed point is a dark cross, unstable is a light cross.

a period 6 orbit which was born mysteriously. The explanation is that there is a tangent bifurcation. Zero curves and fixed points of the sixth composition map are shown in Fig. 34, for values of a just below, and just above, the critical value. The appearance of six pairs of fixed points is clearly seen.

In terms of zero curves, the mechanism is exactly parallel to that discussed in the last lecture. In that case, the condition $f_n'(x^*) = +1$ leads to tangency of the curves $y = f_n(x)$ and $y = x$ when $r = s$; in the present case the condition that $\lambda_+ = +1$ leads to tangency of the curves $\phi_n(x, y; \mu) = 0$ and $\psi_n(x, y; \mu) = 0$ when $a = a^*$. If there are no other coincidences, the curves either cut twice, or not at all, as a varies slightly from the critical value, leading to the tangent bifurcation. As in the one-dimensional case, it is possible to show that one of the orbits is stable, the other unstable, but I will not attempt that here.

Basin boundaries

On looking at pictures of attractors and their basins, it is natural to ask what determines the boundaries between basins. An immediate clue may be found in Fig. 35. The left hand picture shows the strange attractor of Fig. 26 (right) and its basin, but with the addition of the two fixed points of the map. The right hand picture is similar, except that now $a = 1.07$, at which value there are coexisting attractors. I have zoomed in on a small area of Fig. 29, and displayed the fixed points of the sixth composition map. The common feature is that unstable fixed points sit on the basin boundary in both cases.

The rôle of the unstable fixed points is quite profound. Under iteration of the system, points which are exactly on the boundary of the basin must remain on the boundary, for otherwise they are inside one of the basins. The conclusion is that these points are in fact attracting for points exactly on the boundary! This is because the direction of instability is involved in taking iterations away from the boundary, implying that the stable direction is along the boundary itself.

Death of the orbit

Using scChaos for Java, it is possible to follow the fate of the period 6 orbit with increasing a. First it period-doubles to chaos, to become a six-part chaotic attractor. With increasing a, it continues to increase in size, until it collides with the basin boundary at the unstable fixed points. This is an example of a crisis, and as with the example at the end of the last lecture, it is brought about by the unstable periodic orbit which was born in the same tangent bifurcation.

5   Fractals

I have mentioned *fractals* from time to time; in this concluding lecture, I want to say a little more about their relation to chaotic dynamics. A precise definition of fractals, robust enough for general use, does not seem to exist. Indeed, Alligood, Sauer and Yorke [2] write

> ...Scientists know a fractal when they see one, but there
> is no universally accepted definition ...

The word itself was coined by Mandelbrot [19], the root is from the Latin *fractus* (broken). Mandelbrot's definition is that the *Hausdorff dimension* should exceed the *geometric dimension*, which usually implies a fraction; there is however no requirement for this.

I follow Falconer [9] by referring to an object as fractal if it exhibits the following properties:

(i) It has fine structure, meaning that there is always more detail to be seen at arbitrarily small scales.

(ii) It is too irregular to be described using traditional geometry.

(iii) It has some form of exact or approximate *self-similarity*.

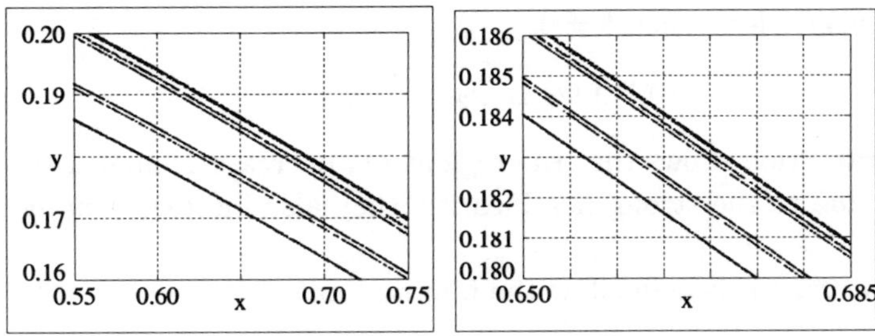

Fig. 36   Self-similarity of the Hénon attractor. The right-hand picture is a detail of the left one; itself a detail from Fig. 26. The repetition of structure at increasingly fine scales is apparent.

(iv) The fractal dimension is not the natural geometric dimension.

Falconer also points out that fractals are often defined in simple ways; as we have already seen.

5.1   *The Hénon attractor*

Let's revisit the Hénon attractor originally shown in Fig. 26; some magnified detail is shown in Fig. 36. Although the attractor is not globally self-similar, there is more here than the simple straightening of curves under repeated magnification. Each of the two pictures consists of groups of *thick lines*; the self similarity is evident in the fact that the structure repeats itself under magnification — it is infinitely complex.

For the Hénon map, it is possible to define simple geometrical *trapping regions*. These are regions which sit in the basin of attraction, but with the further property that they are mapped into themselves under a single iteration. An example which is a quadrilateral is given in [25], p664. Denote it by Q; it is two-dimensional. Denote also the Hénon map by T; it is a smooth function of the coordinates (x,y), so the image $Q_1 = T(Q)$ is also a two-dimensional region of the plane, bent like a boomerang. This argument may be applied repeatedly, to give a sequence of regions Q_n, each two-dimensional:

$$Q \xrightarrow{T} Q_1 \xrightarrow{T} Q_2 \cdots \xrightarrow{T} Q_n \xrightarrow{T} \cdots$$

Moreover, because of the trapping property, together with the fact that the map is *invertible*[34] when $b \neq 0$,

$$\mathcal{Q} \supset \mathcal{Q}_1 \supset \mathcal{Q}_2 \cdots \supset \mathcal{Q}_n \supset \cdots$$

Area contraction shows that the regions $\mathcal{Q}_n$ decrease geometrically in area as well; due to the strong nonlinearity, they also increase rapidly in complexity.

The attractor is defined as the infinite limit

$$\mathcal{Q}_H = \lim_{n \to \infty} \mathcal{Q}_n = \bigcap_{n=0}^{\infty} \mathcal{Q}_n, \tag{5.1}$$

and this is the object approximated by finite computation. Obviously $\mathcal{Q}_H$ already complies with points (i) and (iii) on Falconer's list. Shortly, I shall discuss the fractal dimension of the attractor, which has the typical value $d \approx 1.28$, this is in accord with item (iv). The rapid increase of complexity of the successive $\mathcal{Q}_n$, due to the repeated stretching and bending, accords with point (ii) as well. The conclusion is that $\mathcal{Q}_H$ is a fractal.

5.2   *Fractal dimension*

Definitions of fractal dimension generally rely on scaling properties. For example, the capacity dimension comes from consideration of how to place a sufficient number of small boxes so as to contain the whole set, and how the minimum number required to achieve this increases as their size is decreased.

Given a set A of points in n-dimensional Euclidean space, let $N(\epsilon)$ be the minimum number of n-dimensional cubes of side ϵ needed to cover (contain) every point of A. For an infinite set, it is reasonable to expect that $N(\epsilon) \to \infty$, $\epsilon \to 0$, but in many cases there is the stronger scaling relation

$$N(\epsilon) \sim K\epsilon^{-d_C}, \qquad \epsilon \to 0. \tag{5.2}$$

The exponent d_C is the *capacity dimension*, this gives the intuitively obvious results for smooth curves ($d_C = 1$) or smooth surfaces ($d_C = 2$), and it may

---

[34]Unimodal one-dimensional maps are not invertible, which is the essential source of their chaotic behaviour. Despite this, there are invertible one-dimensional maps of interest in non-linear dynamics, one of which is the *circle map*. Generalised Hénon maps are trivially invertible, as are the Poincaré section maps of driven non-linear oscillators.

Length	0.016	0.008	0.004	0.002	0.001	0.0005	0.00025
10^5	810	1870	4336	9793	21856	41335	62672
$2 \cdot 10^5$	814	1879	4385	10030	23593	51164	91944
$5 \cdot 10^5$	814	1887	4417	10175	24409	57832	124586
10^6	814	1890	4428	10220	24692	59689	137366
$2 \cdot 10^6$	814	1892	4437	10244	24832	60572	142705
$5 \cdot 10^6$	814	1894	4437	10268	24923	61143	145458
10^7	814	1894	4439	10274	24963	61371	146404
$2 \cdot 10^7$	814	1894	4440	10280	24987	61499	146970
$5 \cdot 10^7$	814	1894	4440	10282	25008	61575	147387
10^8	814	1894	4440	10284	25016	61605	147541

Table 1.6   Box counts, Hénon map, $a = 1.4$, $b = 0.3$. The boxes are squares of side ϵ given at the top of each column; the length of the orbit is given in the first column. Initial 10^3 points discarded.

readily be applied to some simple fractals, such as the *middle third Cantor set*,[35] for which

$$d_C = \ln 2 / \ln 3 \approx 0.631.$$

The Hénon attractor

Computing capacity dimension is a difficult numerical task even for a relatively simple system such as the Hénon attractor. From Eq. 5.1 one sees that ideally, a box count $N_{n,\epsilon}$ should be computed for each $\mathcal{Q}_n$, using squares of side ϵ. From this data N_ϵ is obtained from the limit $n \to \infty$, after which the required exponent d_C may be extracted from the limit $\epsilon \to 0$. However there is no practical way to do this.

An alternative, more practical, method is to follow a particular orbit[36] for a large number of iterations, and count how many squares of side ϵ are visited by it. Denote this number by $N_{n,\epsilon}$, where n is the sample length,

---

[35]I don't discuss its construction here, but it is one of the staple examples of a fractal.

[36]Because the orbit is chaotic, the computed shadow should be ergodic.

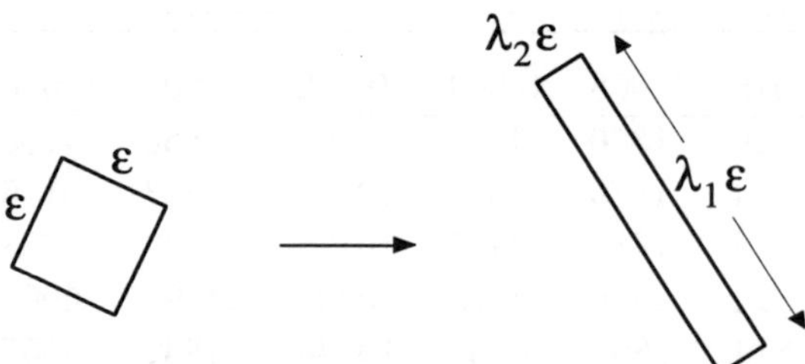

Fig. 37   Using Lyapunov exponents to approximate the effect of a two-dimensional map on a small square of side ϵ.

then

$$N_\epsilon = \lim_{n \to \infty} N_{n,\epsilon}.$$

The infinite limit poses a serious problem; if n is fixed while $\epsilon \to 0$, then the count will *saturate*, that is, $N_{n,\epsilon} \to n$. This would give a dimension of zero! Some data is shown in table 1.6. Reading down each column shows how extraordinarily slow convergence is. Simply using the last count in each column, which is for $n = 10^8$, gives the sequence of estimates for d_C:

$$d_C \approx 1.218, \; 1.229, \; 1.212, \; 1.282, \; 1.300, \; \ldots$$

Obviously the dimension is not an integer, but the data also shows just how bad the infinite limit problem is. For example, when $\epsilon = 0.00025$, of the last $5 \cdot 10^7$ iterations only 154 of them fall into boxes not previously counted!

Lyapunov dimension

The Lyapunov dimension is popular, because it is easy to estimate and directly related to an important measure of dynamical behaviour. To motivate the idea, consider first a periodic attractor; its Lyapunov exponents L_1 and L_2 are both negative and all nearby initial points are attracted to a finite set of points. Clearly the dimension of such an attractor is zero. For a strange attractor, $L_1 > 0$ and $L_2 < 0$, which conveys the fact that the attractor is compressed in one direction and expanded in the other. One can base a simple argument on this, using approximate self-similarity, which leads to a corresponding dimension.

Let $N(\epsilon)$ be the number of squares of side ϵ required to cover the attractor. After one iteration these squares have undergone an uneven change of scale. Choose the orientation of each small square so that one side is stretched and the other shrunk, to give a rectangle of aspect ratio λ_1/λ_2 (see Fig. 37). The λ_i are the exponentials of the L_i, since we want multiplicative factors. One iteration of the map replaces the old covering of N squares by a new covering of $(\lambda_1/\lambda_2)N$ squares having reduced edge size $\lambda_2\epsilon$. This gives a scaling relation, approximate because the actual expansion/contraction factors are in reality position-dependent,

$$\frac{\lambda_1}{\lambda_2}N(\epsilon) \sim N(\lambda_2\epsilon), \qquad \epsilon \to 0.$$

Assume, as usual in making definitions of fractal dimension, the scaling behaviour

$$N(\epsilon) \sim K\epsilon^{-d_L}.$$

Substituting this formula into the relationship between $N(\epsilon)$ and $N(\lambda_2\epsilon)$, and taking logarithms, gives

$$\ln \lambda_1 - \ln \lambda_2 + \ln K - d_L \ln \epsilon = \ln K - d_L(\ln \epsilon + \ln \lambda_2).$$

Solving for d_L and using the fact that $L_i = \ln(\lambda_i)$, the result is

$$d_L = 1 - \frac{L_1}{L_2}. \tag{5.3}$$

This is the *Lyapunov dimension* of the strange attractor, in the case that $L_1 > 0$. If $L_1 \leq 0$ the Lyapunov dimension is zero; the same argument works to show that $N(\epsilon) = N(\lambda_2\epsilon)$.

Hénon map

With $a = 1.4$, $b = 0.3$, simple numerical computation involving orbits whose length is only 10^4–10^6 gives

$$L_1 \approx 0.39, \qquad L_2 \approx -1.59,$$

from which

$$d_L \approx 1 + 0.39/1.59 \approx 1.25.$$

Apart from the fact that we do not require excessive orbit lengths to get good estimates, there is the advantage that this does not require repeated computations for different values of ϵ.

5.3 *Strange attractor of the driven pendulum*

The Poincaré section of the driven pendulum was discussed in the first lecture. Essentially, it is a map of the phase plane, each point evolving according to the equation 1.1. This is better expressed as a pair of first-order equations for the phase plane variables θ and θ':[37]

$$\theta' = \omega,$$
$$\omega' = -c\omega - f(\theta) + g(t). \tag{5.4}$$

The period map

Taking snap-shots of where the system is in the phase plane will always produce a set of points. What is so special about the Poincaré section is the relation to the frequency of the external disturbance. It comes from an important property of the differential equations, namely

if $\big(\theta(t), \omega(t)\big)$ is a solution, then so is $\big(\theta(t + 2\pi/\Omega), \omega(t + 2\pi/\Omega)\big)$.

This follows from the fact that if we change the time variable by $t \to t' = t + 2\pi/\Omega$, nothing changes in the equations.

In general, the two solutions will not be the same. To investigate the connection, let's call them $(\theta_0(t), \omega_0(t))$ and $(\theta_1(t), \omega_1(t))$; they are related by

$$\theta_1(t) = \theta_0(t + 2\pi/\Omega), \qquad \omega_1(t) = \omega_0(t + 2\pi/\Omega).$$

The first pair, $(\theta_0(t), \omega_0(t))$, is the unique solution with initial values

$$\theta_0(t_0) = \theta_0, \qquad \omega_0(t_0) = \omega_0,$$

and after one time step they have the values

$$\theta_1 = \theta_0(t_0 + 2\pi/\Omega), \qquad \omega_1 = \omega_0(t_0 + 2\pi/\Omega).$$

---

[37] The latter is given a new name: ω.

These are the initial values for the solution $(\theta_1(t), \omega_1(t))$, starting from $t = t_0$; the two solutions are part of the same (infinite) orbit, one period out of step.

Phase of section

An initial time t_0 is equivalently specified by an initial *phase of section*, $\phi_0 = \Omega t_0$, which is only determined to within an integer multiple of 2π. An initial point (θ_0, ω_0), starting at initial phase ϕ_0, arrives at a point

$$
\begin{aligned}
(\theta_1, \omega_1) &= \big(\theta_0(t_0 + 2\pi/\Omega), \omega_0(t_0 + 2\pi/\Omega)\big) \\
&= \big(\theta_1(t_0), \omega_1(t_0)\big),
\end{aligned}
$$

after one period of the driving force. The time has been incremented by $2\pi/\Omega$, but the phase remains the same. This is a two-dimensional map of the phase plane to itself, in which the phase appears as a parameter. The vital point is that this two-dimensional map contains complete information about the dynamics of the system.

Lyapunov exponents

It is easy to show that the map just defined is dissipative, with constant area reduction factor of

$$
\exp(-2\pi c/\Omega). \tag{5.5}
$$

Taking into account the time rescaling of Eq. 1.1, this gives the relation

$$
L_1 + L_2 = -c
$$

between the Lyapunov exponents.

There are two possibilities for the Lyapunov dimension of an attractor of this system:

(i) A periodic attractor, which is a one-dimensional curve in three-dimensional space, or a zero-dimensional set of points in the phase plane. This has both Lyapunov exponents negative.

(ii) A strange attractor, which has $L_1 > 0$, $L_2 < 0$, and Lyapunov dimension

$$
d_L = 1 - L_1/L_2 > 1.
$$

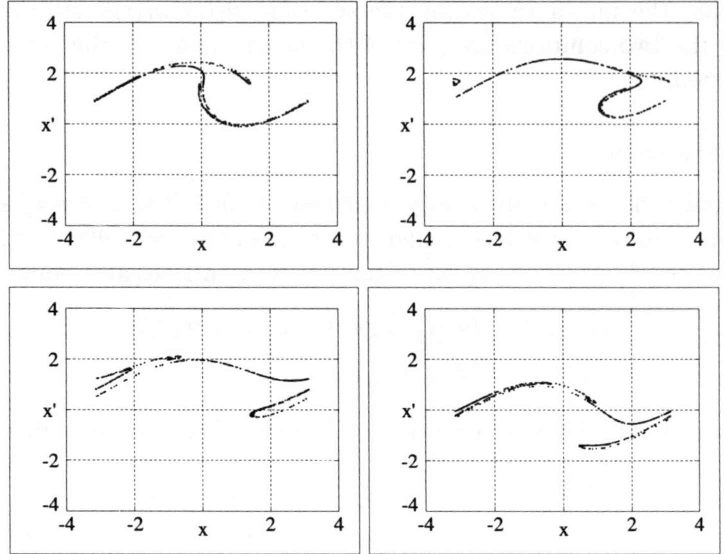

Fig. 38   Poincaré sections, driven pendulum $c = 0.5$, $k = 1.5$. Sample size 10^3 points, initial 10^2 points discarded. The phase of section varies from $0°$ (top left) in increments of $45°$.

For the dimension of the complete orbit in three-dimensional space, just add one; the dimension is greater than two.

The missing possibility is for an attractor with $d_L = 1$, that is, an attracting solution which traces out a smooth surface in three dimensions; this would appear in the Poincaré section as a smooth curve.

Producing the fractal

For the driven pendulum, we can readily observe the stretching and folding mechanism in the Poincaré sections. Some typical pictures, generated by scChaos for Java, are shown in figures 38 and 39. The stretching and folding, as the phase ϕ_0 is increased, is clearly seen, as is the symmetry of the attractor.

Estimated values of the Lyapunov exponents, calculated on the same orbit data (sample size 10^3, initial discard 10^2), were

$$L_1 \approx 0.118, \qquad L_2 \approx -0.618.$$

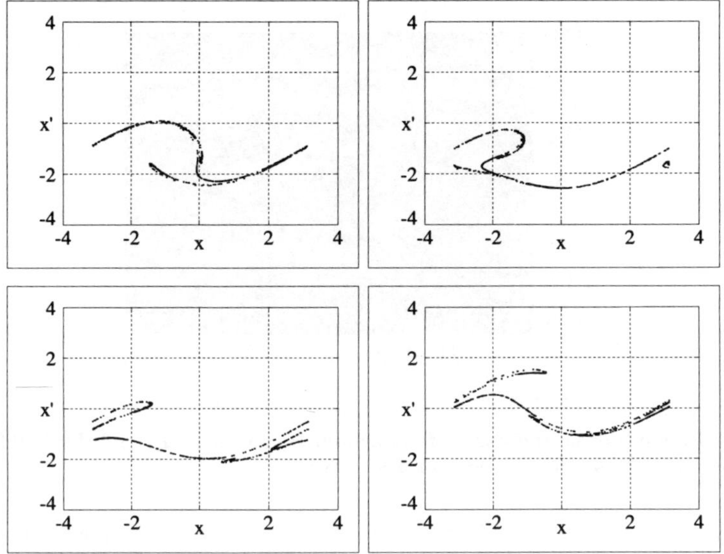

Fig. 39   Continuation of Fig. 38. Phases from 180° in increments of 45°. Sections at ϕ_0 and $\phi_0 + 180°$ are mirror images, indicating the symmetry of the attractor under $\phi_0 \to \phi_o + \pi$.

Note that the sum of the two is $-c = -0.5$ (to the quoted accuracy); any failure would be of the numerics rather than the theory. From these values

$$d_L \approx 1 + 0.118/0.618 \approx 1.19.$$

Increasing the sample size to 2×10^3 changes these estimates to

$$L_1 \approx 0.116, \qquad L_2 \approx -0.616, \qquad d_L \approx 1.19,$$

which indicates the difficulty of accurate numerical estimation. Taken together with the Fourier spectrum of the same data (not shown), this leaves little doubt that the observed behaviour is chaotic, the attractor strange.

5.4   *Fractal basins*

To motivate what follows, I show in Fig. 40 the attractor and its basin of the Hénon map, when the sign of b is changed compared with the well-known Hénon attractor. Far from being chaotic, the system has only attained its first period doubling. However, the basin of attraction appears to be

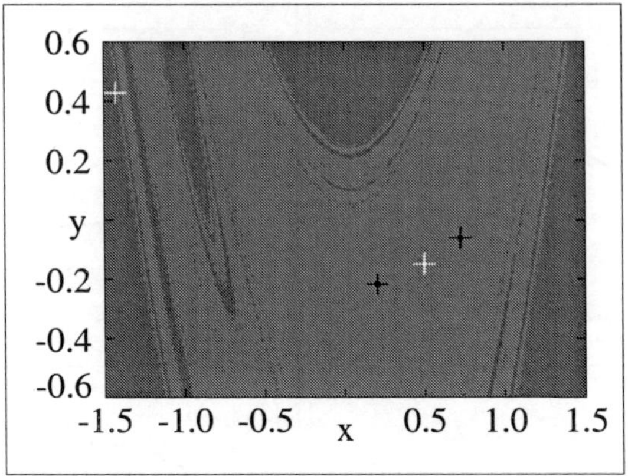

Fig. 40   Period 2 orbit and basin of attraction, Hénon map, $a = 1.4$, $b = -0.3$. Fixed points of map and second composition also shown.

fractal. This is associated with the change of sign for b, which forces a change of sign of λ_- for the fixed point on the basin boundary.

Linear feedback

A unimodal one-dimensional map f will typically have some range of parameter for which there is a non-zero stable fixed point. Suppose we wish to extend that stable behaviour to larger parameter values by incorporating a *feedback loop*. Just beyond the critical value s, iterations of f will tend to alternate; this is measured by the values of $x_k - x_{k-1}$. We can feed back a multiple as a form of linear stabilisation or destabilisation. The result is the discrete dynamical system

$$x_{k+1} = f(x_k) - b(x_k - x_{k-1}),$$

which I shall call a controlled map. It is better viewed as a two-dimensional map which produces the pair (x_{k+1}, x_k) from the pair (x_k, x_{k-1}) at a single iteration. It is in fact a generalised Hénon map:

$$x_{k+1} = f(x_k) - bx_k + y_k.$$
$$y_{k+1} = bx_k,$$
(5.6)

with $h(x) = f(x) - bx$. Stabilisation corresponds to negative b, in which case the term $-b(x_k - x_{k-1})$ corrects the tendency to oscillate about the fixed point.

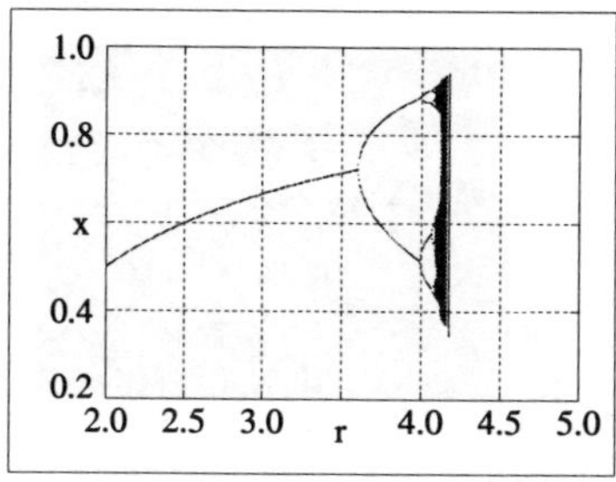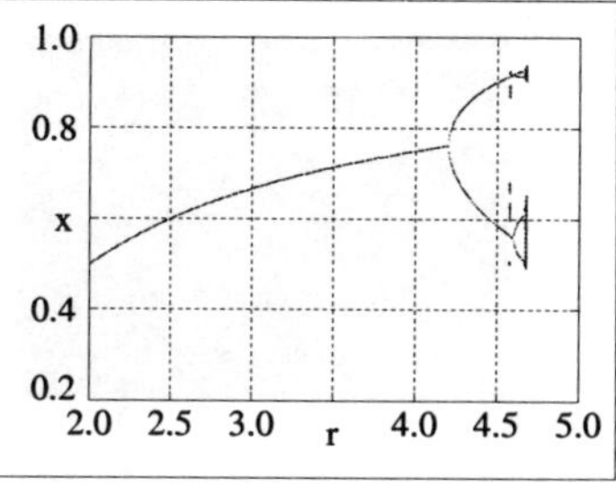

Fig. 41   Final state diagrams, controlled logistic map. $b = -0.3$ (left), $b = -0.6$ (right). Initial state $(0.8, -0.4)$, sample size 10^3 points, initial 10^3 points discarded.

Fixed points

Since $y^* = bx^*$, the fixed point equations 4.6 reduce to the single condition

$$x^* = f(x^*) - bx^* + y^* = f(x^*),$$

so the fixed points are not affected by the feedback. This is to be expected, since the purpose of feedback is to hold the original one-dimensional system at its fixed points.

The stability of these points is of vital interest. Equation 4.15, for the Hénon map, is little changed, becoming

$$\begin{pmatrix} \xi_1 \\ \eta_1 \end{pmatrix} = \begin{pmatrix} f'(x^*) - b & 1 \\ b & 0 \end{pmatrix} \begin{pmatrix} \xi_0 \\ \eta_0 \end{pmatrix}, \tag{5.7}$$

The characteristic equation 4.19, for the eigenvalues λ becomes

$$\lambda^2 - (f'(x^*) - b)\lambda - b = 0,$$

The point (x^*, y^*) is stable until the parameter is increased to the value at which period doubling occurs. Following the method of the previous lecture, this is determined by the fact that $\lambda_+ + \lambda_- = -1 + b$; substituting $\lambda_+ + \lambda_- = f'(x^*) - b$, gives the condition as

$$f'(x^*) = -1 + 2b. \tag{5.8}$$

This shows the effect of control in extending stability beyond the original period doubling; for negative b the derivative has to become more negative for this to occur.

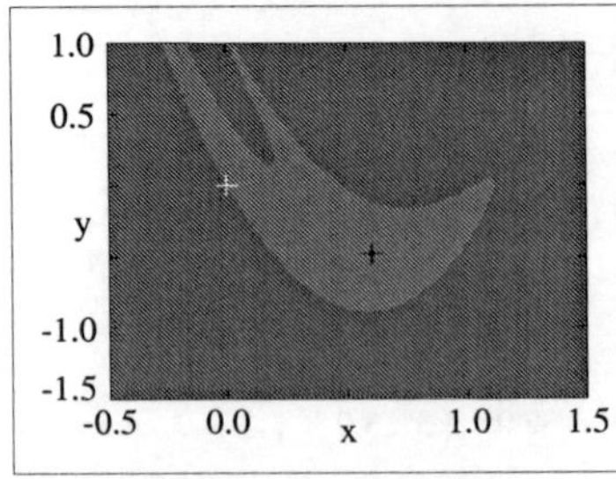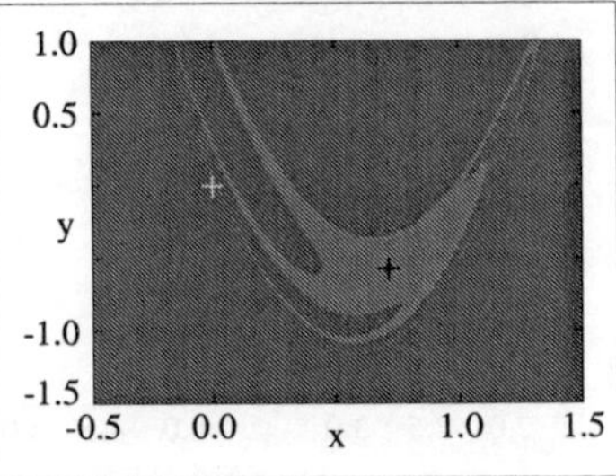

Fig. 42   Erosion of basin boundaries, controlled logistic map. $r = 2.5$, $b = -0.8$ (left), $r = 3.5$, $b = -0.8$ (right).

Controlled logistic map

In the case that f is the logistic map, I call the new map the *controlled logistic map*,

$$x_{k+1} = rx_k(1 - x_k) - bx_k + y_k,$$
$$y_{k+1} = bx_k. \tag{5.9}$$

The function $h(x) = rx(1 - x) - bx$ is a downward parabola, just like the standard Hénon map.[38] For the fixed point $x_1^* = 1 - 1/r$, $f'(x_1^*) = 2 - r$, so period doubling occurs when

$$r_1 = 3 - 2b. \tag{5.10}$$

Two final state diagrams are shown in Fig. 41, for two negative values of b, quite clearly exhibit the stabilisation.

We also know that there is a transcritical bifurcation of the logistic map when s $= 1$, the fixed point $x_0^* = 0$ becoming unstable, caused by the fact that $f'(x_0^*) = +1$. Setting this fixed point into equation 5.7, it is simple to show that $\lambda_+ = +1$, $\lambda_- = -b$, in this case, so the bifurcation is not affected by the feedback. Since $|b| < 1$, the critical value for period doubling, given in Eq. 5.10, is always greater than s $= 1$, however the range of parameter values within which x_1^* is stable becomes smaller as b becomes more positive. This explains the fact that the first period doubling of the standard Hénon map happens rather quickly.

---

[38]The relation between them requires the parameter change $a = (r - b)(r + b - 2)/4$, together with linear transformation of x and y.

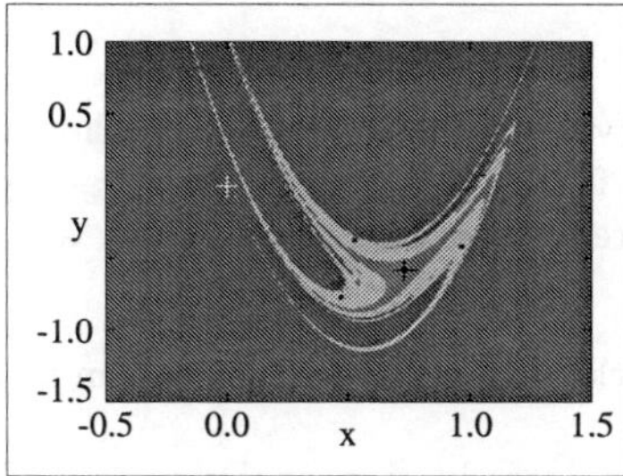 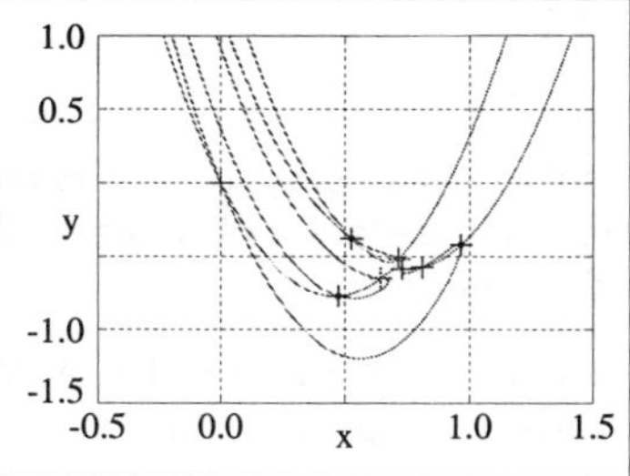

Fig. 43   Fractal invasion. $a = 3.7$, $b = -0.8$, coexisting periodic attractors, periods 1 and 3 (left). Zero curves and fixed points, third composition, showing tangent bifurcation (right).

Fractal basin erosion

In the one-dimensional case, the system is confined to an interval, one end of which is at the fixed point $x_0^* = 0$, which determines the basin boundary of the controlled map via the fixed point $(0,0)$. An interesting development is shown in Fig. 42. Here the feedback has been increased to 80%. Setting $r = 2.5$, below the value for period doubling of the uncontrolled map, the whole interval between x_0 and x_1 (along the line $y = bx$) is within the basin of attraction. When $r = 3.5$, however, for which value feedback is needed in order to maintain control, there has been an obvious *fractal invasion* of the basin, affecting even the previously clear space between the two fixed points. A more drastic invasion soon takes place, and is shown in Fig. 43. With $r = 3.7$, well inside the range for stability, we see a co-existing period 3 attractor; its (fractal) basin has taken over much of the former basin of the fixed point (left). The mechanism of the tangent bifurcation which produces this new orbit is shown on the right; it experiences a crisis between $r = 3.7$ and $r = 4.0$, whereupon its basin joins the basin of infinity.

The phenomenon of a sudden rapid decrease of the safe basin, with increase in an operating parameter, has been called the *Dover cliff phenomenon* by one writer.[39] A good place to commence further reading on the control of chaos is the collection of articles in Ref. [24]; it is, of course, an area of importance and ongoing research.

---

[39] J. M. T. Thompson, "Chaos and fractal basin boundaries in engineering", [22], p201.

References

Because these lectures are an elementary introduction, most of the references are to books which may be consulted for a more extensive treatment. The only exceptions are a small number of references to papers which are directly referenced in the text.

K. T. Alligood, T. D. Sauer and J. A. Yorke. Chaos: An Introduction to Dynamical Systems.Springer, 1997

D. K. Arrowsmith and C. M. Place. An introduction to Dynamical Systems. Cambridge University Press, 1990

G. L. Baker and J. P. Gollub. Chaotic Dynamics, an Introduction. Cambridge University Press, 1990

J. Barrow-Green. Poincaré and the Three-body problem (History of Mathematics, volume 11). American Mathematical Society, 1997

B. Davies. Exploring Chaos: theory and experiment. Perseus Press, 1999

B. Davies. Graphical visualisation of bifurcation theory for two-dimensional discrete systems. American Journal of Physics, submitted

R. L. Devaney. Chaotic Dynamical Systems, 2nd ed. Addison-Wesley, 1989

K. Falconer. Fractal Geometry: Mathematical Foundations and Applications. Wiley, 1990

Mitchell J. Feigenbaum. Quantitative universality for a class of nonlinear transformations. Journal of Statistical Physics,19,25–52 (1978)

J. Gleick. Chaos: Making a New Science. Viking Press, NY, 1987

M. C. Gutzwiller. Chaos in Classical and Quantum Mechanics. Springer, 1990

M. Hénon. A two-dimensional mapping with a strange attractor. Communications in Mathematical Physics, 50, 69–77 (1976)

R. C. Hilborn. Chaos and nonlinear dynamics: an introduction for scientists and engineers. Oxford University Press, 1994

Philip Holmes. A nonlinear oscillator with a strange attractor. Philosophical Transactions of the Royal Society of LondonA292 419–448 (1979)

Philip Holmes. Poincaré, celestial mechanics, dynamical systems theory and chaos. Physics Reports, 193, 137–163 (1990)

Edward N. Lorenz. Deterministic nonperiodic flow. Journal of Atmospheric Science, 20, 130–141 (1963).

E. N. Lorenz. The Essence of Chaos. University of Washington Press, USA, 1993

B. B. Mandelbrot. The Fractal Geometry of Nature. W. H. Freeman, NY, 1983

M. Martelli. Discrete dynamical systems and chaos. Longman Group UK, 1992

F. C. Moon. Chaotic Vibrations. Wiley, 1987

T. Mullin (ed.) The nature of Chaos. Clarendon Press, Oxford, 1993

E. Ott. Chaos in Dynamical Systems. Cambridge University Press, 1993

E. Ott, T. Sauer and J. A. Yorke. Coping with Chaos. Wiley, 1994

H. O. Peitgen, H. Jürgens and D. Saupe. Chaos and Fractals: New Frontiers of Science. Springer-Verlag, NY, 1992

I. Peterson. Newton's Clock: Chaos in the Solar System. W. H. Freeman, NY, 1993

N. Rasband. Chaotic Dynamics Nonlinear Systems. Wiley, 1990

D. Ruelle and F. Takens. On the nature of turbulence. Communications in Mathematical Physics, 20, 167–192 (1971)

H. G. Schuster. Deterministic Chaos: An Introduction. 2nd. ed., VCH Publishers, Weinheim, Germany, 1988

M. Yamaguti, M. Hata and J. Kigami. Mathematics of Fractals. (Translations of Mathematical Monographs, vol. 167) Americal Mathematical Society, 1997

Chapter 2

Hunting Mathematical Butterflies

Nalini Joshi[1]

School of Mathematics and Statistics,
University of Sydney, NSW 2006, Australia

Abstract.   Complex systems as diverse as the weather and the solar system are modelled by non-linear differential equations that have elusive, unstable solutions. An infinitesimally small change in the state of the system at one place can lead to a vast change in its behaviour far away. Such extreme sensitivity is often taken to be a sign of chaos, but it also occurs in completely ordered, integrable systems. In these lecture notes, I will explain some of the mathematical methods for finding and describing such solutions of differential equations in asymptotic limits.

1   Introduction

In complex systems, infinitesimally small disturbances can have vast consequences. In meteorological modelling, the effect of a butterfly's wings[2] is often used as a vivid metaphor of this phenomenon. The motion of the butterfly's wings triggers a tiny disturbance that grows dramatically to change large scale atmospheric motions.

The butterfly represents an extremely unstable solution of the differential equations modelling the weather. But such solutions exist in many differential equations, including ones that are ordered or *integrable* and have no chaos whatsoever.

The complexity of such systems is captured in many mathematical models by differential equations with irregular singularities. To see the

---

[1]e-mail: `nalini@maths.usyd.edu.au`

[2]E. Lorenz, "Predictability: Does the flap of a butterfly's wings in Brazil set off a tornado in Texas?" quoted in *Chaos and Nonlinear Dynamics* by R. C. Hilborn (Oxford University Press, 1994).

butterfly-like solutions, we need to know about the techniques of asymptotics, especially those techniques that show us how to describe solutions near irregular singularities.

The focus of these lecture notes lies in asymptotic analysis of solutions of differential equations in the limit as we approach such a singularity of the equation.

1.1   *Differential Equations with Singularities*

Definition 2.1   Consider the second-order linear ODE

$$y'' + p(x)y' + q(x)y = 0 \tag{1.1}$$

near a point l.

(1) If both $p(x)$ and $q(x)$ are analytic at $x = l$, then l is called an *ordinary* point.

(2) If one of $p(x)$ or $q(x)$ fails to be analytic at $x = l$, but both $(x - l)p(x)$ or $(x - l)^2 q(x)$ are analytic at l, then l is called a *regular* singular point.

(3) If one of $(x - l)p(x)$ or $(x - l)^2 q(x)$ is not analytic at $x = l$ then l is called an *irregular* singular point.

The definition can be applied to the point at infinity by first transforming to $x = 1/t$ and analysing the resultant ODE near $t = 0$.

Example 1.1.1   Consider the linear ODE

$$y'' - y' \left(1 - \frac{1}{x}\right) = 0. \tag{1.2}$$

This equation has a regular singular point at $x = 0$ and an irregular singular point at infinity. All other points in the complex plane are ordinary points.

A theorem of L. Fuchs (see [3]) shows that if l is a regular singular point then at least one solution of Equation (1.1) has the form of a Fröbenius (also called Fuchsian) series

$$y(x) = \sum_{n=0}^{\infty} a_n (x - l)^{n+\alpha}$$

convergent in some domain containing l, where α, called the inidicial exponent, is to be found from the ODE. (The second solution is similar, but may involve $\log(x - l)$.) If no solution has this form, then the point l must be an irregular singular point.

1.2   *Asymptotics*

To understand what happens near irregular singularities, we first need some notation and terminology from asymptotics.

Definition 2.2   Suppose f and g are two (complex-valued) functions of a complex variable x. Assume that x is approaching l along a path γ in a domain containing l or having it as a limit point and assume that $g(x)$ is bounded below along γ. To compare f and g in this limit, we use the following notations.

(1) f is said to be *much, much less than* g (or g is much, much greater than f) in the limit or

$$f(x) \ll g(x) \text{ as } x \underset{\gamma}{\to} l$$

   iff

$$\lim_{x \to l} \frac{f(x)}{g(x)} = 0$$

(2) f is *asymptotic to* g or

$$f(x) \sim g(x) \text{ as } x \underset{\gamma}{\to} l$$

   iff

$$\lim_{x \to l} \frac{f(x) - g(x)}{g(x)} = 0 \iff \lim_{x \to l} \frac{f(x)}{g(x)} = 1$$

(3) f is *of the order of* g i.e.

$$f(x) = O\big(g(x)\big) \text{ as } x \underset{\gamma}{\to} l$$

   iff

$$\exists\, M \text{ s.t. } \left| \frac{f(x)}{g(x)} \right| < M \text{ as } x \underset{\gamma}{\to} l.$$

As an example, consider

$$\sinh x = \frac{e^x - e^{-x}}{2},$$

in the limit as $x \to +\infty$ along the positive real axis. Then we have

$$\sinh x - \frac{e^x}{2} = -\frac{e^{-x}}{2}$$

$$\Leftrightarrow \quad \frac{\sinh x - e^x/2}{e^x/2} = -e^{-2x}$$

$$\Leftrightarrow \quad \lim_{x \to +\infty} \frac{\sinh x - e^x/2}{e^x/2} = 0.$$

This proves that

$$\sinh x \sim \frac{e^x}{2} \text{ as } x \underset{\mathbb{R}}{\to} +\infty$$

Exercise 2.1   Prove the following asymptotic results.

(1) $\sinh x \sim - e^{-x}/2$ as $x \underset{\mathbb{R}}{\to} -\infty$

(2) $\sinh x \ll xe^x$ as $x \underset{\mathbb{R}}{\to} +\infty$

(3) $\sinh x \sin x = O(e^x)$ as $x \underset{\mathbb{R}}{\to} +\infty$

Note that the same function may have different asymptotic representations as $x \to \infty$ along different paths (or directions). Note also that in the last exercise above $\sinh x \sin x$ has no limit as $x \to +\infty$, however, it does have an asymptotic bound. This shows the usefulness of the large-O notation even when the limit may not exist.

A function f can also be compared with reference functions in the limit as $x \to l$. A natural set of reference functions is the set of (integer) powers of $x - l$. These lead to (standard) asymptotic series.

Definition 2.3 f is said to be *asymptotic to a series* $\sum_{n=0}^{\infty} a_n(x - l)^n$ as $x \to l$ if and only if for each integer $N \geq 0$ we have

$$\lim_{x \to l} \frac{f(x) - \sum_{n=0}^{N} a_n(x - l)^n}{(x - l)^N} = 0.$$

In this case, we write

$$f(x) \sim \sum_{n=0}^{\infty} a_n(x - l)^n, \text{ as } x \to l.$$

If the set of reference functions includes other functions such as logarithms and exponentials, the asymptotic series is often referred to as a *generalised* asymptotic series. (See [17] for examples and possible paradoxes that may arise in choosing inappropriate reference functions.)

Some straightforward examples of asymptotic series, valid as $x \to l$, are given by Taylor expansions of analytic functions in a disk centred at l.

Exercise 2.2   Show that the Taylor expansion of $\exp(x)$ in a neighbourhood of $x = 0$ is an asymptotic series as $x \to 0$:

$$e^x \sim \sum_{n=0}^{\infty} \frac{x^n}{n!}, \text{ as } x \to 0.$$

However, not all expansions are asymptotic series in a given limit. The Laurent expansion of a function $f(x)$ with an isolated singularity at $x = l$, convergent in an annulus centred at that point, is not necessarily an asymptotic series of $f(x)$ as $x \to l$.

Example 1.2.1   The Laurent expansion of $\exp(1/x)$

$$e^{1/x} = \sum_{n=0}^{\infty} \frac{1}{n! x^n} \tag{1.3}$$

which is convergent in an annulus around $x = 0$ is *not* an asymptotic expansion of $\exp(1/x)$ as $x \to 0$. To see this, consider

$$
\begin{aligned}
N! \, x^N \left(\exp(1/x) - \sum_{n=0}^{N} (n! x^n)^{-1} \right) &= \sum_{n=N+1}^{\infty} \frac{N!}{n! x^{n-N}} \\
&= \frac{N!}{(N+1)! \, x} + \frac{N!}{(N+2)! \, x^2} + \cdots \\
&\to \infty, \text{ as } x \to 0
\end{aligned}
$$

Exercise 2.3   Show that (1.3) *is* an asymptotic series as $|x| \to \infty$.

Fröbenius expansions [3] of solutions of ODEs near a regular singular point $x = l$ are also asymptotic series in the limit $x \to l$.

Exercise 2.4   Consider the ODE

$$xw''(x) - w(x) = 0,$$

near $x = 0$. Show that $x = 0$ is a regular singular point. Show that there exists at least one solution given by

$$w(x) = \sum_{n=0}^{\infty} a_n x^{n+1}, \quad a_0 \neq 0,$$

where

$$a_n = \frac{1}{(n!)^2 \, (n+1)}.$$

This (convergent) Fröbenius series is also an asymptotic series as $x \to 0$, because

$$\frac{w(x) - \sum_{n=0}^{N} a_n \, x^{n+1}}{x^{N+1}} =: x \, g(x)$$

where $g(x)$ is analytic at $x = 0$.

Taylor, Laurent and Fröbenius series expansions are all convergent series (in an appropriate domain). However, not all asymptotic series are convergent.

Exercise 2.5   Show that $x = 0$ is an *irregular* singular point of the ODE

$$x^2 w''(x) + (3\,x - 1)w'(x) + w(x) = 0.$$

By differentiating, show that it is solved by the formal series

$$\sum_{n=0}^{\infty} n! \, x^n.$$

Show that this series is divergent everywhere except at $x = 0$. (Consider the limit of the ratio of successive coefficients as $n \to \infty$.) However, it is an asymptotic series in the limit $x \to 0$.

Divergent asymptotic series such as in Example 2.5 are *formal solutions* of the differential equation of interest. A major question then arises: do these formal solutions give any information about actual or true solutions of the differential equation? Wasow [18] proved a very general theorem that shows that the answer is yes for a large class of differential equations and formal solutions.

Theorem 1.2.1   *Wasow[18] (Theorem 12.1) Let S be an open sector of the complex x-plane with vertex at the origin and a positive central angle* *not exceeding $\pi/(q+1)$ (where $q \geq 0$ is integer). Let $f(x,w)$ satisfy the following properties.*

[1] *$f(x,w)$ is a polynomial in the components w_j of w, $j = 1,\ldots,r$, with coefficients that are holomorphic in x in the region*

$$0 < x_0 \leq |x| < \infty, \quad x \mathrm{in} S(x_0 \text{ a constant}).$$

[2] *The coefficients of the polynomial $f(x,w)$ have asymptotic series in powers of x^{-1}, as $x \to \infty$, in S.*

[3] *If $f_j(x,w)$ denotes the components of $f(x,w)$ then all eigenvalues λ_j, $j = 1,\ldots,r$, of the Jacobian matrix $\lim_{x\to\infty \mathrm{in} S}\left(\partial f_j/\partial w_k \big|_{w=0}\right)$ are different from zero.*

[4] *The differential equation*

$$x^{-q}w' = f(x,w) \tag{1.4}$$

is formally satisfied by a power series of the form

$$\sum_{k=1}^{\infty} b_k x^{-k}.$$

Then there exists, for sufficiently large x in S, a solution $w = \phi(x)$ of Eq. (1.4) such that, in every proper sub-sector of S,

$$\phi(x) \sim \sum_{k=1}^{\infty} b_k x^{-k}, \quad x \to \infty.$$

We show how to apply this theorem to Equation (1.2) in §2. The theorem shows that there exists a solution of this ODE with asymptotic series

$$y(x) \sim \frac{e^x}{x} \sum_{n=0}^{\infty} \frac{n!}{x^n} \text{ as } x \to \infty, \tag{1.5}$$

in the right half-plane.

However, note that Theorem 1.2.1 does not give a *unique* solution. It simply points out that there exists a true solution asymptotic to the formal series. In the case of Equation (1.2), one solution with asymptotic behaviour (1.5) is given by the exponential integral

$$Ei(x) = \int_{-\infty}^{x} \frac{e^t}{t} dt, \tag{1.6}$$

where the dashed integral sign means a principal value integral. (Check this by differentiation.) This is a well known special function [1] that appears in many physical applications. But there is a whole family of possible such true solutions. These are defined by

$$y(x) := \fint_c^x \frac{e^t}{t}\, dt,$$

where c is any nonzero number. The difference $Ei(x) - y(x)$ is a constant

$$\kappa = \int_{-\infty}^c \frac{e^t}{t}\, dt,$$

which is smaller than every term in the series (1.5). Series that contain such hidden terms are said to suffer from *asymptotics beyond all orders* and the term is said to be *hidden beyond all orders* in the limit.

Hidden terms lead to the subtle and difficult *Stokes' phenomenon* as we travel around in the complex plane. There are special directions in $\mathbb{C}$ where dominant terms (see Definition 2.4) change places with subdominant terms in an asymptotic approximation. Stokes said in 1902:

> *The inferior term enters as it were into a mist, is hidden for a little from view, and comes out with its coefficient changed.*

For exponential integrals like $y(x)$, the positive real line is such a special direction, i.e., a *Stokes' line*. See §2.4 for a more detailed description. There are other directions called *anti-Stokes' lines* where the previously dominant and subdominant terms become approximately equal in size. The series (1.5) becomes oscillatory along the imaginary axis and the constant K that was hidden along the positive real axis becomes comparable in size to $\exp(x)/x$, when x is purely imaginary.

One more definition is necessary before we go onto study non-linear ODEs in asymptotic limits. In studying the asymptotics of solutions of (differential or other) equations, we are often led to a limiting form of the equation in which small terms are neglected. Such forms are called *dominant balances*.

Definition 2.4   An asymptotic limiting form of an equation in which only the largest terms remain is called a *dominant balance*. If the set of largest terms is maximal for the original equation the balance is called *maximal*. Otherwise, it is called *sub-maximal*.

Example 1.2.2 The ODE

$$u' + \frac{u}{3z} = u^2 + 1$$

has the maximal dominant balance

$$u' \sim u^2 + 1,$$

and a sub-maximal balance

$$u^2 \sim -1$$

as $z \to \infty$.

This example is studied in more detail in §3.

1.3 *Non-linear Differential Equations and Chaos*

Nonlinear differential equations give rise to a wide range of behaviours from beautifully ordered and predictable dynamics to amazingly complicated, apparently unpredictable, chaotic dynamics. The analysis of these behaviours, whether ordered or chaotic, is made more complicated by the fact that different behaviours occur in different regions and there are usually no explicit formulæ that allows us to relate these different behaviours.

Example 1.3.1 Consider the nonlinear differential equation governing the angular displacement $\theta(t)$ of a plane pendulum:

$$\theta''(t) = -\big(1 + F(t)\big)\sin\big(\theta(t)\big). \tag{1.7}$$

When $F(t) \equiv 0$, the motion is completely ordered and predictable [11]. When $F(t)$ is a periodic function (i.e., the point of support of the pendulum is jiggled periodically) its motion becomes chaotic [19].

The analysis can be very subtle when very small perturbations somewhere have very large effects far away, as in the above mentioned *butterfly effect*. The word chaos is used in part because, in regions of chaotic behaviour, it is impossible to predict the future behaviour of a solution without knowing its initial value with infinite precision. Two solutions that are qualitatively close at first may be wildly different at a later point in time. This extreme sensitivity of solutions to small changes in initial conditions is symbolically represented by the fluttering of the butterfly in the butterfly effect. The butterfly sets off an instability that has great consequences, such as the land-fall of a typhoon, far away on the earth.

It is commonly believed that this type of instability is a trademark of chaos. Consider Devaney's definition [9] of chaos:

Definition 2.5   A continuous function f on a metric space X is *chaotic* if

(1) f is topologically transitive on X,
(2) the periodic points of f are dense in X, and
(3) f has sensitive dependence on initial conditions.

The iteration of f on X defines a dynamical system. The sequence of image points $f^n(x_0)$ for some $x_0 \text{in} X$ defines the orbit of x_0. The first property means that the collection of orbits for all $x_0 \text{in} X$ cannot be decomposed into two non-intersecting subsets. The second property means that every subset of X contains a periodic fixed point of the iterations of f. The third property implies the instability that is described by the butterfly effect. However, Banks *et al* [2] showed that if f has properties (1) and (2) on X then it must necessarily have property (3). In other words, we could drop property (3) from the definition. In fact such sensitivity to initial conditions can exist in the solutions of systems that are not chaotic at all.

Nevertheless, there is a major mathematical difficulty in the analysis of such extremely sensitive, highly unstable solutions. They can usually only be described explicitly when the independent variable goes to a limiting value. The resulting approximation of the solution is, in general, given by divergent asymptotic series. (A sketch of the theory of asymptotics is given in §2.) The divergence of such series makes it difficult to identify uniquely the solution (or "mathematical butterfly") of interest and to deduce any of their properties far away from the limit.

For linear ODEs, the above problem can usually be overcome by explicit representations of the solutions, such as integral representations. An example is given in §2. For nonlinear ODEs, however, there is no such representation. The problem is made harder by the fact that their solutions cannot usually be written down explicitly in terms of known functions.

Example 1.3.2   The very simple-looking equation

$$y''(x) = 6\,y(x)^2 + x \tag{1.8}$$

has solutions that are highly transcendental. It has been proved that its solutions cannot be described in terms of common operations (addition, multiplication, composition, integration etc) acting on classically known functions (such as polynomials, rationals, $\exp(x)$, or functions such as Bessel functions $J_\nu(x)$).

Such ODEs can have extremely unstable solutions, or "butterflies." For most nonlinear equations, such solutions are impossible to describe explicitly, yet, their behaviours are important for physical applications. For example, the ODE (1.8), known as the first Painlevé equation, appears in the two-dimensional model of quantum gravity [10]. In these lectures, I will describe some mathematical methods for capturing the qualitative behaviours of solutions suffering from asymptotics beyond all orders, including solutions that play the role of mathematical butterflies in asymptotic limits.

2   Linear Asymptotics

In this section, we consider linear examples of asymptotic analysis in more detail.

Example 2.0.3   Recall the linear ODE (1.2).

$$y'' - y' \left(1 - \frac{1}{x}\right) = 0.$$

One question of importance is to find the behaviour of the function as $x \to +\infty$. The desired information can be found by a careful integration by parts. (Care is needed because the real path of integration has a small gap around $t = 0$ when x is real and positive. A rigorous, careful analysis can be found in the book by Olver [17].)

I will illustrate the main point by assuming that x is complex with $\mathrm{Re}\,x > 0$ and small $\mathrm{Im}\,x > 0$, in which case the path of integration can be assumed to be a straight line γ connecting $-\infty$ to x minus the contribution from a semi-circular path around the origin. The first two steps of the integration-by-parts procedure then are

$$
\begin{aligned}
Ei(x) + i\pi &= \frac{e^x}{x} + \int_\gamma \frac{e^t}{t^2}\,dt \\[2mm]
&= \frac{e^x}{x} + \frac{e^x}{x^2} + 2\int_\gamma \frac{e^t}{t^3}\,dt
\end{aligned}
$$

At the nth-step this gives

$$Ei(x) + i\pi = \frac{e^x}{x}\left\{1 + \frac{1}{x} + \frac{2}{x^2} + \ldots + \frac{x}{e^x}n!\int_\gamma \frac{e^t}{t^{n+1}}dt\right\}$$

Noticing that e^x/x is much much larger than a constant as $\mathrm{Re}\,x \to \infty$, we get

$$Ei(x) \sim \frac{e^x}{x}\sum_{n=0}^{\infty}\frac{n!}{x^n} \text{ as } x \to \infty. \tag{2.1}$$

Notice that (2.1) is divergent.

Convergence of a series is equivalent to the convergence of the sequence of N-th partial sums, for each fixed x in some domain, as $N \to \infty$. However, asymptoticity of a series is equivalent to the vanishing of the relative difference between the function and the N-th partial sum, for each *fixed N*, as $x \to l$.

Example 2.0.4   Consider how Theorem 1.2.1 may be applied to Equation (1.2). Let the formal solution (1.5) be denoted as $y_f(x)$. Perturb the solution as $y(x) = y_f + v(x)$. Then the perturbation v satisfies the same equation as y, giving rise to the leading order solution $v = ae^x/x + b$, for arbitrary constants a and b.

However, to be a valid perturbation we must have $v \ll y_f$. This implies $a = 0$. Moreover, it implies that x must lie strictly in the interior of the half-plane $\mathrm{Re}(x) > 0$. Otherwise, near the imaginary axis, e^x becomes oscillatory and the condition $v \ll y_f$ no longer holds. This gives us the sector S of angle not exceeding π (actually the right half-plane).

To have a decaying formal solution, as required by the theorem, we need to transform to $w_1(x) = xe^{-x}y(x) - 1$. Then Equation (1.2) becomes

$$w_1'' + 3\,w_1'(1 - 1/x) + (w_1 + 1)(2 - 2/x + 3/x^2) = 0.$$

Rewrite this as a system:

$$\frac{d}{dx}\begin{pmatrix} w_1 \\ w_2 \end{pmatrix} = \begin{pmatrix} w_2 \\ -3\,w_2(1 - 1/x) - (w_1 + 1)(2 - 2/x + 3/x^2) \end{pmatrix}$$

where $w_1' = w_2$. Letting $q = 0$ and $f = (f_1, f_2)^t$, where $f_1 = w_2$, $f_2 = -3\,w_2(1-1/x) - (w_1+1)(2-2/x+3/x^2)$, the Jacobian matrix in Theorem

1.2.1 is then

$$\lim_{x\to\infty,\, x\,\mathrm{in}\,S}\left(\partial f_j/\partial w_k \,\big|_{w=0}\right) = \begin{pmatrix} 0 & 1 \\ -2 & -3 \end{pmatrix}$$

which has eigenvalues $\lambda_1 = -2$, $\lambda_2 = -1$. All the conditions of Wasow's theorem are satisfied and so we obtain a true solution that is asymptotic to the formal series (1.5) in the right half-plane as $x \to \infty$.

Note that it is easy to make the wrong transformations! Show that if we took $w = xe^{-x}y(x)$, leaving the leading order term unity for w, one of the eigenvalues λ_j above would turn out to be zero.

2.1 *Accuracy of Asymptotic Approximations*

Asymptotic series provide extremely accurate approximations of functions, up to a point. Despite its divergence, the series (1.5) gives a highly accurate approximation to $Ei(x)$ for large x. Let

$$F(x) := xe^{-x}Ei(x).$$

MAPLE (to 20 digits) gives

$$F(100) \underset{20\,\mathrm{digits}}{=} 1.0102062527748357112$$

$$\sum_{n=0}^{99} \frac{n!}{100^n} \underset{20\,\mathrm{digits}}{=} 1.0102062527748357112$$

Notice that the approximation agrees with the calculated value of F to 20 digits! However, this accuracy varies with the number of terms in the partial sum:

$$\sum_{n=0}^{260} \frac{n!}{100^n} \underset{20\,\mathrm{digits}}{=} 1.0108296120102260952$$

$$\sum_{n=0}^{281} \frac{n!}{100^n} \underset{20\,\mathrm{digits}}{=} 732496.06921461904157$$

Contrary to the case for convergent series, a larger number of terms does not imply greater accuracy. For each given x, there exists a number $N = N(x)$, such that the N-th partial sum provides an optimal approximation [3].

2.2   *The Problem of Identification*

Unfortunately, such accuracy does not resolve analytic difficulties. There is a problem of identification of solutions caused by the divergence of the asymptotic series. We will explain this idea in detail by using the ODE (1.2) as our main example in this sub-section.

As explained in §1, Eqn (1.2) possesses other solutions. Recall that a one-parameter family of such solutions with the same asymptotic series (1.5) is given by

$$y(x) = \int_c^x \frac{e^t}{t} dt,$$

where c is an arbitrary constant. Such solutions differ from $Ei(x)$ only by a constant

$$Ei(x) - y(x) = \int_{-\infty}^c \frac{e^t}{t} dt = \kappa.$$

For large positive $\mathrm{Re}(x)$, this constant is smaller than every term of the series (1.5). In other words, for larger and larger x, we would have to know the solution with better and better accuracy to be able to tell whether it is $Ei(x)$ or not. This phenomenon is also called *asymptotics beyond all orders*.

The value of κ plays the role of the butterfly in this example. If it changes, then the solution changes. The changed solution will be very different from other solutions of the same ODE somewhere else in the x-plane.

However, κ cannot be identified from the series (1.5). To identify κ, we would need to either apply the definition of an asymptotic series infinitely often or sum the series. Neither is possible (with conventional tools).

Moreover, if the function $y(x)$ is analytically continued from the first quadrant $(0 < \arg(x) < \pi/2)$ to the fourth quadrant $(-\pi/2 < \arg(x) < 0)$, the value of κ changes. To see this, we use the technique of Borel summation. Roughly speaking, this is the process of replacing the series (1.5), by mapping to a new series

$$\frac{e^x}{x} \sum_{n=0}^{\infty} \frac{1}{x^n}.$$

To see that such a mapping is possible, consider the representation

$$e^{-x} Ei(x) = \int_{\underset{\mathcal{L}}{0}}^{\infty} q(s) e^{-xs} \, ds \qquad (2.2)$$

where $\mathcal{L}$ is a path (for simplicity, a ray) joining 0 to ∞.

Exercise 2.6 Show by using the Gamma function

$$\Gamma(n+1) = \int_{0}^{\infty} t^n e^{-t} \, dt$$

to replace $n!$ in the formal sum $\sum_{n=0}^{\infty} \frac{n!}{x^n}$, and changing the order of summation and integration, that

$$q(s) = \sum_{n=0}^{\infty} \frac{1}{s^n} = \frac{1}{1-s}.$$

Now consider $Ei(x)$, as $x \to \infty$ along different rays. (i) $\arg(x) = \pi/2$, (ii) $\arg(x) = -\pi/2$. For the representation Equation (2.2) to be convergent, we need $\mathrm{Re}(x\,s) > 0$. So in the case (i), the path $\mathcal{L}$ must lie in the lower half-s-plane, while for case (ii), $\mathcal{L}$ must lie in the upper half-plane. Hence the change in $e^{-x} Ei(x)$ is given by

$$\frac{1}{2\pi i} \oint_{C} \frac{e^{-xs}}{1-s} \, ds = e^{-x},$$

where the closed contour C encloses the pole at $s = 1$. That is, the change in κ is unity.

2.3 *Asymptotic Analysis Near Irregular Singular Points*

Now suppose the linear ODE (1.1) has an irregular singular point at $x = \infty$. There is a well known standard approach for finding its formal asymptotic solutions. Write $y(x) = \exp(S(x))$. We get

$$S'' + (S')^2 + p(x)S' + q(x) = 0.$$

Assume

$$S'' \ll S'^2, \quad x \to \infty,$$

for at least one solution. Otherwise, it can be shown that S is logarithmic or small, which implies an algebraic series for both solutions. We assume this is not the case. Therefore, we get

$$S'^2 \sim -p(x)S' - q(x).$$

In general, this quadratic asymptotic equation gives rise to two asymptotic solutions. (For an n-th order ODE, this would be an n-th degree equation with n asymptotic solutions.)

To be more specific, consider the Airy equation

$$y''(x) = x\,y(x). \tag{2.3}$$

The above approach gives

$$S'^2 = x - S''.$$

Recursive substitution for S'', assuming that it is smaller than x, gives

$$
\begin{aligned}
S' &= x^{1/2}\left(1 - S''/x\right)^{1/2} \\[2mm]
&= x^{1/2}\left(1 - (S''/2x) + O(S''/x)^2)\right) \\[2mm]
&= x^{1/2}\left(1 - x^{-3/2}/4 + O(x^{-3})\right) \\[2mm]
&= x^{1/2} - x^{-1}/4 + O(x^{-5/2})
\end{aligned}
$$

which implies

$$S \sim \frac{2}{3}x^{3/2} - \frac{1}{4}\log x + const.$$

Iteration shows that there are two formal solutions given by linear combinations of

$$
\begin{aligned}
y_+(x) &\sim e^{2x^{3/2}/3}x^{-1/4}\sum a_n x^{-3n/2} \\
y_-(x) &\sim e^{-2x^{3/2}/3}x^{-1/4}\sum (-1)^n a_n x^{-3n/2}
\end{aligned}
$$

as $x \to +\infty$, where $a_0 = 1$ and

$$a_n = \frac{3}{4n}(n - 1/6)(n - 5/6)a_{n-1},$$

for $n \geq 1$.

Exercise 2.7   Show that these asymptotic series formally satisfy the Airy equation. By considering

$$\lim_{n\to\infty}\left|\frac{a_{n-1}}{a_n}\right|$$

show that the radius of convergence of these formal solutions is zero.

That is, they are divergent for all $x \neq 0$. The following exercise gives another classical example.

Exercise 2.8   Carry out an asymptotic analysis near infinity for Bessel's equation

$$x^2 y'' + xy' + (x^2 - \nu^2)y = 0,$$

by transforming to $y = \exp(S(x))$. Show that along the positive real axis, the asymptotic behaviour is given (in general) by

$$y(x) \;=\; \frac{a(x)}{\sqrt{x}}\cos\big(x - (\nu + 1/2)\pi/2\big)$$

$$+\frac{b(x)}{\sqrt{x}}\sin\big(x - (\nu + 1/2)\pi/2\big)$$

(the phase $(\nu + 1/2)\pi/2$ is not essential but is chosen to give the standard Bessel's functions $J_\nu(x)$ and $Y_\nu(x)$) where $a(x)$, $b(x)$ are linear combinations of

$$w_1 \;\sim\; \sum_{n=0}^{\infty}(-1)^n c_{2n}x^{-2n}$$

$$w_2 \;\sim\; \sum_{n=0}^{\infty}(-1)^n c_{2n+1}x^{-2n-1}$$

$$c_n \;=\; \frac{(4\nu^2 - 1^2)(4\nu^2 - 3^2)\dots(4\nu^2 - (2n-1)^2)}{8^n n!}$$

$$c_0 \;=\; 1$$

Show that these asymptotic series are divergent.

2.4   *Stokes' Phenomena and Connection Problems*

The standard theorems (of existence and analyticity of solutions for ODEs) show that the solutions of Airy's equation must be entire functions (because the equation has no singularities in the finite plane). However, the asymptotic behaviours we found above are multi-valued functions. The resolution of this apparent paradox lies in the fact that these formal solutions represent the asymptotic behaviours of the Airy functions only in sectors of angular width $< 2\pi$. That is, the analytic continuation of the asymptotic behaviours of $Ai(x)$ in some sector (near infinity) is not necessarily its asymptotic behaviour in other sectors.

Let $\theta := \arg(x)$. Along (or near) the positive real x-axis, the boundary condition

$$y(x) \sim \frac{1}{x^{1/4}} \exp\left(-\frac{2}{3}x^{3/2}\right)$$

defines a unique solution. However, the alternative condition

$$y(x) \sim \frac{1}{x^{1/4}} \exp\left(+\frac{2}{3}x^{3/2}\right)$$

does *not* specify a solution uniquely, because it leaves the coefficient of the small solution $y_-(x)$ undefined. In other words, this behaviour suffers from asymptotics beyond all orders. Moreover, it suffers from Stokes phenomenon.

However, this problem no longer occurs along directions where the two formal solutions have the same order of magnitude as $|x| \to \infty$. In fact, the two solutions are the same size wherever

$$\left|\exp\left(\frac{4}{3}x^{3/2}\right)\right| = 1.$$

This occurs where

$$\frac{4}{3}\cos(3\theta/2) = 0,$$

i.e. along directions where $3\theta/2 = \pi/2 + j\pi$, for integer j. Such directions are called *anti-Stokes* lines.

The directions where one solution is minimal relative to the other one is called a *Stokes* line. The next Stokes line (travelling in an anti-clockwise direction from the positive real axis) occurs along $\theta = 2\pi/3$. Along this direction, it is y_- that is large and y_+ that is hidden beyond all orders.

The connection problem for Airy's equation is to relate all such asymptotic behaviours of a solution valid along different directions of approach to infinity. To solve this problem, we fix our attention on one solution.

Consider the standard solution of Airy's equation $Ai(x)$ defined by initial conditions

$$Ai(0) \quad = \quad 3^{-2/3}/\Gamma(2/3) \tag{2.4}$$
$$Ai'(0) \quad = \quad -3^{-1/3}/\Gamma(1/3). \tag{2.5}$$

It is well known that $Ai(x)$ has an integral representation given by

$$Ai(x) = \frac{1}{2\pi} \int_{\Gamma_1} dk e^{ikx+ik^3/3}. \tag{2.6}$$

where Γ_1 is a path starting from infinity in the sector $2\pi/3 < \theta < \pi$ and ending at infinity in the sector $0 < \theta < \pi/3$. (The second solution $Bi(x)$, linearly independent to $Ai(x)$, also has an integral representation.) The next exercise shows that such integral representations describe all solutions for appropriate choices of path.

Exercise 2.9 Let S_j be a sector of the complex k-plane described by $k \text{ in } [2j\pi/3, (2j+1)\pi/3]$ for $j = 0,1,2$. For any integer j modulus 3, let Γ be a path starting at infinity in S_j and ending in S_{j-1}.

(i) Show that the integral

$$y(x) = \frac{1}{2\pi} \int_{\Gamma} dk e^{ikx+ik^3/3}$$

satisfies the Airy equation.

(ii) Show that there are only two linearly independent choices of path.

The Airy functions also have a symmetry property under rotation of x by a cube root of unity.

Exercise 2.10 Let $\omega = \exp(-2\pi i/3)$. Show by using the integral representation of $Ai(x)$ that

$$Ai(x) = -\omega Ai(\omega x) - \omega^2 Ai(\omega^2 x). \tag{2.7}$$

This property together with Eqn(2.6) can be used to solve the Airy connection problem.

First, we show that $Ai(x)$ is asymptotic to a multiple of y_- in an extended sector centred on the positive real x-axis.

Theorem 2.4.1

$$Ai(x) \sim \frac{1}{\sqrt{\pi}} x^{-1/4} e^{-2x^{3/2}/3}, \quad |x| \gg 1, |\arg x| < \pi.$$

The proof is outlined in the following exercise. (The definition of saddle points and method of steepest descents can be found, e.g., in [3].)

Exercise 2.11   Consider the integral representation of $Ai(x)$.

(i) Show that saddle points of $\rho(k) = k + k^3/(3x)$ are given by $k_s = \pm i\sqrt{x}$.

(ii) Deform the path Γ_1 to pass through the saddle at $i\sqrt{x}$ (transversely to the imaginary k-axis). Show that this yields the leading-order behaviour stated in the theorem above.

(iii) Note that the saddle points move with x. The calculation is valid until the saddle points hit the boundary of the regions in which Γ_1 must lie. Show that this occurs when $|\arg x| = \pi$.

This classical result simultaneously gives the local asymptotic behaviour of $Ai(x)$ and connects it across the anti-Stokes lines $\arg x = \pm\pi/3$ and Stokes lines $\arg x = \pm 2\pi/3$, by effectively tracking $Ai(x)$ along a large circular arc in the complex x-plane. To connect to the one remaining direction $|\arg x| = \pi$ we use the relation (2.7).

Exercise 2.12   Use Eqn (2.7) and Theorem 2.4.1 to find the asymptotic behaviour of $Ai(x)$ as $x \to -\infty$ along the negative real axis.

This result completes the solution of the connection problem for $Ai(x)$ near infinity.

3   Nonlinear Asymptotics

Nonlinear asymptotics here means the asymptotic study of nonlinear ordinary differential equations (ODEs) by methods that rely only on the equations themselves.

3.1 *A First-Order Model*

Our first example is a first-order nonlinear model studied by Boutroux [4]. Consider the Riccati equation

$$y'(x) = y^2 + x, \tag{3.1}$$

as $x \to \infty$. This equation can be linearised to (a version of) the Airy equation.

Exercise 2.13 Show that the solution $y(x)$ of Eqn (3.1) is given by $y = -w'(x)/w(x)$ where $w(x)$ solves

$$w''(x) + xw(x) = 0. \tag{3.2}$$

(This becomes the standard Airy equation under the transformation $x \mapsto -x$.) Asymptotic properties of $w(x)$ could be derived from the asymptotic behaviours of the Airy functions $Ai(-x)$, $Bi(-x)$. However, instead of using such results for Airy functions, we will start from scratch and develop direct methods for the nonlinear equation as given.

To make the largest terms of this nonlinear equation explicit, consider the transformation

$$y = \sqrt{x}u(z), \quad z = \frac{2}{3}x^{3/2}.$$

This maps Eqn (3.1) to

$$u'(z) + \frac{1}{3z}u = u^2 + 1. \tag{3.3}$$

Boutroux considered a larger class given by

$$u' + 2p\frac{u}{z} = u^2 + 1, \quad |z| \to \infty, \tag{3.4}$$

for some constant p. This Riccati equation can be linearised through the transformation

$$u(z) = -\frac{\psi'(z)}{\psi(z)}$$

to

$$\psi'' + 2p\frac{\psi'}{z} + \psi = 0$$

which can in turn be transformed via

$$\psi(z) = z^{1/2-p} w(z)$$

to

$$w'' + \frac{w'}{z} + \left(1 - \frac{(p-1/2)^2}{z^2} \right) w = 0,$$

which is Bessel's equation.

It is well known that Riccati equations, such as (3.1), (3.4) have the Painlevé property [13]. So $u(z)$ is meromorphic except at $z = 0$, or ∞. This shows that $y(x)$ and, therefore, $u(z)$ can only have poles as movable singularities. Recall (see section 1.2) that, as $z \to \infty$, the maximal dominant balance of Eqn(3.3) is

$$u' \sim u^2 + 1.$$

Given $\epsilon \neq 0$, η, we take $|z_0| > 1/|\epsilon|$ and define a solution by the initial condition $u(z_0) = \eta$. This suggests, and we show below, that for generic values of η, the leading-order behaviour of $u(z)$, as $|z| \to \infty$, is given by $\eta + \tan(z - z_0)$.

Before we state and prove this generic result, recall that the function tan is implicitly defined by

$$\int_\eta^u \frac{dv}{v^2 + 1} = z - z_0$$

where $\tan(z_0) = \eta$. Its period π is given by

$$\oint_C \frac{dv}{v^2 + 1}$$

where C is a closed contour enclosing one of the roots of $v^2 + 1$ in the v-plane. In order to define such integrals, the path of the integral must avoid the roots of $v^2 + 1$ where it becomes singular.

Definition 2.6   The following are called *generic conditions*. Assume $\epsilon \neq 0$, $B > 0$, z_0 are given with $|\epsilon| < 1/2$, $|z_0| > 1/|\epsilon|$, $|\epsilon B| < 1/2$.

(i) z lies in the domain

$$\mathcal{Z} := \{ z \mid |z - z_0| < B, |z| > 1/|\epsilon| \}.$$

(ii) γ is a path starting at z_0 in $\mathcal{Z}$ of length $l < 2\pi B$, s.t. $u(z)$, $z$inγ, lies in

$$\mathcal{U} := \{u \mid |u| < |\log \epsilon|, |u^2 + 1| > |\epsilon|^{1/7}\}.$$

Definition 2.7 Suppose ϵ, η, are given that satisfy the generic conditions. Assume $u(z_0) = \eta$. Then the solution defined by this initial condition is called *a generic solution*.

Theorem 3.1.1 *Suppose the generic conditions are satisfied and u is an associated generic solution. Then, this solution satisfies*

$$|\arctan(u - \eta) - z + z_0| < \frac{2\pi B}{3}|\epsilon|^{6/7}|\log \epsilon|. \tag{3.5}$$

Proof 2.8 Let S be defined by

$$S := \int_\gamma \frac{u}{3z(u^2 + 1)}\, dz. \tag{3.6}$$

Dividing Eqn(3.3) by $u^2 + 1$ and integrating gives

$$\int_\Gamma \frac{du}{u^2 + 1} = z - z_0 - S,$$

where Γ is the image of γ under u. Here S is upper-bounded by

$$\begin{aligned} |S| &\leq 2\pi B \frac{|\epsilon|}{3} \frac{|\log \epsilon|}{|\epsilon|^{1/7}} \\ &= \frac{2\pi B}{3}|\epsilon|^{6/7}|\log \epsilon| \end{aligned}$$

by the generic conditions.

By taking Γ to be a closed contour starting at 0 and enclosing one of the points $\pm i$, this theorem extends to give the asymptotic spacing between successive zeroes of u. If z_0, z_1 are two such zeroes then we get

$$|z_1 - z_0 \mp \pi| < \frac{2\pi B}{3}|\epsilon|^{6/7}|\log \epsilon|. \tag{3.7}$$

where the choice of sign in the left side is dependent on the orientation of Γ.

Note that the power of $|\epsilon|$ in Eqn(3.5) or (3.7) depends on how closely we skirt around the roots of $u^2 + 1$. If $|u^2 + 1|$ is lower bounded by a constant (independent of ϵ) on γ this error can be made proportional to $|\epsilon \log \epsilon|$.

The generic assumptions leave five possible gaps in the (asymptotic) description of the solution space, namely the limits where the path Γ (or equivalently γ) becomes very long, its end-point becomes very large or becomes very close to $\pm i$ and where η becomes unboundedly large or arbitrarily close to $\pm i$. For further details on the asymptotic estimates in these cases see [14]. We summarise the differences here.

The case of very long Γ The result is the same as in the above theorem with the addition of $N\pi$ inside the modulus signs on the left side of (3.5).

The case of large η This is equivalent to the case of large end-point of Γ (modulo $N\pi$ being added to the left side of (3.5)). The same theorem holds then for v with η replaced by η_1 where $v(z_0) = \eta_1 = 1/\eta$ and v lies in a ball of radius $1/|\log \epsilon|$ around the origin. Moreover, the estimate (3.7) holds for the spacing between successive poles of u.

Nearly degenerate case This is the case when u^2 is close to -1 but not arbitrarily close. The asymptotic behaviour of u is an almost degenerate form of the generic leading-order behaviour, obtained as a limit as $\tan(z - z_0) \to \pm i$. To leading-order u oscillates around the value $\pm i$.

Degenerate case This is the case when u^2 is arbitrarily close to -1. In this case, the solution is asymptotic to a power series, in powers of $1/z$, with no apparent parameter. It can be shown that no free parameter appears explicitly at any order of its asymptotic series.

Finally, we can show also that the asymptotic behaviours form a connected component of the solution surface. So this description is complete.

These results also show that the degenerate behaviours are free of poles or zeroes in their region of validity. Transforming back to the x-plane shows that, in general, for $|x| \gg 1$, the solution $y(x)$ has three (half-)lines of poles, aligned asymptotically with the rays of angle 0, $2\pi/3$, $4\pi/3$.

3.2   *The First Painlevé Equation*

Consider the first Painlevé equation P_I (1.8). Boutroux [4] showed that it can be mapped to

$$u_{zz} = 6u^2 + 1 - \frac{u_z}{z} + \frac{4}{25}\frac{u}{z^2} \tag{3.8}$$

under the transformation

$$y(x) = \sqrt{x}\, u(z), \ z = 4x^{5/4}/5,$$

Such changes of variables can be found by asymptotic arguments (see page 320 of [16]).

The maximal dominant balance of Equation (3.8) is solved by elliptic functions. Boutroux integrated the equation by multiplying first by u_z and integrating as though only the dominant terms

$$u'' \approx 6u^2 + 1$$

were present (where primes now denote derivatives with respect to z). However, this leads to problems in estimating the contribution of u_z^2/z in the small terms of the integrated equation.

Exercise 2.14   Show that multiplying Equation (3.8) by u' and integrating leads to

$$\frac{u'^2}{2} = 2u^3 + u + c - \int^z \left(\frac{u'^2}{z} + \frac{4}{25} \frac{uu'}{z^2} \right) dz$$

where c is a constant of integration.

Because u'^2 appears on the right as well as the left of this integrated equation, it requires more argument to show that u'^2 must be upper (or lower) bounded if $2u^3 + u + c$ is. Boutroux did not provide such an argument.

However, we can keep the term u'/z with the dominant terms and integrate the resultant asymptotic equation:

$$u'' + \frac{u'}{z} \approx 6u^2 + 1.$$

The two terms on the left have integrating factor z. To integrate the polynomial in u on the right we need the integrating factor u'. We combine the two and multiply the equation by $z^2 u'$ to get

$$\left((zu')^2/2 \right)' = \left(2z^2 u^3 + z^2 u \right)' - 4zu^3 - 2zu + \frac{2}{25}(u^2)'$$

Integration yields

$$z^2 u'^2 = z^2(4u^3 + 2u) + 2K - 4 \int_{z_0}^z z(2u^3 + u)dz + \frac{4}{25} u^2 \qquad (3.9)$$

where K is a constant of integration. To fix a solution, assume that initial values are given at a (nonzero) point z_0 by

$$u(z_0) = \eta, u'(z_0) = \eta'.$$

Note that K is related to these values respectively by

$$K = \frac{z_0^2}{2}\left(\eta'^2 - 4\eta^3 - 2\eta\right) - 2\eta^2/25,$$

If $|z_0|$ and $|z|$ were (equally) large, each term in the dominant balance of Equation (3.9) would be of $O(z_0^2, z^2)$. Since this will be the case in our treatment, we redefine

$$E := \frac{K}{z_0^2},$$

and rewrite the above equation as

$$\begin{aligned}
u'^2 &= 4u^3 + 2u + 2E + 2E\left(\frac{z_0^2}{z^2} - 1\right) \\
&\quad - \frac{1}{z^2}\int_{z_0}^{z} 4z(2u^3 + u)dz + \frac{4u^2}{25z^2},
\end{aligned} \tag{3.10}$$

To show that the solution has asymptotic behaviours given by elliptic functions, we need to first assume some conditions on the initial values and the domain in which we integrate the function. In particular, we assume that the local domain in which we integrate the equation lies in a fixed local patch near infinity. We also need some notation.

Definition 2.9   Define

$$P(u) := 4u^3 + 2u + 2E.$$

In general, $P(u)$ has three distinct roots, which we will denote by u_i, $i = 1, 2, 3$, in the complex plane.

It is well known [1] that

$$\int_{\eta}^{u} \frac{dv}{\sqrt{P(v)}} = z - z_0$$

defines the inverse of a Weierstrass elliptic function (conventionally denoted for standard initial values by $\wp(z)$), with periods

$$\omega_j = \oint_{C_j} \frac{dv}{\sqrt{P(v)}},$$

where C_j, $j = 1, 2$, are two linearly independent closed contours each enclosing two points in the set $\{u_1, u_2, u_3, \infty\}$.

There exist special values of E for which two roots of $P(u)$ coincide. To find these, consider

$$\frac{dP}{du} = 12u^2 + 2 = 0$$

$$\Rightarrow u = d_k := (-1/6)^{1/2}, \ k = 1, 2.$$

If such points are also zeroes of $P(u)$ then we must have

$$E = D_k := -2d_k(2d_k^2 + 1)$$

$$= -2(-1/6)^{1/2}(-1/3 + 1) = -\frac{4}{3}(-1/6)^{1/2}, \ k = 1, 2.$$

For each such value of E, $P(u)$ has a double root. Note, however, that $P(u)$ cannot have a triple root, because $P''(u) = 24u$ cannot vanish simultaneously with $P'(u)$. Now we estimate the spacing between the roots u_i when E is close to a degenerate value D_k for some k.

Definition 2.10 Suppose ϵ, B, z_0, η, η' are given numbers satisfying the conditions below.

(i) $0 < |\epsilon|$ is given such that

$$
\begin{aligned}
|\epsilon| &< \min(1/e, 1/B) \\
|\log \epsilon| &> \max(72\pi, 2/(5\sqrt{eB})) \\
8B|\epsilon|^{6/7}|\log \epsilon|^4 &< 1/2
\end{aligned}
$$

(ii) $|z_0| > 1/\epsilon$ and z lies in the domain $\mathcal{Z}$, defined by

$$\mathcal{Z} := \left\{ z \ \middle| \ |z - z_0| < B, |z| > 1/\epsilon \right\}.$$

(iii) $|\eta|$, $|\eta'|$ are bounded above by $|\log \epsilon|$, $|\eta'| > |\epsilon|^{1/7}$, and E defined by

$$2E = \eta'^2 - 4\eta^3 - 2\eta - \frac{4\eta^2}{25z_0^2}$$

satisfies $|E - D_k| > 2|\epsilon|^{2/7}$, $|E| < |\log \epsilon|^4$. Moreover, $u(z)$ satisfies the initial conditions $u(z_0) = \eta$, $u'(z_0) = \eta'$.

(iv) γ is a path joining z_0 to z in $\mathcal{Z}$ of length $l < 2\pi B$ such that its image Γ under u lies in the domain

$$\mathcal{U} := \left\{ u \ \middle| \ |u| < |\log \epsilon|, |P(u)| > |\epsilon|^{1/7} \right\}.$$

These conditions are called *generic conditions* and the solution $u(z)$ defined in (iii) above is called a *generic solution*.

Note that ϵ_0 can always be found so that the upperbounds on $|\epsilon|$ are satisfied for $0 < |\epsilon| < \epsilon_0$ and that a nonempty connected domain $\mathcal{U}$ and a path γ in $\mathcal{Z}$ exist because all solutions of P_I are known to be meromorphic.

It can be shown that the roots u_i of $P(u)$ are separated by a distance of order $O\big(|\epsilon|^{1/7}\big)$. The first of the results of the main theorem below shows that $|u'|$ is lowerbounded by a nonzero number on γ (in fact on $\mathcal{Z}$) if $|P(u)|$ is so. Therefore, $u(z)$ is invertible (by the inverse function theorem). In particular, the mapping $\gamma \mapsto \Gamma$ is invertible.

Our main result is

Theorem 3.2.1   *Under these conditions, $\exists$ positive ϵ_0 such that $\forall\ 0 < |\epsilon| \leq \epsilon_0$, the generic solution satisfies*

$$\big|u'^2 - P(u)\big| < 8B|\epsilon||\log \epsilon|^4,$$

and

$$\left|\int_{\eta\ \mathrm{r}}^{u} \frac{dv}{\sqrt{P(v)}} - (z - z_0)\right| < 16\pi B^2 |\epsilon|^{6/7}|\log \epsilon|^4,$$

in $\mathcal{Z}$. Moreover, if z_0 and z_{10} are two successive points in $\mathcal{Z}$ where $u = \eta$, then for $j = 1$ or 2,

$$|(z_{10} - z_0) - \omega_j| < 16\pi B^2 |\epsilon|^{6/7}|\log \epsilon|^4.$$

The proof of this main result and other cases can be found in [14].

3.3   *Tronquée Solutions*

P_I has other solutions that are not given by elliptic-function-type asymptotic behaviours. Here we state some results for the solutions that are asymptotic to algebraic power series in x as $|x| \to \infty$ in sectors of angular width $4\pi/5$. For simplicity of notation, we first map $x \mapsto -x$ and consider P_I in the form

$$y'' = 6y^2 - x.$$

The details and proofs for the results we describe here can be found in [15].

Proposition 2.11   *For any branch of $x^{1/2}$, the formal series*

$$y_f = \frac{x^{1/2}}{\sqrt{6}} \sum_{k=0}^{\infty} \frac{a_k}{(x^{1/2})^{5k}}, \tag{3.11}$$

where $a_0 = 1$ and the coefficients a_{k+1}, $k \geq 0$, are given by the recurrence relation

$$a_{k+1} = \frac{25k^2 - 1}{8\sqrt{6}}\, a_k - \frac{1}{2} \sum_{m=1}^{k} a_m a_{k+1-m}, \ k \geq 0, \tag{3.12}$$

is a formal solution of P_I.

Proposition 2.12   tronquée *solutions For any sector of angle less than $4\pi/5$, there exists a solution of P_I whose asymptotic behaviour as $|x| \to \infty$, in this sector, is given by the asymptotic series* (3.11).

Definition 2.13   The solutions defined by Proposition 2.12 are called the *tronquée* solutions (or *intégrales tronquées*).

Boutroux was the first person to study these solutions. He called them *tronquée* because the lines of poles of the generic solutions are "truncated" in the case of these more special solutions.

Exercise 2.15   Show by perturbing around a *tronquée* solution, that there exist formal exponential small terms hidden beyond all orders of the series (3.11). Moreover, show that these terms contain a free parameter.

So the true solutions given by Proposition 2.12 form a one-parameter family of solutions. Since the free parameter is hidden beyond all orders, in the interior of the sectors associated with these solutions, it is difficult to distinguish such solutions from each other. Nevertheless, there is a unique distinguished solution for each sector which is asymptotic to the formal series (3.11) in a wider sector.

Theorem 3.3.1   tritronquée *solution There exists a unique solution $Y(x)$ of P_I which has asymptotic expansion $y_f(x)$ in the sector $|\arg(x)| \leq 4\pi/5$ with $x^{1/2}$ denoting the analytic continuation of $-\sqrt{x}$.*

This surprising result is proved in [15] by starting with tronquée solutions in two overlapping sectors of width $4\pi/5$. However, the properties of the tritronquée solution $Y(x)$ in the finite plane are much harder to deduce. One of the fruitful ideas is to construct $Y(x)$ as the limit of a sequence of true solutions that are asymptotic to the leading order term $\sqrt{x/6}$. Details, including numerical estimates, can be found in [15]. See §4.3 below for a summary of this method of constructing $Y(x)$.

4   Optimal Asymptotics

At a given point z (in the domain of validity), it is well known that an asymptotic series such as (1.5) has a smallest term. This is illustrated by the numerical results displayed in §2.1. A highly accurate approximation of a function asymptotic to the series can be obtained by truncating the series just before the smallest term (see pp. 16 of [6] for illuminating examples). Such an approximation is called optimal approximation (see pp. 94 of [3]) and associated methods are called optimal asymptotics.

Much higher accuracies can be obtained by expanding the remainder, then truncating that at its smallest term, expanding *its* remainder and doing so again. Such techniques were developed by Berry and Howls [7; 8] to obtain highly accurate approximations of functions with formal series suffering from asymptotics beyond all orders. Furthermore, they have shown through such techniques that the Stokes phenomenon is not a discontinuous jump in asymptotic description across Stokes lines, but actually a smooth, rapid transition, with a generic form described by an error function. However, these major developments have been used primarily to improve the accuracy of approximations of functions, rather than to define true functions asymptotic to a given formal series. Related ideas are explored in [5].

Here we construct a unique true solution of Equation (1.2) that is asymptotic to the formal series (1.5) along the Stokes line by using the ideas of optimal asymptotics. Suppose N is a given positive integer. Let x_N be a point in the right half-plane where the Nth-term of the series (1.5) is the smallest term. We can find x_N, for large N, by considering the ratio of successive terms

$$\frac{(n+1)!}{x^{n+1}} \bigg/ \frac{n!}{x^n} = \frac{n+1}{x}.$$

Since the smallest term is approximately equal to the terms on either side of the minimum in the sum, we get the *optimal point* $x_N = N + 1$. (Note that these points lie precisely on the Stokes line.) Conversely, for any x in the right half-plane, we can define an optimal index $N = [|x|] - 1$ where $[z]$ means the integer part of z.

Now define the *optimal partial sums*

$$Y^{[N]}(x) := \frac{e^x}{x} \sum_{n=0}^{N-1} \frac{n!}{x^n}.$$

We use the fact that these provide the best approximation to the true solution y. They are optimal in the sense that they are partial sums of the series (1.5) truncated just before its smallest term at x_N. This motivates the construction of a sequence $y_N(z)$ of solutions given locally by optimal partial sums. We construct a true solution asymptotic to (1.5) by taking the limit of this sequence as $N \to \infty$.

4.1   *Construction of Optimal Sequence*

In this section, we explain how to construct a sequence $y_N(x)$ of solutions of Equation (1.2) with the appropriate properties by starting with the optimal partial sums $Y^{[N]}(x)$ at the corresponding optimal points x_N. This sequence will be referred to as an *optimal sequence*.

To construct appropriate solutions, we consider the sum of an optimal partial sum $Y^{[N]}(x)$ and a remainder-like function $\Delta_N(x)$. Take $x_N = N$ and define a solution $y_N(x)$ by

$$\begin{cases} y_N'' - y_N' \left(1 - 1/x\right) &= 0 \\ y_N(x_N) &= Y^{[N]}(x_N) \\ y_N'(x_N) &= Y^{[N]'}(x_N) \end{cases}$$

Proposition 2.14   *The truncated sums satisfy*

$$Y^{[N]''} - Y^{[N]'} \left(1 - \frac{1}{x}\right) = e^x \frac{N!N}{x^{N+2}}.$$

Proof 2.15   The proof is by straightforward calculation. Note that

$$Y^{[N]'} = e^x \sum_{n=0}^{N-1} \frac{n!}{x^{n+2}} \left(x - (n+1)\right).$$

Hence

$$Y^{[N]''} - Y^{[N]'}\left(1 - \frac{1}{x}\right)$$

$$= e^x \sum_{n=0}^{N-1} \frac{n!}{x^{n+3}} \left[\left(x^2 - 2(n+1)x + (n+1)(n+2)\right) \right.$$

$$\left. - \left(x^2 - (n+1)x - x + (n+1)\right)\right]$$

$$= e^x \sum_{n=0}^{N-1} \frac{1}{x^{n+3}} \left(-n\,n!\,x + (n+1)\,(n+1)!\right)$$

$$= -\frac{e^x}{x^2} \left(\sum_{k=0}^{N-1} \frac{k\,k!}{x^k} - \sum_{j=1}^{N} \frac{j\,j!}{x^j}\right)$$

$$= -\frac{e^x}{x^2} \left(0 - \frac{N\,N!}{x^N}\right)$$

as required.

To study how well these partial sums approximate the solutions away from the initial point x_N, we need to consider

$$\Delta_N(x) := y_N(x) - Y^{[N]}(x).$$

Corollary 4.1

$$\Delta_N'' - \Delta_N'\left(1 - \frac{1}{x}\right) = -e^x \frac{N!N}{x^{N+2}} \tag{4.1}$$

Proof 2.16   The result follows from the fact that

$$\Delta_N'' - \Delta_N'\left(1 - \frac{1}{x}\right) = -Y^{[N]''} + Y^{[N]'}\left(1 - \frac{1}{x}\right).$$

Proposition 2.17

$$\Delta_{N+1}(x_N) \sim -\sqrt{2\pi}N^{-1/2}e^{-1}, \text{ as } N \to +\infty.$$

Proof 2.18   We integrate Equation (4.1) by using the solutions of its homogeneous version as integrating factors. First, multiply the ODE (4.1) by xe^{-x}. Noting that

$$\left(x\,e^{-x}\Delta_N'(x)\right)' = x\,e^{-x}\left[\Delta_N'' - \Delta_N'\left(1 - \frac{1}{x}\right)\right]$$

and using the initial condition $\Delta_N'(x_N) = 0$ we get

$$x\,e^{-x}\Delta_N'(x) = \frac{N!}{x^N} - \frac{N!}{N^N} \tag{4.2}$$

Now multiplying by e^x/x, integrating once more and using the initial condition $\Delta_N(x_N) = 0$, we get

$$\Delta_N(x) = N! \int_N^x e^t \left(\frac{1}{t^{N+1}} - \frac{1}{N^N\,t} \right) dt. \tag{4.3}$$

We use this integral representation, with $N \mapsto N+1$, to estimate $\Delta_{N+1}(x_N)$. The first integral on the right then is

$$
\begin{aligned}
-\int_N^{N+1} dt\, \frac{e^t}{t^{N+2}} \;&=\; -\int_0^1 ds\, \frac{e^{N+s}}{(N+s)^{N+2}} \\
&=\; -\frac{1}{N^{N+2}} \int_0^1 ds\, \frac{e^{N+s}}{(1+s/N)^{N+2}} \\
&=\; -\frac{e^N}{N^{N+2}} \int_0^1 ds\,(1 + O(1/N))
\end{aligned}
$$

where the $O(1/N)$ term in the last integrand can be upperbounded by $5/N$ (for $N \geq 1$), since

$$(N+2)\log(1+s/N) = (N+2)\left(\frac{s}{N} - \left(\frac{s}{N}\right)^2 \frac{1}{2(1+\theta/N)^2} \right)$$

for some $\theta \text{in} [0,1]$, which implies

$$\frac{1}{(1+s/N)^{N+2}} < e^{-s}\, e^{3/(2N)} < e^{-s}\left(1 + \frac{3\,e}{2N} \right).$$

The second integral on the right in Equation (4.3) can be expanded as

$$
\begin{aligned}
\int_N^{N+1} dt\, \frac{e^t}{t} \;&=\; \left[\frac{e^t}{t} \right]_N^{N+1} + \int_N^{N+1} dt\, \frac{e^t}{t^2} \\
&=\; e^N \left(\frac{e}{N+1} - \frac{1}{N} + O(1/N^2) \right)
\end{aligned}
$$

where the $O(1/N^2)$ term on the right can be upperbounded by $1/N^2$ (for

$N \geq 1$). Putting these together, we get

$$\Delta_{N+1}(N) \; = \; e^N \, (N+1)! \left(\frac{e}{(N+1)^{N+2}} - \frac{1}{N(N+1)^{N+1}} - \frac{1}{N^{N+2}} \right) \cdot$$
$$\cdot \, (1 + O(1/N))$$

We use Stirling's formula [1]

$$\Gamma(z+1) = (z+1)^{z+1/2} \, e^{-z-1} \sqrt{2\pi} \, (1 + O(1/z)), \quad \text{as } z \to \infty,$$

along with $\Gamma(N+2) = (N+1)!$ and $(1+1/N)^{N+p} = e(1 + O(1/N))$ for $N \gg 1$, $p \, \text{in} \, \mathbb{R}$, to get

$$\Delta_{N+1}(N) \; = \; e^N \, (N+2)^{N+3/2} \, e^{-N-2} \sqrt{2\pi}$$
$$\cdot \left(\frac{e}{(N+1)^{N+2}} - \frac{1}{N(N+1)^{N+1}} - \frac{1}{N^{N+2}} \right)$$
$$\cdot \, (1 + O(1/N))$$
$$= \; e^{-2} \sqrt{2\pi} \left(1 + \frac{1}{N+1} \right)^{N+3/2}$$
$$\cdot \left(\frac{e}{(N+1)^{1/2}} - \frac{1}{N^{1/2}} - \frac{(1+1/N)^{N+3/2}}{N^{1/2}} \right)$$
$$\cdot \, (1 + O(1/N))$$
$$= \; -e^{-1} \sqrt{2\pi} \, N^{-1/2} \, (1 + O(1/N))$$

as required.

To prove convergence of the optimal sequence of solutions, we need to study the differences between successive members of this sequence. Define

$$\hat{y}_N(x) \; := \; y_{N+1}(x) - y_N(x)$$
$$= \; Y^{[N+1]}(x) + \Delta_{N+1}(x) - \left(Y^{[N]}(x) + \Delta_N(x) \right)$$
$$= \; \frac{e^x \, N!}{x^{N+1}} + \Delta_{N+1}(x) - \Delta_N(x). \tag{4.4}$$

Since $\hat{y}_N(x)$ satisfies Equation (1.2), we can write it as

$$\hat{y}_N(x) = \alpha_N \int_1^x dt \frac{e^t}{t} + \beta_N \tag{4.5}$$

for some constants α_N and β_N. Proposition 2.17 allows us to estimate $y_N(x)$ at x_N, and hence find estimates for α_N and β_N.

Proposition 2.19

$$
\begin{aligned}
\hat{y}_N(x_N) &= \sqrt{2\pi}\, N^{-1/2} \left(1 - e^{-1}\right)\left(1 + O(1/N)\right) \\
\hat{y}'_N(x_N) &= \sqrt{2\pi}\, N^{-1/2} \left(1 - e^{-1}\right)\left(1 + O(1/N)\right)
\end{aligned}
$$

Proof 2.20   First note that

$$
\hat{y}_N(x_N) = \frac{e^N\, N!}{N^{N+1}} + \Delta_{N+1}(N)
$$

Hence by Proposition 2.17, we get

$$
\begin{aligned}
\hat{y}_N(x_N) &= \frac{e^N\, N!}{N^{N+1}} - e^{-1}\sqrt{2\pi}\, N^{-1/2}\left(1 + O(1/N)\right) \\
&= \sqrt{2\pi}\, N^{-1/2}\left(1 - e^{-1}\right)\left(1 + O(1/N)\right)
\end{aligned}
$$

where we have used Stirling's formula to obtain the last estimate. Using Equation (4.2)

$$
\Delta'_{N+1}(x) = \frac{e^x}{x}\left(\frac{(N+1)!}{x^{N+1}} - \frac{(N+1)!}{(N+1)^{N+1}}\right)
$$

and the definition (4.4) of $\hat{y}_N$, along with $\Delta'_N(x_N) = 0$, we obtain

$$
\begin{aligned}
\hat{y}'_N(x_N) &= \frac{e^N\, N!}{N^{N+1}} - \frac{e^N\, N!(N+1)}{N^{N+2}} \\
&\qquad + \frac{e^N}{N}\left(\frac{(N+1)!}{N^{N+1}} - \frac{(N+1)!}{(N+1)^{N+1}}\right) \\
&= \frac{e^N\, N!}{N^{N+1}}\left(1 - \frac{N+1}{N} + \frac{N+1}{N} - \frac{1}{(1+1/N)^N}\right) \\
&= \frac{e^N\, N!}{N^{N+1}}\left(1 - e^{-1}\right)\left(1 + O(1/N)\right) \\
&= \sqrt{2\pi}\, N^{-1/2}\left(1 - e^{-1}\right)\left(1 + O(1/N)\right)
\end{aligned}
$$

Corollary 2.21

$$
\begin{aligned}
\alpha_N &= \sqrt{2\pi}\, N^{1/2} e^{-N}\left(1 - e^{-1}\right)\left(1 + O(1/N)\right) \\
\beta_N &= O\left(1/N^{3/2}\right)
\end{aligned}
$$

Proof 2.22   From the solution (4.5), we have

$$
\begin{aligned}
\alpha_N &= N e^{-N} \hat{y}'_N(x_N) \\
&= \sqrt{2\pi}\, N^{1/2} e^{-N} \left(1 - e^{-1}\right)\left(1 + O(1/N)\right)
\end{aligned}
$$

and

$$
\begin{aligned}
\beta_N &= \hat{y}_N(x_N) - \alpha_N \int_1^{x_N} dt\, \frac{e^t}{t} \\
&= \sqrt{2\pi}\, N^{-1/2} e^{-N}\left(1 - e^{-1}\right)\left(1 + O(1/N)\right) \\
&\quad - \sqrt{2\pi}\, N^{1/2} e^{-N}\left(1 - e^{-1}\right)\left(\frac{e^N}{N} + O\left(e^N/N^2\right)\right) \\
&= O\left(1/N^{3/2}\right)
\end{aligned}
$$

4.2   *Natural Sum*

These results show that $\hat{y}_N(x)$ is uniformly summable for all $x \neq 0$. That is, the telescoping sum

$$
\begin{aligned}
\sum_{N=2}^{\infty} \hat{y}_N(x) &= y_3(x) - y_2(x) + y_4(x) - y_3(x) + \ldots \\
&= -y_2(x) + \lim_{N\to\infty} y_N(x)
\end{aligned}
$$

exists. Therefore, $\{y_N(x)\}$ convergees uniformly. Uniformity implies analyticity of the limit function $y(x)$. Also, since $y_N(x)$ satisfies the ODE (1.2), so does $y(x)$. Moreover, given a large positive x and small $\epsilon > 0$, we have $N \gg 1$ such that

$$
|y(x) - y_N(x)| < \epsilon
$$

which implies that

$$
|y(x) - Y^{[N]}(x) - \Delta_N(x)| < \epsilon.
$$

That is,

$$
\frac{|y(x) - Y^{[N]}(x)|}{x^N e^x} < \frac{\epsilon + (N!/N^N)e^x/x}{x^N e^x} \xrightarrow[x\to\infty]{} 0
$$

In other words, we have constructed a unique solution of (1.2) that is asymptotic to the series (1.5). We call this solution the *natural sum* of the series.

4.3   *Tritronquée Solution of* P_I

The idea of a natural sum can also be extended to non-linear ODEs such as P_I. In [15], however, instead of using increasing partial sums of the formal series (3.11), we used the simplest partial sum, i.e. the first term of the series.

Definition 2.23   For $x_0 > 0$, the *tangent solutions* y_{x_0} of P_I are those with initial data $y_{x_0}(x_0) = -\sqrt{x_0/6}$, $y_{x_0}'(x_0) = -1/(2\sqrt{6x_0})$.

Proposition 2.24   *For any* $x \geq 0$,

$$y_{x_0}(x) \underset{x_0 \to +\infty}{\to} Y(x).$$

To complete our analogy, the tronquée solutions are the mathematical butterflies. We have caught one unique one, i.e. the tritronquée solution, along the positive real axis. By using this solution we can identify the others by considering their differences. Asymptotically, their differences are the terms that are exponentially small in the limit as $x \to \infty$.

References

M. Abramowitz and I. Stegun (eds). *Handbook of Mathematical Functions*. Dover, New York, 1972.

J. Banks, J. Brooks, G. Cairns, G. Davis, and P. Stacey. On Devaney's definition of chaos. *Amer. Math. Monthly* **99** (1992), 332–334.

C. Bender and S. Orszag. *Advanced Mathematical Methods for Scientists and Engineers*. McGraw-Hill, New York, 1978.

P. Boutroux. Recherches sur les transcendantes de M. Painlevé et l'étude asymptotique des équations différentielles du second ordre. *Ann. École Norm.* **30**, 1913, 265–375.

O. Costin and M. D. Kruskal. Optimal uniform asymptotics and rigorous asymptotics beyond all orders for a class of ODEs. *Proc. Roy. Soc. Lond. A*, 452:1057–1085, 1996.

E.T. Copson. *Asymptotic Expansions*. Cambridge University Press, London, 1965.

M. V. Berry. Uniform asymptotic smoothing of Stokes' discontinuities. *Proc. Roy. Soc. Lond. A*, 422:7–21, 1989.

M. V. Berry and C. Howls. Hyperasymptotics. *Proc. Roy. Soc. Lond. A*, 430:653–668, 1990.

R. Devaney. *Introduction to Chaotic Dynamical Systems*. Second edition. Addison-Wesley Publishing Co. Menlo Park, California, 1989.

D. J. Gross and A. A. Migdal. Non-perturbative two-dimensional quantum gravity. *Phys. Rev. Letts.* **64**, 1990, 127–130.

H. Goldstein. *Classical Mechanics.* Second edition. Addison-Wesley Series in Physics. Addison-Wesley Publishing Co., Reading, Mass., 1980.

V. Hakim and K. Mallick. Exponentially small splitting of separatrices, matching in the complex plane and Borel summation. *Nonlinearity* **6**, 1993, 57–70.

E. Ince. *Ordinary Differential Equations.* Dover, New York, 1954.

N. Joshi. Asymptotic Studies of the Painlevé Equations. in *The Painlevé Property: One Century Later* ed. R. Conte.

N. Joshi and A. V. Kitaev. On Boutroux's tritronquée solutions of the first Painlevé equation. *Stud. Appl. Math.* **107** (2001) 253–291.

N. Joshi and M. D. Kruskal. The Painlevé connection problem: an asymptotic approach I. *Stud. Appl. Math.*, 86:315–376, 1992.

F. W. J. Olver. *Asymptotics and Special Functions.* Academic Press, London, 1992.

W. R. Wasow. *Asymptotic expansions for ordinary differential equations.* Robert E. Krieger, Huntington, N.Y., 1976.

S. Wiggins. *Global Bifurcations and Chaos.* Applied Mathematical Sciences, **73**. Springer-Verlag, New York, 1988.

Acknowledgment

The research on which this material is based is supported by the Australian Research Council.

Chapter 3

The Dynamics of Fermi Acceleration: From Cosmic Rays to Discharge Heating

M.A. Lieberman

Department of Electrical Engineering and Computer Sciences, University of California, Berkeley, CA 94720

Abstract. The heating of electrons by time-varying fields is fundamental to the operation of radio frequency (rf) and microwave discharges. Ohmic heating, in which the phase of the electron oscillation motion in the field is randomized locally by interparticle collisions, can dominate at high pressures. Phase randomization can also occur due to electron thermal motion in spatially inhomogeneous rf fields, even in the absence of collisions, leading to collisionless or stochastic heating, which can dominate at low pressures. Electrons are heated collisionlessly by repeated interaction with fields that are localized within a sheath, skin depth layer, or resonance layer inside the discharge. A simple heating model of a ball bouncing elastically back and forth between a fixed and an oscillating wall was proposed originally by Fermi to explain the origin of cosmic rays. This model of Fermi acceleration can be used as a paradigm to describe collisionless heating and phase randomization in capacitive, inductive, and electron cyclotron resonance (ECR) discharges. Mapping models for Fermi acceleration are introduced and related to the fundamentals of Hamiltonian dynamics. The Fokker-Planck description of the heating and the effects of phase correlations and dissipation are described. The collisionless heating rates are determined in capacitive and inductive discharges and compared with self- consistent (kinetic) calculations where available. Experimental measurements and computer simulations are reviewed and compared to theoretical calculations. Incomplete phase randomization and adiabatic barriers can modify the heating in low pressure ECR discharges.

1   Introduction

The heating of electrons by time-varying fields is fundamental to the operation of radio frequency (rf) and microwave plasma discharges. In a *uniform* oscillating electric field $\mathbf{E}(t) = \mathrm{Re}\,\mathbf{E}_0\,e^{j\omega t}$, a single electron has a coherent velocity of motion that lags the phase of the electric field force $-e\mathbf{E}$ by $90°$. Hence the time-average power transfered from the field to the electron is zero. Electron collisions with other particles destroy the phase coherence of the motion, leading to a net transfer of power. For an ensemble of n electrons per unit volume, it is usual to introduce the macroscopic current density $\mathbf{J} = en\mathbf{u}$, with $\mathbf{u}$ the macroscopic electron velocity, and to relate the amplitudes of $\mathbf{J}$ and $\mathbf{E}$ through a local conductivity: $\mathbf{J}_0 = \sigma_p\mathbf{E}_0$, where $\sigma_p = e^2 n/m(\nu_m + j\omega)$ is the plasma conductivity and ν_m is the electron collision frequency for momentum transfer. In this "fluid" approach, the average electron velocity $\mathbf{u}$ still oscillates coherently but lags the electric field by less than $90°$, leading to an ohmic power transfer per unit volume:

$$p_{\mathrm{ohm}} = \frac{1}{2}\mathrm{Re}\,\mathbf{J}_0 \cdot \mathbf{E}_0^* = \frac{1}{2}|\mathbf{E}_0|^2\,\mathrm{Re}\,(\sigma_p) = \frac{1}{2}|\mathbf{J}_0|^2\,\mathrm{Re}\,(\sigma_p^{-1}).$$

Although the average velocity is coherent with the field, the fundamental mechanism that converts electric field energy to thermal energy is the breaking of the phase-coherent motion of individual electrons by collisions: the total force (electric field force plus that due to collisions) acting on an individual electron becomes spatially non-uniform and non-periodic in time.

These observations suggest that a spatially *non-uniform* electric field by itself might lead to electron heating, even in the absence of interparticle collisions, provided that the electrons have thermal velocities sufficient to sample the field inhomogeneity. This phenomenon has been well-known in plasma physics since Landau [1] demonstrated the collisionless damping of an electrostatic wave in a warm plasma, and is variously referred to in the literature as *collisionless, noncollisional, stochastic, transit time*, or *anomalous* heating or dissipation. Since that time, collisionless dissipation has been studied extensively in fusion and space plasma physics. With the increased emphasis on industrial applications of low pressure gas discharges, it has also become evident that collisionless dissipation phenomena are fundamental to rf and microwave discharges [2, 3].

In almost all discharges, the spatial variation of the time-varying field is strongly non-uniform, with a low field in the bulk of the plasma and one or more highly localized field regions (rf sheath, skin depth layer, etc), usually near the plasma boundaries. An electron, being confined for hundreds to thousands of bounce times by the dc ambipolar and boundary sheath potential in the discharge, interacts repeatedly with the high field regions, but interacts only weakly during its drift through the plasma bulk. This suggests a dynamical model to investigate the energy transfer and loss of phase coherence: a ball bounces elastically back and forth between a fixed and an oscillating wall. This model was first introduced by Fermi [4] to explain the origin of cosmic rays. The process in which the ball repeatedly interacts with the oscillating wall, resulting in phase randomization and stochastic heating, is known as *Fermi acceleration.* This process has been studied extensively as a paradigm in dynamics. We adapt the Fermi acceleration model as our fundamental approach for understanding collisionless heating in weakly ionized gas discharges.

In Section 2 we describe the model of Fermi acceleration. We motivate its introduction to explain the origin of cosmic rays, and introduce mapping models to describe the dynamics. We show the relation of mapping models to conventional Hamiltonian dynamics [5, 6]. We introduce a Fokker-Planck formalism to describe the collisionless heating in the presence of complete phase randomization, and we describe the effects of partial phase randomization and dissipation. In Section 3, we review collisionless heating in capacitive rf discharges. We introduce a simple Fermi acceleration model for a homogeneous sheath to determine the collisionless heating rate, and we make comparisons to experiments and to fluid and particle-in-cell (PIC) simulations. In Section 4 we review collisionless heating in inductive rf discharges. We introduce the classical and anomalous skin effects and a Fermi acceleration model of the collisionless (anomalous) heating. We compare this to a self-consistent kinetic model and describe recent experiments that identify effects due to collisionless heating, such as the existence of regions of negative electron power absorption within the discharge bulk due to space dispersion caused by electron thermal motion. In Section 5, we review briefly some features of collisionless heating in electron cyclotron resonance (ECR) discharges. We introduce a Fermi acceleration heating model and show some comparisons to experiments. Both the model and the experiments suggest that incomplete phase randomization can reduce the heating rate and lead to an adiabatic barrier to the heating. In Section 6 we summarize our conclusions.

2   Fermi acceleration

A. Cosmic Rays: Discovery and Properties

In the morning of August 7, 1912, Austrian physicist Viktor Hess ascended to over five kilometers in a balloon gondola as "an observer for atmospheric electricity." During the journey, he made careful measurements of the rate of discharging of three electroscopes, and he noted a several-fold increase in the rate of discharging as the balloon rose in altitude. In his publication in Physikalische Zeitschrift in November 1912, Hess suggested that the results of his observations were best explained "by a radiation of great penetrating power entering our atmosphere from above." Further flights confirmed these findings, and the American physicist Robert Millikan, although initially skeptical of the extraterrestrial origin, introduced the name *cosmic rays.*

It is now generally agreed that the majority of cosmic rays have a galactic origin. The cosmic ray flux is isotropic and of order 1 $\mathrm{cm}^{-2}\mathrm{s}^{-1}$, the energy density is approximately 1 $\mathrm{eV/cm}^3$, and the lifetime is approximately 10^7 years. Cosmic rays are mostly protons, but are rich in heavy nuclei compared to solar abundences. The particle energy ranges from $W \sim 10^8$–10^{20} eV, with a power law distribution $\mathcal{N}(W) \propto W^{-(2-2.5)}$.

Cosmic rays are believed to originate from supernovas such as the well-studied Crab nebula, which is the remnant of a supernova in 1054 A.D. With one galactic supernova every fifty years within a galactic disk volume of 10^{67} cm^3 creating 10^{43} J of fast particles, the energy balance is

$$\frac{10^{43}\ \mathrm{J}}{50\ \mathrm{yrs}} \approx \frac{e \times 1\ \mathrm{eV/cm}^3 \times 10^{67}\ \mathrm{cm}^3}{10^7\ \mathrm{yrs}}.$$

Measurements of radiation from supernova remnants clearly show the presence of synchrotron radiation, demonstrating the existence of high-energy ($> 10^{11}$ eV) electrons. Exactly how the fast particles are formed and accelerated is not well understood. Early theories emphasized acceleration across high voltages or by means of shock waves.

B. Fermi's Proposal

In 1949, Fermi put forth the idea that "cosmic rays are originated and accelerated primarily in the interstellar space of the galaxy by collisions against moving magnetic fields". He went on to assert the basic acceleration mechanism as follows:

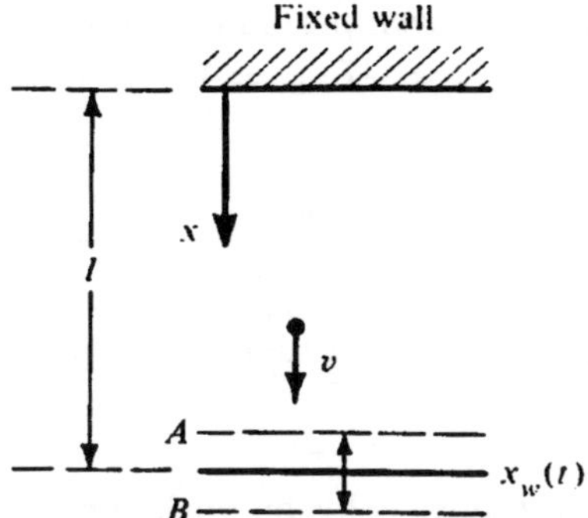

Fig. 1   Fermi acceleration: a particle bounces between a fixed and an oscillating wall.

It may happen that a region of high field intensity moves toward the cosmic-ray particle which collides against it. In this case, the particle will gain energy in the collision. Conversely, it may happen that the region of high field intensity moves away from the particle. Since the particle is much faster, it will overtake the irregularity of the field and be reflected backwards, in this case with loss of energy. The net result will be average gain, primarily for the reason that head-on collisions are more frequent than overtaking collisions because the relative velocity is larger in the former case.

Fermi noted that this idea naturally leads to a power law energy distribution, but that it failed to explain in a straightforward way the heavy nuclei observed in the primary cosmic radiation. It is now believed that cosmic rays below 10^{15} eV are produced within our galaxy by Fermi acceleration of particles within the shock waves of supernova remnants.

C. Fermi Maps as Hamiltonian Dynamical Systems

The Fermi problem of a particle bouncing between a fixed and an oscillating wall is a classical model of Hamiltonian dynamics [5] and is illustrated in Fig. 1. This model of energy gain by repeated collisions of a particle with an oscillating wall has been examined numerically in various approximations.

To find the exact mapping dynamics for this system, we introduce a fixed surface of section as some $x = $ const. Defining $u_n = v_n/2\omega a$ to be the normalized velocity, $\theta_n = \omega t$ to be the phase of the moving wall at the nth collision with the fixed surface at $x = 0$, then a difference equation for the motion of the particle can be determined in terms of a wall motion

$x_w(t) = aF(\psi)$, where F is an even periodic function of the phase $\psi = \omega t$, with period 2π and with $F_{\max} = -F_{\min} = 1$. We obtain, in implicit form the equations of motion

$$
\begin{aligned}
u_{n+1} &= u_n + F'(\psi_c), & \text{(2.1a)}\\
\theta_{n+1} &= \psi_c + \frac{[2\pi M - \frac{1}{2}F(\psi_c)]}{u_{n+1}}, & \text{(2.1b)}\\
\psi_c &= \theta_n - \frac{1}{2}\frac{F(\psi_c)}{u_n}. & \text{(2.1c)}
\end{aligned}
$$

Here ψ_c is the phase at the next collision with the moving wall, after the nth collision with the fixed surface $x = 0$, $M = l/2\pi a$, with l the distance between the walls, and $F' = dF/d\psi$ is the velocity impulse given to the ball. It is well known [13] that measuring the position from the fixed wall as x, conjugate to the velocity v, then the phase θ is a time-like variable conjugate to the energy-like variable $w = u^2$. That is, in the extended phase space $(v, x, -w, t)$ for this Hamiltonian system, the choice of a surface $x = 0$ gives an area-preserving mapping for the remaining pair $(-w, t)$. As we show in Section 2D, this implies that a stochastic orbit has a uniform invariant distribution over the accessible (w, θ) phase space. Hence, assuming all phases are accessible, the energy w has a uniform invariant distribution.

Because of its complicated form, (2.1) is not convenient for analytical study. Substituting $w = u^2$, assuming a sinusoidal wall motion in (2.1), and expanding to first order in F' (and F), we obtain

$$
w_{n+1} = w_n + 2\sqrt{w_{n+1}}\,\sin\theta_n, \tag{2.2a}
$$

$$
\theta_{n+1} = \theta_n + \frac{2\pi M}{\sqrt{w_{n+1}}} + \frac{\cos\theta_n}{\sqrt{w_{n+1}}}. \tag{2.2b}
$$

A still simpler, non-implicit form can be constructed if the sinusoidally oscillating wall imparts momentum to the ball, according to the wall velocity, without the wall changing its position in space. The problem defined in this manner has many of the features of the more physical problem. In this simplified form the mapping is

$$
\begin{aligned}
u_{n+1} &= |u_n + \sin\psi_n|, & \text{(2.3a)}\\
\psi_{n+1} &= \psi_n + \frac{2\pi M}{u_{n+1}} \pmod{2\pi}. & \text{(2.3b)}
\end{aligned}
$$

The mapping in (2.3) serves as an approximation (with suitably defined variables) to many physical systems in which the transit time between kicks is inversely proportional to a velocity. The absolute-value signs in (2.3) correspond to the velocity reversal, at low velocities $u < 1$, which appears in the exact equations (2.1). The absolute value has no effect on the region $u > 1$, which is the primary region of interest. For the simplified problem, a proper canonical set of variables are the ball velocity and phase just before the nth impact with the moving wall. The normalized velocity u then has a uniform invariant distribution, as will be seen in Section 2D.

Let us examine the relation between mappings such as (2.1)–(2.3) and the conventional formulation of Hamiltonian dynamics [5]. Hamilton's equations for a non-dissipative dynamical system are

$$\frac{dp_i}{dt} = -\frac{\partial H}{\partial q_i}, \qquad \frac{dq_i}{dt} = \frac{\partial H}{\partial p_i}, \qquad \frac{dH}{dt} = -\frac{\partial H}{\partial t}, \tag{2.4}$$

where $H(p_1, \ldots p_N, q_1, \ldots q_N, t)$ is the Hamiltonian, p_i and q_i are the generalized (or "canonical") momenta and coordinates, and N is the number of degrees of freedom. Generally H has the interpretation of the total energy of the system (sum of kinetic and potential energies). We say a system is *autonomous* if H is explicitly independent of time t. A non-autonomous system of degree N can be transformed into an autonomous system of degree $N + 1$ by introducing an extended phase space with the additional momentum variable $-H$ and the additional coordinate variable t. We can view the motion of an autonomous H in the $2N$-dimensional phase space of the system. A set of initial conditions $(\mathbf{p}_0, \mathbf{q}_0)$ evolves in time according to (2.4), generating a Hamiltonian flow in the phase space. Hamiltonian flows can be shown to conserve the phase space volume (Liouville's theorem).

We are often interested in periodic or quasiperiodic motions for Hamiltonian systems. Autonomous systems with $N = 1$, described by $H(p, q)$ with a single p and q, are generally integrable; i.e., a periodic solution for the motion can be found. An example is the motion of a pendulum described by the Hamiltonian $H = p^2/2 - \cos\phi$, where the canonical momentum p is the (appropriately normalized) angular momentum and the canonical coordinate ϕ is the angle of the pendulum bob from the vertical. The general procedure for solving the motion is to find a "canonical" transformation (which preserves the form of Hamilton's equations) to new canonical variables, the new momentum variable I (the "action") and the new coordinate variable θ (the "angle"), such that the transformed Hamiltonian H' is a function of I alone, independent of θ. In the transformed coordinates, Hamilton's equations can be trivially integrated to yield the periodic solution $I(t) = I_0$ and $\theta(t) = \omega_0 t + \theta_0$, where $\omega_0 = (\partial H'/\partial I)_{I_0}$ is the frequency of the periodic motion. The action can be expressed in terms of the old variables as $I = (2\pi)^{-1} \oint p\, dq$, which specifies the required transformation [5]. In the case of a pendulum, the periodic solutions correspond to either back and forth "librations" of the bob with $\phi_{\max} < \pi$, or "rotations" about the axis for large energies where ϕ continually increases with time. The special case ("separatrix") where the motion evolves in time from the bob initially hanging straight down ($\phi = 0$) to straight up ($\phi = \pi$) separates the librations and rotations.

Some non-autonomous Hamiltonians with $N \geq 1$ and some autonomous Hamiltonians with $N \geq 2$ are integrable. In these cases we can do the action-angle transformation on each degree of freedom separately to write $H_0' = H_0'(\mathbf{I})$, where $\mathbf{I}$ is an N-vector of actions, with corresponding frequencies $\omega = \partial H_0'/\partial \mathbf{I}$ and angles $\boldsymbol{\theta}$. In general the frequencies are incommeasurate, corresponding to a quasiperiodic integrable motion. The special case where two or more frequencies are fractionally related (i.e., $\omega_1/\omega_2 = k/l$ with k and l integers) is called a *resonance*.

Most such systems are not integrable in this way. However, we can try to understand the dynamics of "near-integrable" systems having Hamiltonians of the form

$$H(\mathbf{p},\, \mathbf{q}) = H_0(\mathbf{p},\, \mathbf{q}) + \epsilon H_1(\mathbf{p},\, \mathbf{q}),$$

or, transforming to the action-angle varibles of H_0,

$$H'(\mathbf{I},\, \boldsymbol{\theta}) = H_0'(\mathbf{I}) + \epsilon H_1'(\mathbf{I},\, \boldsymbol{\theta}),$$

where H_0 is integrable and ϵH_1 is the "perturbed" part of the Hamiltonian. An example of a near-integrable system is the driven pendulum

$$H = p^2/2 - \cos\phi - \epsilon \cos(\phi - \Omega t). \tag{2.5}$$

It is often convenient to view the long-term motion of a Hamiltonian system in phase space by choosing some fixed surface in the phase space, such as $p_1 = \text{const}$ or $q_1 = \text{const}$, and recording the successive "piercings" of this surface (in the same sense of direction) as the motion evolves in time. This is sometimes called a *Poincare surface of section*. Either the new $(\mathbf{I},\, \boldsymbol{theta})$ or the old $(\mathbf{p},\, \mathbf{q})$ variables can be used. As shown in Fig. 2,

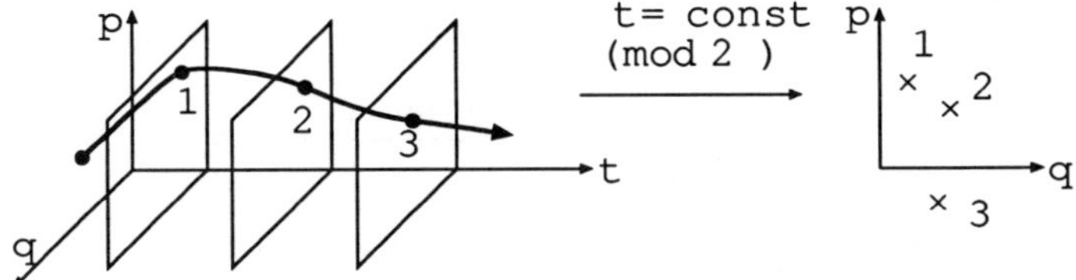

Fig. 2   Surface of section for an $N = 1$ nonautonomous Hamiltonian $H(p, q, t)$.

for a non-autonomous Hamiltonian $H(p, q, t)$ with $N = 1$ that is periodic in t with period $2\pi/\Omega$, we can view the motion in the two-dimensional surface of section $\Omega t = $ const (mod 2π). For autonomous Hamiltonians, similar surfaces of section exist having dimension $2N - 1$.

It is well-known that near-integrable systems have intermingled regions of regular and chaotic motion [5]. The regular ("integrable" or "KAM") trajectories depend discontinuously on the initial conditions. The chaotic trajectories lie arbitrarily close to every point in the phase space or surface of section. As we will see, the Fermi map also has this structure and can be represented as a non-autonomous Hamiltonian system with $N = 1$. We can see this by explicitly writing the Fermi map in Hamiltonian form, with the iteration number n playing the role of the time. Introducing the periodic delta function

$$\delta_1(n) = \sum_{k=-\infty}^{\infty} \delta(n - k) = 1 + 2\sum_{k=1}^{\infty} \cos 2\pi kn, \tag{2.6}$$

then we can write the difference equations (2.3) in the form of two differential equations

$$\frac{du}{dn} = \epsilon\, \delta_1(n)\, \sin\theta, \tag{2.7a}$$

$$\frac{d\theta}{dn} = \frac{2\pi M}{u}, \tag{2.7b}$$

where u_n and θ_n are $u(n)$ and $\theta(n)$ just before "time" n. We have introduced ϵ into (2.7a) to identify the perturbation. Equations (2.7) are in the form of Hamilton's equation with the non-autonomous $N = 1$ Hamiltonian

$$H = 2\pi M \ln u + \epsilon\, \delta_1(n)\, \cos\theta. \tag{2.8}$$

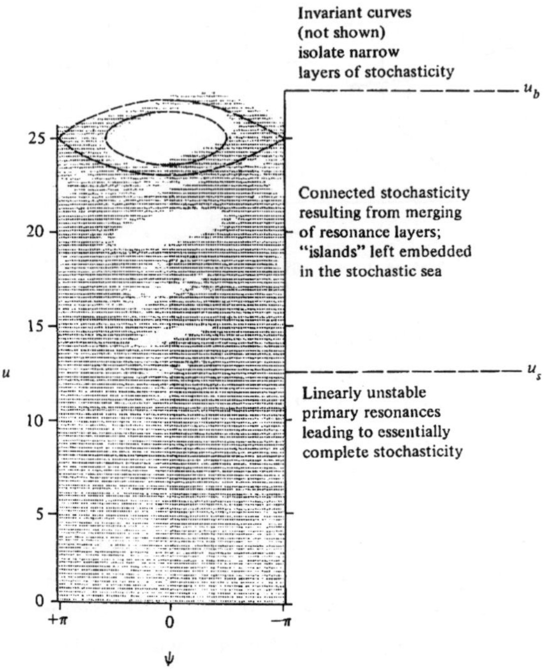

Fig. 3   Surface of section for the Fermi problem, showing occupation of phase space cells for 623,000 iterations of a single initial condition. Dashed curves are calculated from secular perturbation theory (after [5]).

Here u is the action, θ is the angle variable, n is the time, and the second term on the right hand side is the perturbed part of the Hamiltonian.

Transformations of the type (2.1)–(2.3) can be examined numerically for many thousands of iterations, thus allowing both detailed knowledge of the structural behavior and statistical properties of the dynamical system to be determined. Figure 3 shows the u–ψ surface for the simplified Fermi map (2.3) with $M = 100$ for 623,000 wall collisions of a single trajectory, with an initial condition at low velocity $u_0 \approx 1$. The surface has been divided into 200×200 cells, with a blank indicating no occupation of that cell. We find that the phase plane consists of three regions:

(1) a region for large u, $u > u_b = \sqrt{2\pi M}$, in which invariant adiabatic curves predominate and isolate narrow layers of stochasticity near the separatrices of the various resonances;

(2) an interconnected stochastic region for intermediate values of u, $u_s \approx$

$\frac{1}{2}u_b < u < u_b$, in which adiabatic islands near linearly stable periodic solutions are embedded in a stochastic sea; and

(3) a predominantly stochastic region for small u, $u < u_s$, in which all primary periodic solutions appear to be unstable.

Both regions (2) and (3) exhibit *strong* or *global stochasticity* of the motion. In the latter region, although some correlation exists between successive iterations, over most of the region it is possible to approximate the dynamics by assuming a *random phase approximation* for the phase coordinate, thus describing the momentum coordinate by a diffusion equation. We explore this question more fully in the next subsection.

D. The Fokker-Planck Equation

In regions of the phase space that are stochastic or mostly stochastic with small isolated adiabatic islands, it may be possible to describe the evolution of the distribution function in action space (or velocity space) alone. This is, in fact, the problem of most practical interest. In the Fermi acceleration problem, for example, the motivation was to find a possible mechanism for heating of cosmic rays. The variations in the phases of the particles with respect to their accelerating fields are of little interest except as they are required for determining the heating rates and the final energy distribution.

Let us consider in what sense the evolution of the distribution function $f(u,n)$ can be described by a stochastic process in the action u alone. Clearly we must confine our attention to a globally stochastic region of the phase space in which adiabatic islands do not exist or occupy negligible phase space volume. In such a region, it may be possible to express the evolution of $f(u,n)$, the distribution in u alone, in terms of a Markov process in u:

$$f(u, n + \Delta n) = \int f(u - \Delta u, n)W_t(u - \Delta u, n, \Delta u, \Delta n)\, d(\Delta u), \qquad (2.9)$$

where $W_t(u, n, \Delta u, \Delta n)$, the transition probability, is the probability that an ensemble of phase points having an action u at a "time" n suffers an increment in action Δu after a "time" Δn. If we make the additional assumption that there exists an intermediate time scale $\Delta n \gg 1$ such that $\Delta u \ll (f^{-1}df/du)^{-1}$, then we can expand the first argument of the integrand fW_t in (2.9) to second order in Δu to obtain the Fokker-Planck equation

$$\frac{\partial f}{\partial n} = -\frac{\partial}{\partial u}(Bf) + \frac{1}{2}\frac{\partial^2}{\partial u^2}(Df). \tag{2.10}$$

For Hamiltonian systems, the friction coefficient B and the diffusion coefficient D are related [5] as

$$B = \frac{1}{2}\frac{dD}{du}, \tag{2.11}$$

allowing (2.10) to be written in the form of a diffusion equation

$$\frac{\partial f}{\partial n} = \frac{\partial}{\partial u}\left(\frac{D}{2}\frac{\partial f}{\partial u}\right). \tag{2.12}$$

Assuming that phase randomization occurs on the time scale Δn, then we can average Δu over a uniform distribution of phases to obtain the so-called *quasilinear diffusion coefficient*

$$D(u) = \frac{1}{2\pi}\int_0^{2\pi} d\psi[\Delta u(\psi)]^2. \tag{2.13}$$

$B(u)$ is then obtained directly from (2.11).

For the simplified Fermi map (2.3) with sinusoidal velocity, for which $\Delta u = \sin\psi$, we obtain $D = \frac{1}{2}$ and $B = 0$. Hence the Fokker-Planck equation for the velocity distribution is

$$\frac{\partial f}{\partial n} = \frac{1}{4}\frac{\partial^2 f}{\partial u^2}. \tag{2.14}$$

Similarly, for the Fermi map (2.2), we obtain

$$\bar{D} = \frac{1}{2\pi}\int_0^{2\pi} 4w\cos^2\theta\, d\theta = 2w \tag{2.15}$$

and the Fokker-Planck equation for the energy distribution g is, from (2.12)

$$\frac{\partial g}{\partial n} = \frac{\partial}{\partial w}\left(w\frac{\partial g}{\partial w}\right). \tag{2.16}$$

To obtain a steady-state solution to the Fokker-Planck equation, we assume perfectly reflecting barriers at $u = 0$ and $u = u_b$. Setting $\partial/\partial n = 0$ in (2.14) and taking the net flux to be zero, we obtain a uniform invariant distribution in velocity $f(u) = $ const for the simplified map. For the

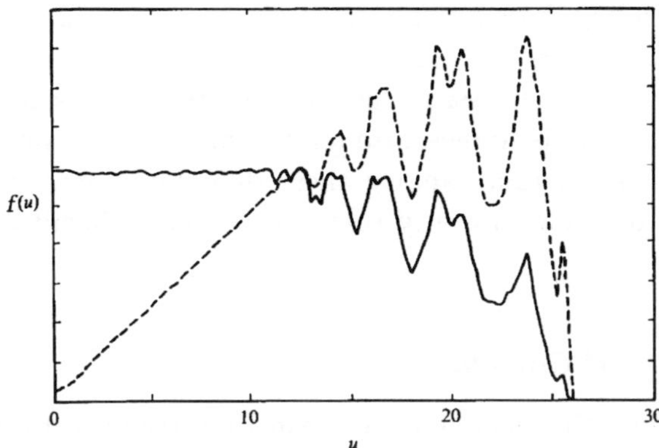

Fig. 4   Comparison of velocity distribution $f(u)$ [here $P(u)$] for the simplified Fermi map (2.3) [solid line] and the exact Fermi map (2.1) [dashed line] (after [7]).

map (2.2), we obtain similarly a uniform invariant distribution in energy $g(w) = \text{const}$. Introducing the velocity distribution $f(u)$ for (2.2) through

$$f(u)\,du = g(w)\,dw \tag{2.17}$$

and using $dw = u\,du$, we see that $f(u) = \text{const} \times u$ for (2.2). In Fig. 4 we compare the numerically calculated distributions for $M = 100$ and 5×10^6 interactions with these predictions. In the region below $u_s = (\pi M/2)^{1/2} \approx 12.5$, the predictions are verified. Above u_s, the distributions both fall off due to the presence of islands and higher-order correlations in the phase space, with the dips near the island centers.

We can also solve the transient Fokker-Planck equation. For the simplified Fermi mapping (2.3), with initial conditions of a δ-function at $u = 0^+$, we can solve (2.14) to obtain

$$f(u, n) = \frac{2}{(\pi n)^{1/2}}\, \exp\left(-\frac{u^2}{n}\right), \tag{2.18}$$

which yields the distribution function for the transient heating of the particles. This time development only holds, of course, until the particles begin to penetrate into the region with islands, $u > u_s$.

In real discharges there is always a non-zero flux in action space due to particle generation and loss processes. For example, electrons might be born by ionization at low energies and lost to the walls or by inelastic collisions at high energies. Adding these generation and loss terms to (2.16) and solving yields non-constant energy or velocity distributions that typically decrease with increasing energy. Classical (electron-electron) collisions, which are always present in real discharges, also tend to produce Maxwellian electron distributions. The interplay among these different processes determines the distribution function in a way that can be very difficult to determine analytically.

E. The Effects of Correlations

The complete dynamics, including the transition region with adiabatic islands embedded in a stochastic sea, is very complicated and can only be solved numerically. To gain some understanding of the diffusion in the phase space region where correlations are important, it is convenient to first transform the Fermi map to a local map near a resonance. Taking the simplified Fermi map of (2.3), we obtain the so-called *standard mapping* by linearization in action space near a given period-1 fixed point. These fixed points are located at

$$\frac{2\pi M}{u_1} = 2\pi k, \qquad k \text{ integer.} \tag{2.19}$$

Putting $u_n = u_1 + \Delta u_n$ and shifting the angle

$$\theta_n = \psi_n - \pi, \qquad -\pi < \theta_n < \pi,$$

then the mapping equations take the standard form

$$\begin{aligned}
I_{n+1} &= I_n + K \sin \theta_n, \tag{2.20a} \\
\theta_{n+1} &= \theta_n + I_{n+1}, \tag{2.20b}
\end{aligned}$$

where

$$I_n = -\frac{2\pi M \Delta u_n}{u_1^2} \tag{2.21}$$

is the new action and

$$K = \frac{2\pi M}{u_1^2} \tag{2.22}$$

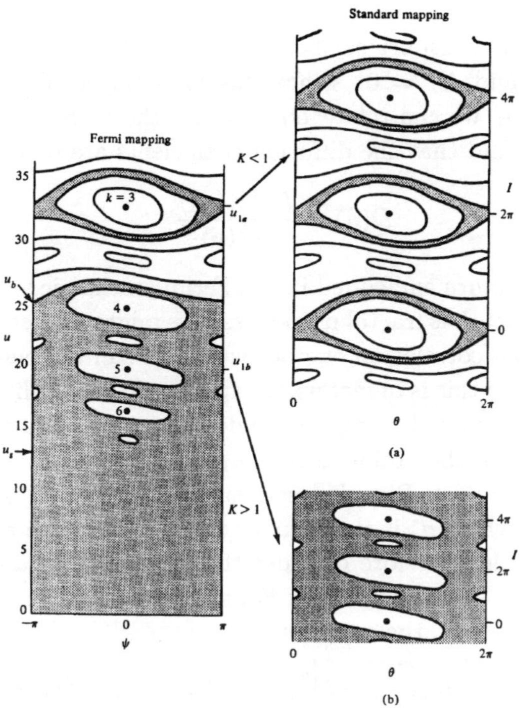

Fig. 5   Local approximation of the Fermi mapping by the standard mapping. (a) Linearization about u_{1a} leading to K small and local stochasticity; (b) linearization about u_{1b} leading to K large and global stochasticity (after [5]).

is the *stochasticity parameter*. We have thus related K to the old action u_1. The conversion from Fermi to standard mapping is illustrated in Fig. 5 for two different values of u_1, leading to two different values of K.

The dynamics of the standard mapping (2.20) can be considered to evolve on a two-torus, with both θ and I taken modulo 2π. The periodicity of the mapping in I gives rise to a special type of periodic orbit (period-1 fixed point) in which I advances by $\pm 2\pi$ every iteration of the mapping. The condition for these so-called *accelerator modes* is that $I_{1l} = 2\pi k$ and $K \sin \theta_{1l} = 2\pi l$, k and l integers, with $l \neq 0$. The accelerator modes are stable provided $|2 \pm K \cos \theta_{1l}| < 2$, which implies that stability windows for period-1 fixed points exist for successively higher values of K as I increases ($\cos \theta_{1l}$ decreases). Remnants of these accelerator modes, called *quasi-accelerator modes*, can exist in the Fermi mapping, leading to enhanced diffusion.

The quasilinear transport coefficients for the standard mapping (2.20) are $D_I = K^2/2$ and $B_I = 0$. Since this mapping locally approximates the Fermi mapping, we can relate D_I to D for the Fermi mapping. Using $\Delta I = -K\Delta u$, we find that the diffusion coefficients are related by

$$D(u) = \frac{D_I(K(u))}{K^2(u)} \tag{2.23}$$

The island structure embedded in the Fermi stochastic sea is exceeding complex, and, in fact, has fractal properties. We might expect this structure to lead to long time correlation of stochastic orbits in the neighborhood of adiabatic orbits, and this is in fact what happens. The quasilinear transport coefficients are determined using the random phase assumption applied to a single step jump in the action $\Delta u_1 = u_1 - u_0$. However, as pointed out in Section 2D, the Fokker-Planck description of the motion is valid only in the limit $n \gg n_c$, where n_c is the number of steps for phase randomization to occur. We should therefore consider the jump $\Delta u_n = u_n - u_0$, where $n > n_c$. This has been done using Fourier techniques for the standard mapping. To order K^{-1}, the result is [5]:

$$D_n = D_{QL}[1 - 2\mathrm{J}_2(K) - 2\mathrm{J}_1^2(K) + 2\mathrm{J}_2^2(K) + 2\mathrm{J}_3^2(K)], \tag{2.24}$$

where $D_{QL} \equiv D_I/2 = K^2/4$ and the J's are Bessel functions. A numerical calculation of D_{50} using 3000 particles is compared with (2.24) in Fig. 6. There is good agreement, except near the first few peaks of D, which are due to the presence of *accelerator modes.*

Deterministic Fermi acceleration mappings are useful tools for understanding the purely dynamical aspects of the phase randomization and heating of particles by periodic fields. However, let us note that for the heating of electrons in weakly ionized gas discharges, the *extrinsic stochasticity* associated with electron-electron, electron-ion, and electron-neutral collisions can play a critical, and in many cases dominating role.

F. The Effects of Dissipation

The phase space structure in near-integrable Hamiltonian systems of regions of persistent chaotic motion densely interwoven with regions of regular motion is not stable under dissipative perturbation. The stable fixed points of period k shown in Fig. 5 become attracting centers (sinks), and all KAM

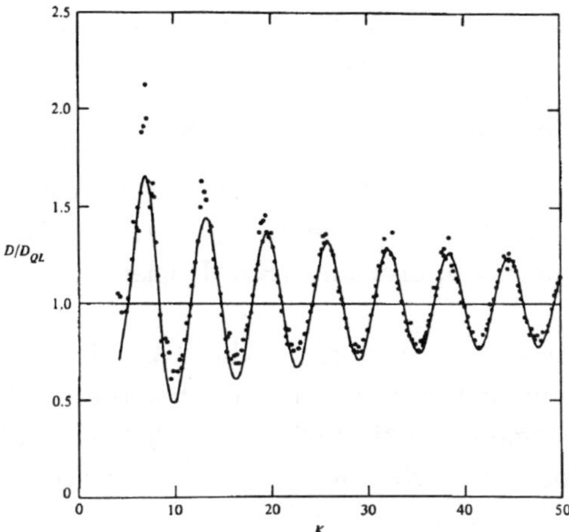

Fig. 6   Plot of D/D_{QL} versus stochasticity parameter K. The dots are the numerically computed values and the solid line is the theoretical result in the large K limit (after [5]).

trajectories are destroyed. Although transient chaotic motion generally exists, the phase point eventually enters an embedded island of period k an is attracted to an island sink; the motion ultimately becomes periodic. The complete destruction of persistent chaos when weak dissipation is added to a near-integrable Hamiltonian system is typical and probably generic behavior. However, above a critical dissipation strength, a new type of attractor ("strange attractor") in the phase plane can make its appearance, on which the motion is persistent and chaotic [5].

To illustrate these effects, we introduce dissipation into the simplified Fermi map (2.3) by assuming that the ball suffers a fractional loss δ in velocity upon collision with the fixed wall. The map is then

$$\bar{u} = (1 - \delta)u_n - \sin\psi_n, \tag{2.25a}$$

$$\bar{\psi} = \psi_n + \frac{2\pi M}{\bar{u}} \quad (\text{mod } 2\pi), \tag{2.25b}$$

$$(\psi_{n+1},\, u_{n+1}) = (\bar{\psi},\, \bar{u})\, \text{sgn}\, \bar{u}. \tag{2.25c}$$

We have shifted the phase ψ by $180°$. The function $\text{sgn}\,\bar{u} = 1$ for $\bar{u} > 0$ and -1 for $\bar{u} < 0$ is introduced to maintain $u_{n+1} \geq 0$ for low velocities, as physically occurs in the exact model, while preserving the continuity of the map near $u = 0$. The Jacobian of the map is $\partial(u_{n+1}, \psi_{n+1})/\partial(u_n, \psi_n) = 1 - \delta$, and thus the map is area preserving (Hamiltonian) for $\delta = 0$.

The primary fixed points of the map are found by setting $u_{n+1} = u_n$ and $\psi_{n+1} = \psi_n \bmod 2\pi$ in (2.25). We obtain

$$[u_k, \psi_k] = [M/k, \sin^{-1}(-u_k\delta)], \qquad (2.26)$$

where k is an integer. There are two fixed points for each k: $\psi_k \approx 0$ or $\psi_k \approx \pi$ for $u_k\delta \ll 1$. $\psi_k \approx \pi$ is stable for $u_k > u_s = (\pi M/2)^{1/2}$; $\psi_k \approx 0$ is always unstable. For $\delta = 0$, invariant (KAM) island orbits surround the stable fixed points.

For weak dissipation, $0 < \delta < \delta_{\mathrm{crit}}$, where δ_{crit} is a critical dissipation strength, the numerical interations show that the fixed points of the Hamiltonian map become attracting centers (sinks), the KAM curves no longer exist, and all persistent chaotic motion is destroyed. However, transient chaotic motion surrounds the sinks. As an example, for $M = 30$ ($\delta_{\mathrm{crit}} \approx 0.02$) and $\delta = 0.003$, an initial phase point chosend randomly at low velocity undergoes transient chaotic motion for about 13,000 interations before it enters an embedded island and becomes trapped in an island sink. In Fig. 7, we plot the cumulative phase-integrated distribution

$$\bar{f}(u) = 100 \int_0^N dn \int_0^{2\pi} d\psi\, f(u, \psi, n),$$

after $N = 50{,}000$ interations, for 100 initial conditions at low velocities chosen randomly. We see evidence of attracting sinks between u_s and u_b near the primary resonances at $k = 3$ and $k = 4$. The density leaving the stochastic region flows into these sinks, forming spikes in the figure.

Numerical studies show that a decaying quasistatic distribution

$$f(u, \psi, n) = f_Q(u)\, \exp(-\bar{\alpha}n) \qquad (2.27)$$

is formed for values of u outside of the "sticky" islands. The distribution f_Q and the decay rate $\bar{\alpha}$ can be found analytically by solving the appropriate Fokker-Planck equation (2.10) for the map, where, to first order in δ, D is the diffusion coefficient for the area-preserving ($\delta = 0$) map, and $B = -u\delta$ is the friction coefficient due to the dissipation. For $u \lesssim u_s$, $D \approx \frac{1}{2}$, the quasilinear value.

Above the critical dissipation strength δ_{crit}, a strange attractor is found, corresponding to persistent (not transient) chaotic motion. For example, for $M = 100$ ($\delta_{\mathrm{crit}} \approx 0.03$) and $\delta = 0.1$, Fig. 8a shows the (ψ, u) surface of section in the range $4 \leq u \leq 6$ after 500,000 iterations of a single initial condition. The leaved structure of the attractor is evident. Figure 8b shows

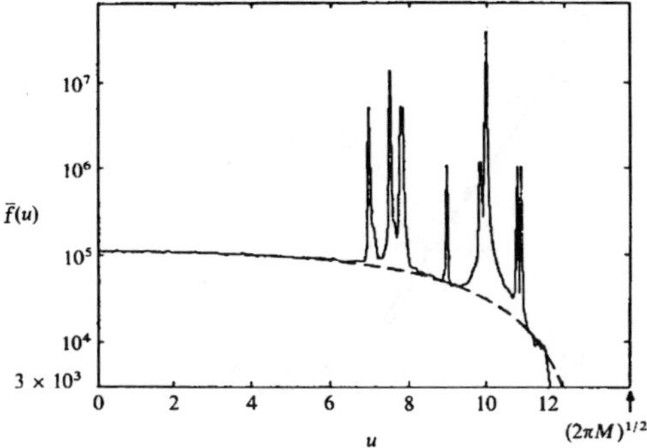

Fig. 7   Cumulative phase-averaged distribution for $M = 30$, $\delta = 0.003$, and $N = 50,000$; the solid curve shows the numerical result and the dashed curve shows the quasistatic theory.

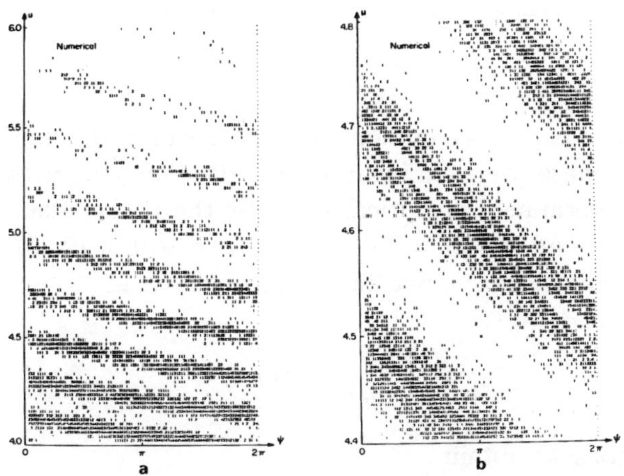

Fig. 8   Phase space u-ψ for the Fermi map, $M = 100$ and $\delta = 0.1$. (a) One particle with 500,000 collisions, of which 5943 occupations appear in the range $4 \leq u < 6$; (b) one particle with 3,000,000 collisions, of which 938 occupations appear in the range $4.4 \leq u < 4.8$.

the expanded region $4.4 \leq u \leq 4.8$; we now see finer structures within the leaves. Here the expanded region has been divided into 100 intervals along u and 100 intervals along ψ, forming 10,000 cells. The map is iterated 3×10^6 times for a single initial condition, and the number inside each cell (not readily seen) is a logarithmic measure of the number of occupations.

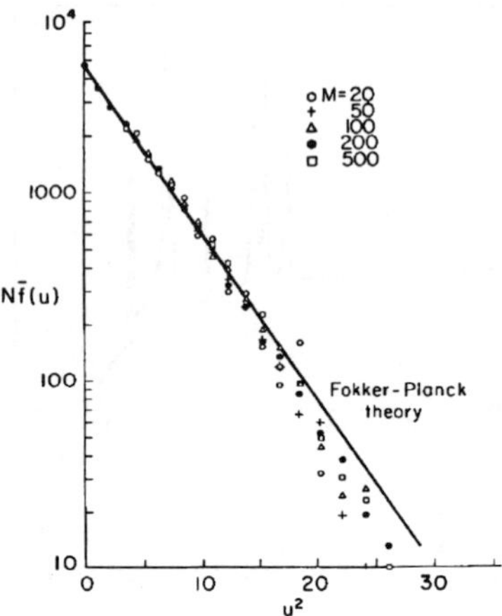

Fig. 9   Comparison of the numerically calculated phase-averaged invariant distribution with the Fokker-Planck solution $\bar{f}(u)$ for $\delta = 0.1$ and various values of M for the Fermi map.

When the number of occupations per cell is summed over the phase ψ at a fixed u, a phase-averaged invariant distribution $\bar{f}(u)$ is obtained numerically. To determine $\bar{f}$ analytically, we use the Fokker-Planck equation (2.10) with $B = -u\delta$ and $D = \frac{1}{2}$. Setting $\partial/\partial n \equiv 0$ in (2.10), we find in the steady state and with no net flux of particles,

$$-B\bar{f} + \frac{1}{2}\frac{d}{du}(D\bar{f}) = 0, \qquad (2.28)$$

which we integrate to obtain

$$\bar{f}(u) = \left(\frac{8\delta}{\pi}\right)^{1/2} \exp(-2\delta u^2). \qquad (2.29)$$

The integration constant has been found from the requirement that $\int_0^\infty \bar{f}\, du = 1$. In Fig. 9, we compare this analytic expression (solid line) with the numerical results for $\delta = 0.1$ and various values of M. The theory and numerical calculations are in good agreement.

Successively better approximations to the attractor can be found by successively back-iterating the map [5]. For example, inverting (2.25) and

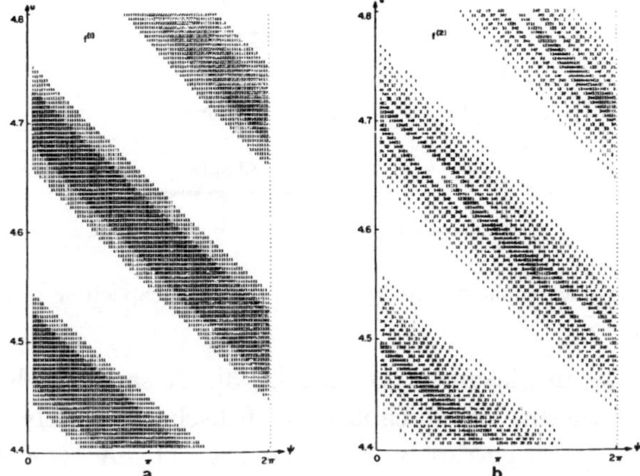

Fig. 10 Analytical calculation of the invariant distribution for dissipative Fermi acceleration; (a) first order result $f^{(1)}(\psi, u)$; (b) second order result $f^{(2)}(\psi, u)$.

using (2.29), we obtain the next order approximation

$$
\begin{aligned}
f^{(1)}(\psi, u) = {} & \frac{1}{1-\delta} \left(\frac{8\delta}{\pi} \right)^{1/2} \\
& \times \exp \left\{ -\frac{2\delta}{(1-\delta)^2} \left[u + \sin \left(\psi - \frac{2\pi M}{u} \right) \right]^2 \right\}.
\end{aligned}
\tag{2.30}
$$

To compare (2.30) with the numerically calculated invariant distribution $f(\psi, u)$, we plot in Fig. 10a the expected occupation numbers using $f^{(1)}$, in the same expanded region of the surface of sectoin as for the numerical calculation of f in Fig. 8b. The band structure seen in the magnified image of $f^{(1)}$ corresponds closely to the numerically determined bands seen in Fig. 8b. To see a still closer correspondence between theory and numerical calculation, we plot the higher order expression $f^{(2)}$ in Fig. 10b; this shows even better agreement with the numerical calculation than $f^{(1)}$.

3 Capacitive RF discharges

We begin with the application of Fermi acceleration to electron heating in capacitive discharges where collisionless dissipation of rf power has been intensively studied over the last few decades, due to the wide application

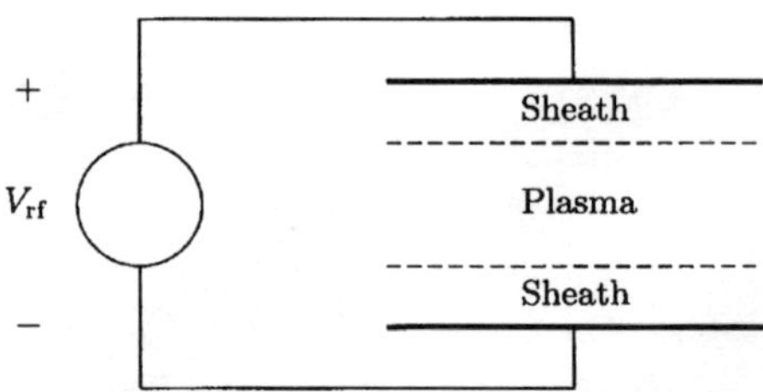

Fig. 11   Sheath-plasma-sheath sandwich structure of a capacitive rf discharge.

of these discharges in plasma processing [2, 3]. A sandwich-like (sheath-plasma-sheath) structure of the capacitive rf discharge widely accepted in analyzing the basic properties is shown in Fig. 11. A typical discharge consists of a vacuum chamber containing two planar electrodes separated by a spacing of order 2–10 cm and driven by an rf power source. The substrates are placed on one electrode, feedstock gases are admitted to flow through the discharge, and effluent gases are removed by the vacuum pump. The typical rf driving voltage is $V_{\rm rf} = 100\text{--}1000$ V, and for etching of thin films pressures are in the range 10–100 mTorr, power densities are 0.1–1 W/cm^2, and the driving frequency is usually 13.56 MHz. Plasma densities are relatively low, 10^9–10^{10} cm^{-3}, and electron temperatures are of order 3 V. For deposition of films, pressures tend to be higher, and frequencies can be lower than 13.56 MHz.

The operation of capacitively driven discharges is reasonably well understood. The mobile plasma electrons, responding to the instantaneous electric fields produced by the rf driving voltage, oscillate back-and-forth within the positive space charge cloud of the ions. The massive ions respond only to the time-averaged electric fields. Electron thermal motion and oscillation of the electron cloud create sheath regions near each electrode that contain net positive charge when averaged over an oscillation period; i.e., the positive charge exceeds the negative charge in the system, with the excess appearing within the sheaths. This excess produces a strong time-averaged electric field within each sheath directed from the plasma to the electrode. Ions flowing out of the bulk plasma near the center of the discharge can be accelerated by the sheath fields to energies of order of $V_{\rm rf}$ as they flow to the substrate, leading to energetic-ion enhanced processes. The positive ions continuously bombard the electrode over an rf cycle, whereas electrons are lost to the electrode only when the oscillating cloud closely approaches the electrode. During that time, the instantaneous sheath potential collapses to near-zero, allowing sufficient electrons to escape to balance the ion charge delivered to the electrode. Except for such brief moments, the instantaneous potential of the discharge must always be positive with respect to any large electrode or wall surface; otherwise the mobile electrons would quickly leak out. The sheath impedance is generally much larger than that of the plasma and plays the essential role in limiting the rf discharge current.

An explicit application of Fermi acceleration to electron heating in rf discharges was by Godyak [7]:

In an oscillating double sheath, the potential distribution, and thus the coordinate of the electron-reflection point depend on the time, and the electron reflection is analogous to that of solid particles from a vibrating wall. On the average particles acquire energy in this case (the Fermi acceleration mechanism).

Godyak went on to determine the electron power deposition for a dc sheath with a small sinusoidally vibrating fluctuation, and put forward the idea that Fermi acceleration might be a major mechanism to sustain a capacitive discharge at low gas pressures.

A. Homogeneous Discharge Model

Let us consider the collisionless power absorption in the simplest model of a capacitive discharge with a homogeneous ion background in the electrode gap and with a high rf sheath voltage. Such a simplified discharge model qualitatively describes the main features of capacitive discharges in practice.

Electrons reflecting from the large decelerating fields of a moving high voltage sheath can be approximated by assuming the reflected velocity is that which occurs in an elastic collision (in the moving reference frame) of a ball with a moving wall

$$u_r = -u + 2u_{es} \tag{3.1}$$

where u and u_r are the incident and reflected electron velocities parallel to the time varying electron sheath velocity u_{es}. If the parallel electron velocity distribution at the sheath edge is $f_{es}(u,t)$, then in a time interval dt and for a speed interval du, the number of electrons per unit area that collide with the sheath is given by $(u - u_{es})f_{es}(u,t)dudt$. This results in a power transfer per unit area,

$$dS_{\text{stoc}} = \frac{1}{2}m(u_r^2 - u^2)(u - u_{es})f_{es}(u, t)\, du. \tag{3.2}$$

Using $u_r = -u + 2u_{es}$ and integrating over all incident velocities, we obtain

$$S_{\text{stoc}} = -2m \int_{u_{es}}^{\infty} u_{es}(u - u_{es})^2 f_{es}(u, t)\, du. \tag{3.3}$$

In the physical problem f_{es} varies with time, as the sheath oscillates, and the problem becomes quite complicated. For the uniform density model we note that

$$\int_{-\infty}^{\infty} f_{es}(u, t)\, du = n_{es}(t) = n, \text{ a constant.} \tag{3.4}$$

Furthermore, for the purpose of understanding the heating mechanism we make the simplifying approximations that $f_{es}(u, t)$ can be approximated by a Maxwellian, ignoring the plasma drift, and that $u_{es} \ll \bar{v}_e = (8e\text{T}_e/\pi m)^{1/2}$, the mean electron speed. These approximations simplify the calculation. Consistent with our approximation, we can set the lower limit in (3.3) to zero. Before performing the average over the distribution function, we substitute

$$u_{es} = u_0 \cos \omega t, \tag{3.5}$$

in (3.3) and average over time. Only the term in $\sin^2 \omega t$ survives giving

$$\bar{S}_{\text{stoc}} = 2mu_0^2 \int_0^{\infty} u f_{es}(u)\, du. \tag{3.6}$$

Now, consistent with our approximation that f_{es} is Maxwellian, we note that the integral gives the usual random flux $\Gamma_e = \frac{1}{4}n\bar{v}_e$, and (3.6) becomes

$$\bar{S}_{\text{stoc}} = \frac{1}{2}mu_0^2 n\bar{v}_e. \tag{3.7}$$

Inside the plasma the rf current I_1 is almost entirely conduction current, such that

$$I_1 = J_1 A = -enu_0 A, \tag{3.8}$$

where A is the cross-sectional area. Substituting (3.8) into (3.7) yields the stochastic electron power in terms of the (assumed) known current. Since we are calculating the power per unit area, we use the current density, to obtain, for a single sheath,

$$\bar{S}_{\text{stoc}} = \frac{1}{2}\frac{m\bar{v}_e}{e^2 n} J_1^2.$$ (3.9)

Within the homogeneous model, the time-average electron heating per unit area due to ohmic heating in the discharge bulk is

$$\bar{S}_{\text{ohm}} = \frac{1}{2}J_1^2 l \text{Re}\,(\sigma_p^{-1}),$$ (3.10)

where l is the length of the bulk region containing electrons and $\sigma_p = e^2 n/m(\nu + j\omega)$ is the plasma conductivity. Substituting σ_p into (3.10), we find

$$\bar{S}_{\text{ohm}} = \frac{1}{2}J_1^2 \frac{m\nu_m l}{e^2 n},$$ (3.11)

where ν_m is the electron-neutral momentum transfer frequency.

Adding (3.9) (for two sheaths) and (3.11), the total time average electron power per unit area is

$$S_e = \frac{1}{2}\frac{m}{e^2 n}(\nu_m l + 2\bar{v}_e)J_1^2.$$ (3.12)

A useful interpretation of this result is to introduce an effective collision frequency

$$\nu_{\text{eff}} = \nu_m + \frac{2\bar{v}_e}{l}.$$ (3.13)

Then we can consider that the stochastic heating introduces an additional, gas pressure-independent collision frequency, the electron bounce frequency $2\bar{v}_e/l$, into the expression for ohmic heating of the discharge.

Although the preceding calculation illustrates simply the mechanism of collisionless power absorption, in a real discharge one has to account for the oscillatory electron drift velocity. If this velocity is independent of position and exactly equals the velocity of the plasma-sheath edge, as given in (3.5), then transforming to an accelerated frame moving with the plasma electrons, we see a stationary electron sheath edge. Therefore no energy is transfered to electrons that collide with the sheath, and there is no stochastic heating. Analytical rf sheath heating models are very sensitive to assumptions about the form of the electron velocity distribution at the moving plasma-sheath edge. In particular, if the electron velocity distribution there is assumed to be symmetric about the time-varying sheath velocity, then there is no heating. However, simulations for an inhomogeneous sheath indicate that the distribution is not symmetric and that sheath heating does occur.

B. Experimental Results

An early experiment to investigate stochastic heating is described in [8]. Electrical and plasma parameters were studied in a parallel-plate capacitive rf discharge symmetrically driven at 40–110 MHz in mercury vapor. The current-voltage characteristic, the rf power, the plasma density and the electron temperature were simultaneously measured in the mercury pressure range between 2×10^{-4} and 1×10^{-1} Torr.

The effective collision frequency $\nu_{\rm eff}$ versus pressure was evaluated from the shape of the measured discharge current-voltage characteristic, and directly by measuring the rf power absorbed by the discharge, from which $\nu_{\rm eff}$ was obtained from the relationship for power absorbed per unit area

$$S_{\rm abs} = \frac{1}{2} \frac{|\tilde{J}_{\rm rf}|^2}{e^2 n} m \nu_{\rm eff} l, \tag{3.14}$$

where $|J_{\rm rf}|$ is the cross-section averaged discharge current density. The measurements were done at relatively low rf voltages, and the power absorption due to ion acceleration in the rf sheaths was neglected. The effective collision frequency found from experiment as a function of mercury pressure is shown in Fig. 12. Both the asymptotic leveling off of $\nu_{\rm eff}$ at low pressure p, characteristic of stochastic heating which is independent of p, and the linear increase of $\nu_{\rm eff}$ with p at high p, characteristic of ohmic heating, are clearly visible. The good agreement of the measurements with $\nu_{\rm eff}$ calculated from the stochastic heating formula is somewhat fortuitous, however, as a uniform sheath rather than a self-consistent sheath was used in the calculation, and the ion power loss S_i was neglected in determining $\nu_{\rm eff}$ from the measurements.

C. Fluid and Particle Simulations

Monte Carlo and particle-in-cell (PIC) simulations of capacitively coupled discharges at low pressure performed in the last decade have confirmed the

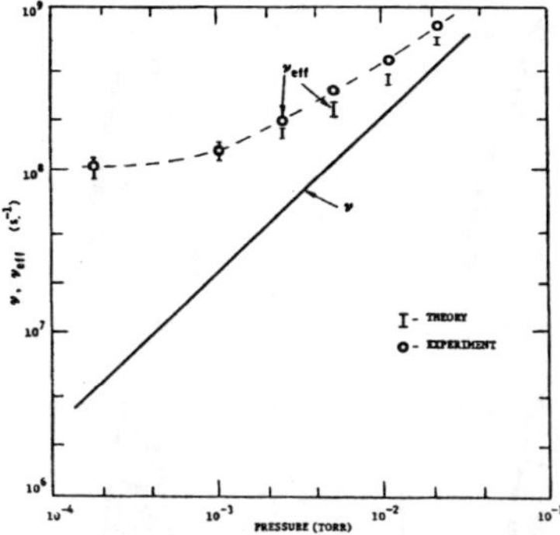

Fig. 12   Effective collision frequency ν_{eff} versus pressure p for a mercury discharge driven at 40.8 MHz. The solid line shows the collision frequency due to ohmic dissipation alone (after [8]).

existence of collisionless electron heating produced by oscillating electrode sheaths. One simulation of discharge behavior [3] was performed at $p = 3$ mTorr (argon) with a spacing of 10 cm between parallel plates, and over a range of rf voltages between 100 and 1000 volts. A two temperature distribution was found, as in the experiments, and the distribution varied in both space and time. It is clear that a deeper understanding of the discharge behavior involves the space and time variations of f_e. Figure 13 shows the one dimensional electron distribution function $f_e(x, v_x, t)$ versus v_x at 15 positions near the sheath region ($x = 0$–3 cm) and at eight different times during the rf cycle. Each plot covers 1/32 of a cycle temporally, and each line in a plot covers a 2 mm thick region spatially. The units on the vertical axis are proportional to f_e. At time 0/32, the sheath is fully expanded, and the two-temperature nature of the discharge near the sheath can be seen as the wide "base" and narrow "peak" of the distribution. As the rf cycle progresses to time 8/32, the distributions within the sheath region at each position display a drift toward the electrode (negative velocity) that is approximately equal to the collapsing sheath velocity. By time 12/32, fast electrons have arrived from the opposite electrode, moving at a velocity of about 4×10^6 m/s (small peak at extreme left of figure). At time 16/32, the sheath is fully collapsed, the drift in the sheath has disappeared, and the fast electron group moving toward the electrode shows a lower velocity as slower electrons arrive from the opposite electrode. As the sheath begins to expand, as shown here at times 18/32 and 20/32, the electrons in the

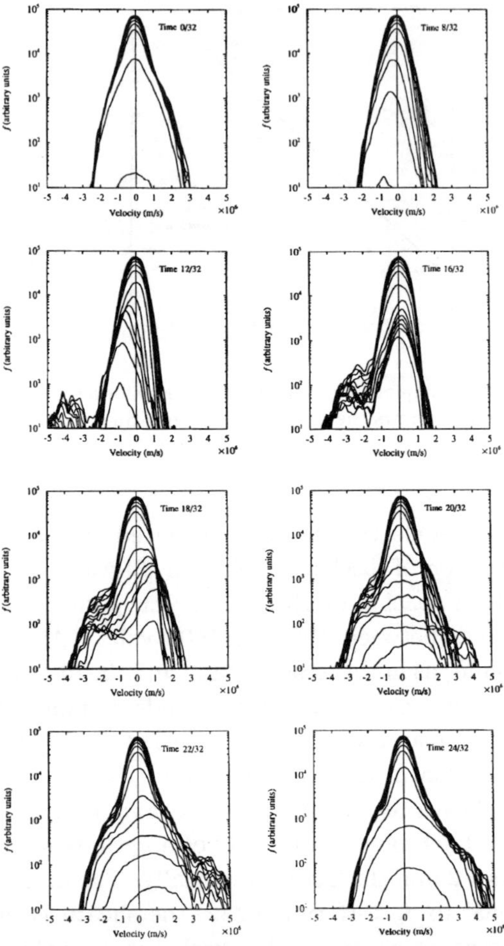

Fig. 13   One-dimensional electron velocity distribution function $f_e(x, v_x, t)$ for a 10 cm electrode spacing in a 3 mTorr argon discharge; each plot covers a time window of 1/32 of an rf cycle. Each line on a plot represents a spatial window of 2 mm (after [3]).

sheath region are strongly heated, and the beginning of an electron beam produced by this expansion can be seen moving away at a positive velocity. As the sheath continues to expand, the drift of the distribution within the sheath away from the electrode can be seen to initially match the sheath velocity (time 22/32) but then decays (time 24/32) to a velocity much slower than when the sheath was collapsing. One consequence of these

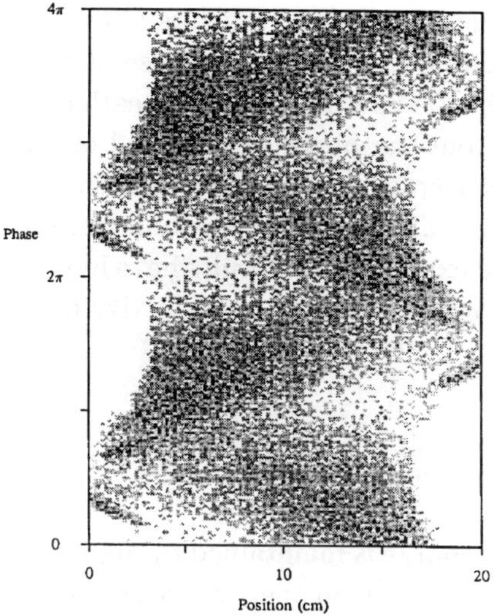

Fig. 14   Spatiotemporal distribution of ionizing collisions collected over 20 rf cycles, for a 10 MHz, 20 mTorr hydrogen discharge ([3]).

complicated dynamics near the sheath edge is that f_e is not symmetric about the velocity of the moving sheath edge during the entire rf cycle.

The existence of more energetic electrons near the plasma edge due to local electron heating increases the ionization there at higher pressures ($\lambda_e \ll l$), tending to flatten the plasma profile. Furthermore, the ionization is not constant, but follows the density variations in space and time of the more energetic electrons. This is shown for a PIC simulation [3] in the plot of Fig. 14, in which the darkness of each square is proportional to the number of ionizing collisions within that square of position and time intervals. Most of the ionization is seen to occur along a path of fastest electrons that are reflected off of the sheath at the phase at which it is most rapidly expanding. There is also somewhat more ionization near the sheaths, an effect that becomes more pronounced at higher pressures where the ionization mean-free-path is shorter, which has been observed in experiments.

The spatial distribution of the electron power absorption has been examined in several PIC simulations. While the absorption is large and positive near and within the rf sheaths, it can become negative within certain regions in the discharge bulk under conditions of strong stochastic heating. This is particularly apparent in simulations at low pressures in a Ramsauer gas, where the ohmic dissipation (which is always positive) is small. Negative power absorption occurs where the phase of the electron current (transfered from the stochastic heating at the rf sheath edge by the electron thermal motion) differs from the phase of the local electric field by more than 90°. There have been no experimental measurements confirming the existence of negative power absorption in the bulk region of capacitive rf discharges, although some experimental results have hinted at the existence of this effect. Negative power absorption in inductive discharges is treated extensively in Section 4.

4   Inductive RF discharges

Plasma in an inductive discharge is maintained by application of rf power to an inductive coil, resulting in electron energy absorption due to the induced rf electric field near the coil. The driving frequency is usually 13.56 MHz, although lower (and higher) frequencies are sometimes used. As shown in Fig. 15ab, planar or cylindrical coils in a low aspect ratio (length/diameter) discharge are generally used for low pressure materials processing. The planar coil is a flat helix wound from near the axis to near the outer radius at one end of the discharge chamber ("electric stovetop" coil shape). For a typical 30 cm chamber diameter, the rf power is typically 100-1000 W.

Another kind of inductive discharge is an rf lamp with an internal coil, as shown in Fig. 15c. The internal coil, usually with a ferrite core, is inserted into a re-entrant cavity inside a glass bulb coated inside with fluorescent powder (phosphor). The bulb is filled with a mixture of inert gases such as argon or krypton at a pressure of hundreds of millitorr, and mercury vapor at a few millitorr. The inductive plasma excited inside the bulb has a very high rf power conversion to the mercury resonance uv radiation (60–70%, mainly at 253 nm), and the uv radiation excites the phosphor to emit visible light. The absence of electrodes provides a highly efficient (4–5 times more than an incandescent bulb) and durable (up to 100,000 hours) light source.

Because the voltage across the exciting coil of an inductive discharge can be as large as several kilovolts, a discharge can also be capacitively driven by the coil. There is generally a capacitively driven discharge at low plasma densities, with a transition to an inductive discharge at high densities. An electrostatic shield placed between the coil and the plasma can reduce the capacitive coupling if desired, while allowing the inductive field to couple

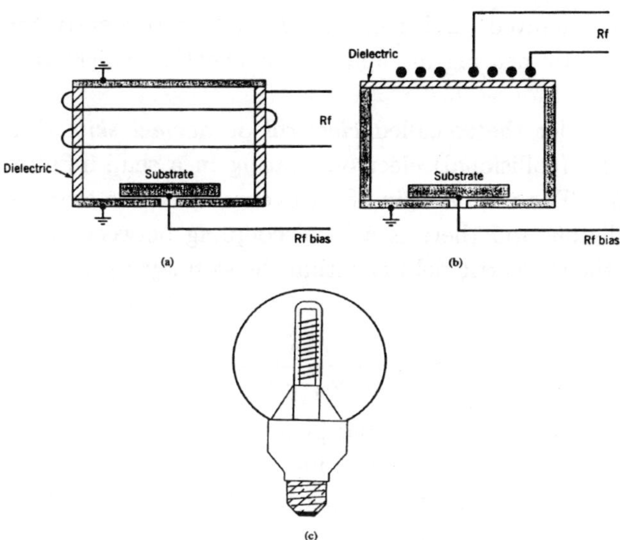

Fig. 15   Schematic of inductively driven discharge in (a) cylindrical, (b) planar, and (c) re-entrant geometries; (a) and (b) are used for materials processing and (c) is used for lighting.

unhindered to the plasma. Inductive discharges for materials processing are sometimes referred to as ICP's (inductively coupled plasmas), TCP's (transformer coupled plasmas), or RFI's (rf inductive plasmas).

The inductive electric field is non-propagating ($\omega \ll \omega_{pe}$) and typically penetrates into the discharge a distance on the order of a plasma skin depth, which is typically 1–3 cm. Hence plasma heating occurs near the dielectric window surface. The dc plasma potential in these discharges is typically of order 30–40 volts with respect to the walls, and the plasma density is typically in the range 10^{11}–10^{12} cm^{-3}. Hence the sheath thickness is of order 0.1–1 mm (a few Debye lengths).

A. Classical and Anomalous Skin Effect

In an inductively coupled plasma, power is transferred from the electric fields to the plasma electrons within a skin depth layer of thickness δ near the plasma surface by collisional (ohmic) dissipation and by a collisionless heating process in which bulk plasma electrons "collide" with the oscillating inductive electric fields within the skin layer. In the latter situation, electrons are accelerated (and decelerated) and subsequently thermalized much like stochastic heating in capacitive rf sheaths, which we treated in Section 3.

We first consider the so-called *classical* or *normal skin effect* accompanied by ohmic (collisional) electron heating in a semi-infinite spatially uniform plasma. The normal skin effect occurs when the electron thermal motion is negligible and there is a local coupling between the rf current density $\mathbf{J}$ and the rf electric field $\mathbf{E}$ within the skin layer given by $\mathbf{J} = \sigma_p \mathbf{E}$, where

$$\sigma_p = \frac{e^2 n}{m(\nu_m + j\omega)} \tag{4.1}$$

is the complex conductivity of a cold plasma. We consider the case when $\omega \ll \omega_{pe}$, which is always true for inductively coupled plasmas. We also assume a Maxwellian EEDF and an energy-independent electron-atom collision frequency $\nu_m(\mathcal{E}) = \text{const.}$

According to Maxwell's equations, the penetration of the transverse electric field into the plasma is described by the complex wave equation

$$\frac{d^2 E_y}{dx^2} = j\omega \mu_0 \sigma_p E_y, \tag{4.2}$$

having solution

$$E_y = E_{y0}\, e^{-x/\delta}\, \cos(\omega t - \beta x), \tag{4.3}$$

where

$$\delta^{-1} = \text{Re}(j\omega\mu_0\sigma_p)^{1/2} \tag{4.4}$$

is the inverse skin depth and

$$\beta = \text{Im}(j\omega\mu_0\sigma_p)^{1/2} \tag{4.5}$$

is the propagation constant. Substituting (4.1) into (4.4), one obtains the general expression for the classical (normal) skin depth,

$$\delta = \frac{\delta_0}{\cos(\epsilon/2)}, \tag{4.6}$$

where

$$\delta_0 = \frac{c}{\omega_{pe}} \left(1 + \frac{\nu_m^2}{\omega^2}\right)^{1/4} \qquad \text{and} \qquad \epsilon = \tan^{-1}(\nu_m/\omega). \tag{4.7}$$

In the collisional limit ($\nu_m \gg \omega$), typical for non-superconducting metals and high pressure plasmas, $\epsilon = \pi/2$ and

$$\delta = \delta_c = \frac{c}{\omega_{pe}}\left(\frac{2\nu_m}{\omega}\right)^{1/2} = \delta_p\left(\frac{2\nu_m}{\omega}\right)^{1/2} \tag{4.8}$$

and the rf energy collisionally dissipates within the skin layer.

In the high frequency limit ($\nu_m \ll \omega$), called the non-dissipative or high frequency skin effect, $\epsilon = 0$ and

$$\delta = \delta_p = \frac{c}{\omega_{pe}}. \tag{4.9}$$

In this case the electrons collisionlessly oscillate within the skin layer with no net energy gain. For an electromagnetic wave incident on the plasma boundary this case corresponds to the total reflection of the wave from the plasma. For discharge maintenance in this case, the wave reflection is not perfect, and a small fraction of the incident wave power is locally and/or nonlocally deposited within the skin layer.

There is a third situation (anomalous skin effect) for which electrons incident on a skin layer of thickness δ_a satisfy the condition $\bar{v}_e/\delta_a \gg \omega, \nu_m$, where δ_a is determined below. In this case the interaction time of the electrons with the skin layer is short compared to the rf period or the collision time. The rf field penetration in this regime was first estimated by Pippard [9] with application to the high frequency skin effect in metals at low temperatures, and was determined self-consistently by Reuter and Sondheimer [10] for metals and by Weibel [11] for a homogeneous plasma half-space with a Maxwellian EEDF. To see the essential scalings, following Pippard, we consider the ordering $\nu_m \gg \omega$ and divide the electrons into two groups, those moving at small angles to the surface which spend most of their time between collisions within the skin layer, and the rest, whose chance of collision is small. We ignore the latter group of ineffective electrons. The velocities of the effective electrons form angles less than δ_a/λ_e and their relative number is of order δ_a/λ_e, where λ_e is the electron-neutral mean free path for momentum transfer. This results in an effective plasma density $n_{\text{eff}} = K_{\text{eff}} n \delta_a/\lambda_e$ and an effective plasma frequency $\omega_{pe\text{eff}} = \omega_{pe}(K_{\text{eff}}\delta_a/\lambda_e)^{1/2}$ within the skin layer, where K_{eff} is a constant of order unity. Substituting these effective quantities into (4.1) and (4.8) yields an expression for the effective collision frequency ν_{eff}, analogous to that introduced in (3.13) for capacitive rf discharges, and for the anomalous skin depth δ_a:

$$\nu_{\text{eff}} = \frac{\bar{v}_e}{K_{\text{eff}}\delta_a} \qquad \text{and} \qquad \delta_a = \left(\frac{2\bar{v}_e c^2}{K_{\text{eff}}\omega\omega_{pe}}\right)^{1/3}.$$

A more careful averaging based on the kinetic theory of the anomalous skin effect [11] gives $K_{\text{eff}} = 4$ and

$$\nu_{\text{eff}} = \frac{\bar{v}_e}{4\delta_a} \qquad \text{and} \qquad \delta_a = \left(\frac{\bar{v}_e c^2}{2\omega\omega_{pe}^2}\right)^{1/3} = \delta_p\left(\frac{\bar{v}_e\omega_{pe}}{2c\omega}\right)^{1/3}. \qquad (4.10)$$

According to (4.10), for a strong anomalous skin effect where $\bar{v}_e\omega_{pe} \gg c\omega$ and $\omega \gg \nu_m$, the penetration of the rf electric field into the plasma is deeper than for the non-dissipative skin effect in the high frequency limit: $\delta_a > \delta_p$.

A general *nonlocality parameter* for the non-local interaction of electrons with the electromagnetic field is [11]:

$$\Lambda = \frac{\pi}{4}\left(\frac{\bar{v}_e\omega_{pe}}{c}\right)^2 \frac{\omega}{(\omega^2 + \nu_m^2)^{3/2}}. \qquad (4.11)$$

Formulae for the classical skin effect are applicable when $\Lambda \ll 1$; for $\Lambda \gtrsim 1$, the anomalous skin effect takes place. Note that Λ is small for both very low and very high frequency and reaches it maximum at $\omega = \nu_m/\sqrt{2}$ where $\Lambda_{\max} \approx (\omega_{pe}\bar{v}_e/\omega c)^2$, which is close to its high frequency $(\omega \gg \nu_m)$ limit. In the opposite case $(\nu_m \gg \omega)$, $\Lambda \approx (\omega_{pe}\bar{v}_e/\omega c)^2(\omega/\nu_m)^3 \ll 1$, and the normal (collisional) skin effect occurs. Figure 16 shows the boundary dividing the classical and the anomalous skin effect $(\Lambda = 1)$ in the space of λ_e/δ_0 and ω/ν_m.

Let us describe some essential features of the anomalous skin effect which are the result of the nonlocal interaction of electrons with the electromagnetic field due to their thermal motion:

(1) The spatial decay of the electromagnetic field into a plasma for the anomalous skin effect is not exponential, as it is in the classical skin effect. Moreover, the decay may be nonmonotonic and may exhibit local maxima and minima in the plasma.

(2) In the regime of the strong anomalous skin effect neither the skin depth δ_a nor the energy dissipation in the skin layer depend on the collision frequency ν_m. In this case the rf energy dissipation process occurs even in the limit $\nu_m \to 0$. The non-collisional dissipation has a simple explanation. For a relatively thin skin layer when $\delta < \bar{v}_e/\omega$ and $\delta < \lambda_e$, the electrons reflecting from the space charge sheath at the plasma-wall boundary cross

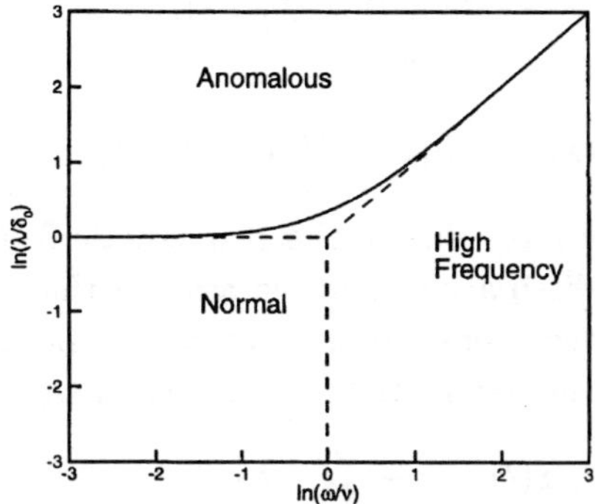

Fig. 16   Plot of $\ln(\lambda_e/\delta_0)$ versus $\ln(\omega/\nu_m)$, showing the regimes of collisional, high frequency, and anomalous skin effect in a semi-infinite plasma; the solid line, corresponding to nonlocality parameter $\Lambda = 1$, is the boundary of the anomalous skin effect (after [2]).

the skin layer in a time less than the rf field period. Hence, on the average the electrons gain energy within the skin layer as in a dc field. This differs from collisionless electron motion in a homogeneous rf field when $\delta \gg \bar{v}_e/\omega$, corresponding to the high frequency limit of the normal skin effect in which electrons gain energy from the field during one quarter cycle and return the energy back to the field during the next quarter cycle.

For both the normal and the anomalous skin effect, there is no electron heating unless some phase mixing (randomization) mechanism breaks the regularity of the electron motion. For collisional heating corresponding to the normal skin effect, randomization occurs locally within the skin layer due to electron-atom (and/or electron-ion) collisions. For the anomalous skin effect the randomization is provided by the electron thermal motion which moves electrons out of the skin layer into the neighboring plasma having no rf field, thus preventing the electrons from returning the energy acquired in the skin layer back to the rf field. For a bounded plasma such as an inductive discharge where electrons can repeatedly interact with the rf field in the skin layer, some randomization mechanism must be present in the bulk plasma to provide an effective electron heating in the skin layer.

B. Fermi Acceleration Model of Collisionless Heating

To determine the heating at low pressures using a Fermi acceleration model, we consider an electron from the bulk plasma incident on the rf electric field within a skin depth layer in slab geometry. We assume a simple model in which the transverse electric field within the slab decays exponentially with distance x from the edge into the slab,

$$E_y(x,t) = E_0 e^{-|x|/\delta} \cos(\omega t + \phi). \qquad (4.12)$$

We also assume that the force due to the rf magnetic field is negligible in determining the power dissipation, and we assume that the collisionality is weak, $\nu_m \ll \bar{v}_e/\delta$; hence there are no electron collisions within the skin layer. Because there are no x-directed forces, we can write

$$\begin{aligned} x(t) \quad &= -v_x t, \quad t < 0, \\ &= v_x t, \quad t > 0, \end{aligned} \qquad (4.13)$$

where the electron reflects from the surface at $t = 0$. Substituting (4.13) into (4.12) yields the transverse electric field seen by the electron,

$$\begin{aligned} E_y(t) \quad &= \operatorname{Re} E_0 e^{(j\omega + v_x/\delta)t + j\phi}, \quad t < 0, \\ &= \operatorname{Re} E_0 e^{(j\omega - v_x/\delta)t + j\phi}, \quad t > 0. \end{aligned} \qquad (4.14)$$

The transverse velocity impulse,

$$\Delta v_y = -\int_{-\infty}^{\infty} dt \, \frac{e E_y(t)}{m}, \qquad (4.15)$$

is calculated by substituting (4.14) into (4.15) and integrating to obtain

$$\Delta v_y = \frac{2 e E_0 \delta}{m} \frac{v_x}{v_x^2 + \omega^2 \delta^2} \cos \phi. \qquad (4.16)$$

The energy change $\Delta \mathcal{E}$, averaged over a uniform distribution of initial electron phases ϕ, is then

$$\begin{aligned} \Delta \mathcal{E} \quad &= \tfrac{1}{2} m \langle (\Delta v_y)^2 \rangle_\phi \\ &= \tfrac{1}{4} m \left(\frac{2 e E_0 \delta}{m} \right)^2 \frac{v_x^2}{(v_x^2 + \omega^2 \delta^2)^2}, \end{aligned} \qquad (4.17)$$

which can be integrated over the particle flux to obtain the stochastic heating power

$$S_{\mathrm{stoc}} = \int_{-\infty}^{\infty} dv_y \int_{-\infty}^{\infty} dv_z \int_{0}^{\infty} dv_x \, f_e v_x \Delta \mathcal{E}(v_x). \qquad (4.18)$$

For a Maxwellian electron distribution f_e, the integrals over v_y and v_z are easily done, and the v_x integral can be evaluated in terms of the exponential integral E_1. For the regime of large non-locality ($\Lambda \gg 1$),

$$\zeta = \frac{4\omega^2 \delta^2}{\pi \bar{v}_e^2} \ll 1, \tag{4.19}$$

we obtain [21]

$$S_{\text{stoc}} \approx \frac{m n_s}{\bar{v}_e} \left(\frac{eE_0 \delta}{m}\right)^2 \frac{1}{\pi} \left[\ln\left(\frac{1}{\zeta}\right) - 1.58\right]. \tag{4.20}$$

We can introduce an effective collision frequency ν_{eff} by equating the stochastic heating (4.20) to an effective collisional heating power flux,

$$\begin{aligned}
S_{\text{ohm}} &= \tfrac{1}{2} \int_0^\infty dx \, \left(E_0 e^{-x/\delta}\right)^2 \frac{e^2 n_s}{m} \frac{\nu_{\text{eff}}}{\nu_{\text{eff}}^2 + \omega^2} \\
&= \tfrac{1}{4} \frac{e^2 n_s \delta}{m} \frac{\nu_{\text{eff}}}{\nu_{\text{eff}}^2 + \omega^2} E_0^2 .
\end{aligned} \tag{4.21}$$

For $\omega \ll \nu_{\text{eff}}$, we have

$$S_{\text{ohm}} = \frac{1}{4} \frac{e^2 n_s \delta}{m \nu_{\text{eff}}} E_0^2,$$

and equating this to (4.20), we determine

$$\nu_{\text{eff}} = \frac{\bar{v}_e}{\delta} \frac{4}{\pi} \left[\ln\left(\frac{1}{\zeta}\right) - 1.58\right], \tag{4.22}$$

With $\zeta \ll 1$, we find $\nu_{\text{eff}} \sim \bar{v}_e/\delta$, in agreement with the simple estimate (4.10).

Although Fermi acceleration models of the velocity impulse lead to simple estimates of the effective collision frequency due to stochastic heating, they are not self-consistent because the form of the spatial variation of the electric field has been assumed. The heating power and effective collision frequency have been determined over the entire range of collisionality from a Fermi acceleration model with an exponentially decaying electric field profile by Vahedi et al [12]. We summarize their results and compare them to a self-consistent model in the next subsection.

C. Self-Consistent Collisionless Heating

The self-consistent analysis of Reuter and Sondheimer [10], using a Fermi-Dirac electron distribution to determine the anomalous skin resistance of a low temperature metal, was first applied by Weibel [11] to a classical plasma having a Maxwellian electron distribution. The rf power absorption due to the skin effect can be characterized by a complex surface impedance

$$Z_s = Z_{s\mathrm{R}} + jZ_{s\mathrm{I}} = E_y(0)/H_z(0). \tag{4.23}$$

The time-average power absorbed by the plasma per unit area can be written in terms of Z_s as:

$$S_{\mathrm{abs}} = \mathrm{Re}\left(\frac{1}{2}E_y(0)H_z^*(0)\right) = \mathrm{Re}\left(\frac{1}{2}Z_s|H_z^*(0)|^2\right). \tag{4.24}$$

For the classical (normal) skin effect, we find

$$Z_s = j\omega\mu_0\delta_p\left[\sqrt{b(1+b)/2} + j\sqrt{b(1-b)/2}\right]^{-1}, \tag{4.25}$$

where $b = (1+\nu_m^2/\omega^2)^{-1/2}$. For $\nu_m \gg \omega$, corresponding to the normal skin effect dominated by collisional heating, we find

$$Z_s = \frac{\omega\mu_0\delta_c}{2}(1+j), \tag{4.26}$$

with δ_c given by (4.8). For the anomalous skin effect (see [11]), we obtain

$$Z_s = \frac{2\omega\mu_0\delta_a}{3}\left(\frac{1}{\sqrt{3}} + j\right), \tag{4.27}$$

with δ_a given by (4.10). In the collisional regime where $\nu_m/\omega \geq 3$, the results of kinetic (warm plasma) theory practically coincide with hydrodynamic (cold plasma) theory, corresponding to the classical skin effect. In the collisionless limit when $\nu_m/\omega \to 0$, the real part of the surface impedance disappears in the classical (cold plasma) skin effect theory (there is no heating in the high frequency limit), but remains a constant, independent of the collision frequency, in the kinetic theory of the skin effect acounting for the electron thermal motion.

Typical behaviors for the effective skin depth and the normalized rf power dissipation in the skin layer are shown in Figs. 17 and 18 respectively, as calculated by Vahedi et al [12]. In these figures the calculated parameters are given as functions of the normalized rf frequency $w = c\omega/\bar{v}_e\omega_{pe}$ for $\lambda_e > \delta$, where $w = \Lambda^{-1/2}$ (Λ is the nonlocality parameter for $\omega^2 \gg \nu_m^2$). The comparisons with self-consistent results for the formula for the effective skin depth (Fig. 17 for $w < 1$) and for the normalized rf power absorption (Fig. 18) show good agreement and justify the simplifications assumed by Vahedi et al [12] for calculating the absorbed power.

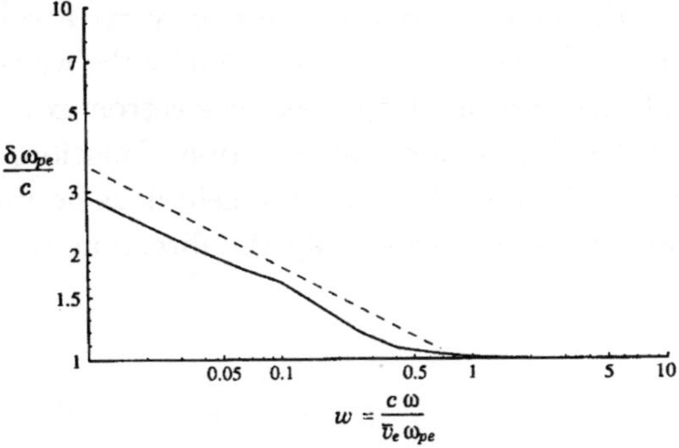

Fig. 17   The normalized effective skin depth δ/δ_p versus the normalized frequency $w = (\omega c)/(\bar{v}_e \omega_{pe})$ for a near-collisionless case of $\nu_m/\omega = 0.008$. The dashed line shows δ_a/δ_p obtained from (4.10) for comparison (after [2]).

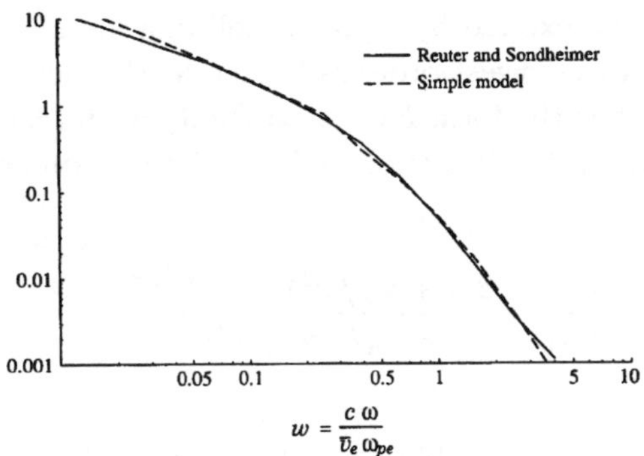

Fig. 18   Normalized input power versus normalized frequency $w = (\omega c)/(\bar{v}_e \omega_{pe})$. The dashed line shows the result obtained from the (non self-consistent) Fermi model for a near-collisionless case of $\nu_m/\omega = 0.008$, line , and the solid line shows the result of the nonlocal theory of Weibel [11] (Reuter and Sondheimer [10] theory using a Maxwellian electron distribution) (after [2]).

One might think that the stochastically heated electrons absorb energy along the direction of the wave electric field; i.e., along the y-direction in the slab model. That this is not the case can be seen by considering the canonical angular momentum $P_y = mv_y - eA_y(x,t)$, where $E_y(x,t)$ is obtained from the vector potential as $E_y = -\partial A_y/\partial t$. Because the Hamiltonian for the motion is independent of y, P_y is conserved. Since A_y vanishes a few skin depths into the plasma, it follows that v_y within the plasma is a constant of the motion. Hence, the transverse acceleration of the electron along y within the skin layer is converted by the wave magnetic field into a longitudinal acceleration along x as the electron exits the layer. Therefore the stochastic heating is along the direction of motion, a classical Fermi acceleration model. Because the magnetic field does no work on the electron, the kick in energy is the same; only the direction of the kick is altered.

D. Experiments

In a typical ICP experiment in a short metal cylindrical chamber with a dielectric window and a planar coil, the axial distribution of the rf field even without plasma is very inhomogeneous. Therefore, the rf field decay during its propagation into the plasma is a combined effect of the chamber geometry and the plasma screening due to the skin effect. For the classical skin effect in a uniform plasma excited by a planar coil in a metallic cylindrical chamber of radius R, for the lowest-order radial mode, the electric field profile next to the window has the form $E_\theta(r, z) = E_{\theta 0} \mathrm{J}_1(3.83 r/R) \exp(-z/\delta)$, where J_1 is the first-order Bessel function and the plasma skin depth is (see, for example, [12]):

$$\delta = \delta_p \left(\frac{2(1 + \nu_m^2/\omega^2)}{a[1 + (1 + \nu_m^2/\omega^2 a^2)^{1/2}]} \right)^{1/2}, \tag{4.28}$$

where

$$a = 1 + \frac{\delta_p^2}{\delta_v^2} \left(1 + \frac{\nu_m^2}{\omega^2} \right) \tag{4.29}$$

and $\delta_v \approx R/3.83$. For the anomalous skin effect, ν_m can approximately be replaced by an effective collision frequency ν_{eff} as in (4.10). It follows from (4.28) that for $\delta_p \gg \delta_v$ (low plasma density), $\delta = \delta_v$; and for $\delta_p \ll \delta_v$ (high plasma density), δ is the classical skin depth given by (4.6).

An experimental study of a low pressure inductive discharge over a wide range of argon gas pressure, rf power and frequency was reported by Godyak [13]. The EEDF and rf field and current density distribution were measured in a near-collisionless regime in a metal chamber with a planar ICP excitation coil using Langmuir and magnetic probes. The chamber diameter was 20 cm and the length was 10.5 cm. From the magnitudes and phases of the measured components of rf magnetic field B_r and B_z within

p (mTorr)	P_{plasma} (W)	n_1 (10^{10} cm^{-3})	ν_m (10^7 s^{-1})	ν_{eff} (10^7 s^{-1})	ν_{stoc} (10^7 s^{-1})	$\nu_{\text{eff}} - \nu_m$ (10^7 s^{-1})
1	25	1.1	0.71	3.0	1.8	2.3
1	100	4.1	0.55	3.9	2.4	3.3
10	50	4.8	4.5	5.3	1.8	0.8
10	150	16	3.0	5.7	2.3	2.7

Table 3.1   Comparison of collisional and collisionless heating frequencies

the skin layer and the plasma, the rf electric field E_θ and current density J_θ were determined from Maxwell's equations. The absolute values of the power density absorbed and the effective electron collision frequency were then directly found according to the relations

$$p_{\text{abs}} = J_\theta E_\theta \cos\psi \qquad \text{and} \qquad \nu_{\text{eff}} = \frac{e^2 n_1 E_\theta \cos\psi}{m J_\theta},$$

where ψ is the measured phase shift between J_θ and E_θ, and n_1 is the local plasma density measured with a Langmuir probe. In Table 1 are shown results of calculations for 6.78 MHz based on measurements within the skin layer at the maximum of the rf current density distribution (approximately 1 cm from the glass window) at a radial position of 4 cm, corresponding to maximum of the rf electric field radial distribution.

As is seen in Table 1, at $p = 1$ mTorr, $\nu_{\text{eff}} \gg \nu_m$, such that collisionless electron heating dominates. At $p = 10$ mTorr, $\nu_{\text{eff}} \approx \nu_m$, and power absorption in the skin layer is predominantly collisional. It is interesting to compare values of ν_{eff} found experimentally at 1 mTorr where $\nu_{\text{eff}} \gg \nu_m$ with a theoretical expression for the collisionless (stochastic) frequency $\nu_{\text{stoc}} = \bar{v}_e/4\delta$ given by Vahedi et al [12] for a one-dimensional model with an exponential rf profile. The values δ were determined as the distance from the glass window where the measured rf electric field decays by one e-folding. The calculated values of ν_{stoc} given in Table 1 appear to be close to the measured values of $\nu_{\text{eff}} - \nu_m$.

The rf power absorption density integrated along the direction of electromagnetic field propagation, $S(z) = \int_0^z E_\theta J_\theta \cos\psi \, dz'$, is shown in Fig. 19 as a function of the distance from the window. Here the measured total absorbed power flux S is compared with that calculated from the measured

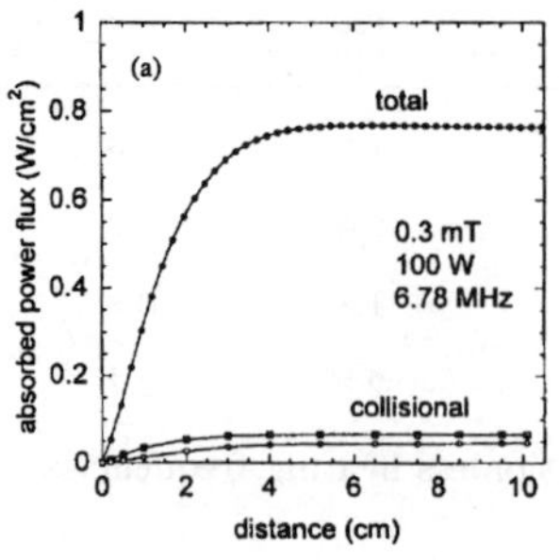
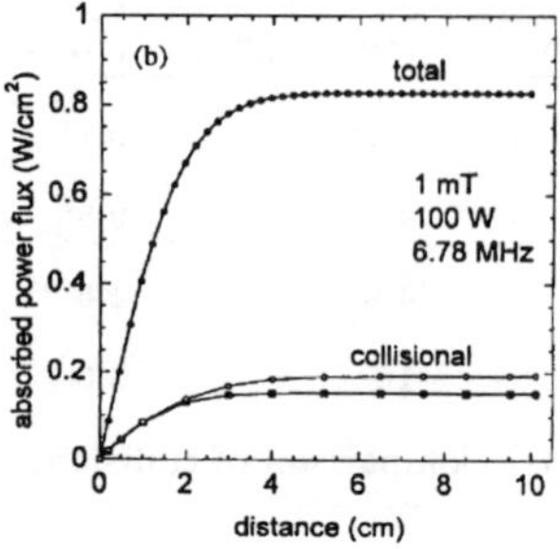

Fig. 19   Absorbed rf power flux versus distance for (a) $p = 0.3$ mTorr and (b) $p = 1$ mTorr (after [2]).

E_θ and J_θ distributions assuming collisional power absorption:

$$S_{\nu E} = \int_0^z E_\theta^2 \, \mathrm{Re}(\sigma_p) dz' \qquad \text{and} \qquad S_{\nu J} = \int_0^z J_\theta^2 \, \mathrm{Re}(\sigma_p^{-1}) dz',$$

with σ_p given in (4.1). The comparison shows that the collisionless process dominates the rf power absorption in these particular cases.

For the normal skin effect, the rf current is defined by the product of the rf electric field and the local value of the cold plasma conductivity, such that $J_\theta E_\theta \cos\psi$ is always positive. For the anomalous skin effect the current far away from the skin layer is that which is translated from the skin layer by the electron thermal motion, and its phase is defined by the transit time $\tau_J \approx z/\bar{v}_e$, while the electric field phase is defined by a delay due to the phase velocity, which depends on frequency and plasma density, $\tau_E \approx z/v_{\mathrm{ph}}$. The different mechanisms of phase delay for current and electric field can result in various phase combinations including those corresponding to a negative power absorption. Apparently, the negative power absorption can exist only in the collisionless regime ($\lambda_e \gtrsim \delta$); otherwise, electron-atom collisions destroy the translational motion of the current, yielding a local coupling between rf current and electric field. The rf power absorption along the plasma is shown in Fig. 20 at various frequencies for $p = 10$ mTorr (where $\lambda_e \approx L$), and a discharge power of 100 W. A transit time effect is seen for the appearance of the first negative power absorption region. The distance d between the middle of the skin layer and the first zero crossing of the power is inversely proportional to the rf frequency, $d \propto f^{-1}$. Note that apparently there is no negative power absorption at the lowest frequency of 3.39 MHz, where the anomalous skin effect (large Λ) and the

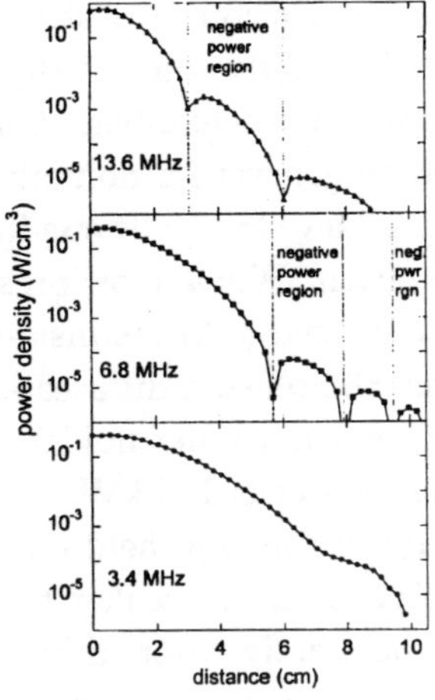
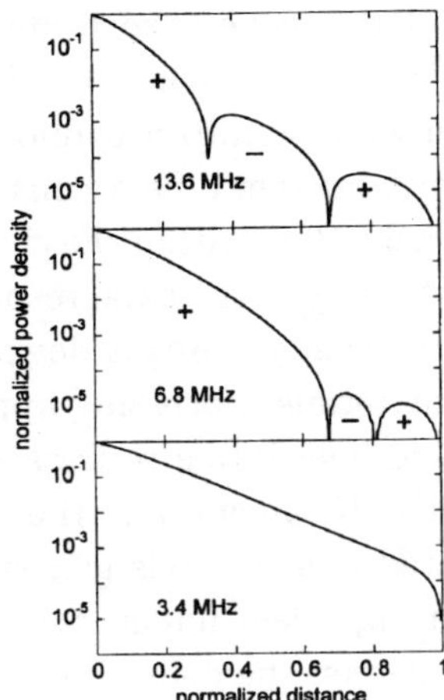

Fig. 20   Effect of frequency variation on the spatial distribution of the rf power density absorbed by the plasma; (a) measured distribution for 10 mTorr and 100 W absorbed plasma power, and (b) distribution calculated from a 1-D model by Kolobov (after [2]).

electron ballistic phenomena are expected to be greatest. This has a simple explanation: for 3.39 MHz, the chamber length L is too small.

An analytical calculation of rf power absorption for parameters of the Godyak [13] experiment has been performed by Kolobov [14]. The result shown in Fig. 20b demonstrates a good agreement with the experiment in Fig. 20a.

5   Electron cyclotron resonance discharges

Waves generated near a plasma surface can propagate into the plasma or along the surface where they can be subsequently absorbed, leading to heating of plasma electrons and excitation of a discharge [3]. The classical example is an electron cyclotron resonance (ECR) discharge, in which a right circularly polarized electromagnetic wave propagates along dc magnetic field lines to a resonance zone, where the wave energy is absorbed by collisionless heating. This is a type of stochastic heating in which electrons receive "kicks" in energy at each passage through the resonance zone; i.e., it is a type of Fermi acceleration.

ECR discharges are generally excited at microwave frequencies (e.g.,

2450 MHz), and the wave absorption requires application of a strong dc magnetic field (875 G at resonance). The power is usually coupled through a dielectric end-window into a cylindrical metal source chamber. One or several concentric magnetic field coils are used to generate a nonuniform, axial magnetic field $B(z)$ within the chamber to achieve the ECR condition, $\omega_{ce}(z_R) \approx \omega$, where z_R is the axial resonance position. When a low pressure gas is introduced, the gas breaks down and the discharge forms inside the chamber. For materials processing applications, the plasma diffuses along the magnetic field lines into a process chamber toward a wafer holder. The source diameter is 15–30 cm, and the microwave power is 1–5 kW.

Two separated magnet coils generate a *magnetic mirror* field configuration having a high field underneath the two coils and a weaker field in the midplane between the coils. Electrons can be axially trapped between the high field regions by the axial magnetic field gradients and repeatedly bounce between the mirror coils (e.g., see [3]). By proper choice of the field strength and profile, there can be two resonance zones symmetrically located with respect to the midplane. This configuration can yield high ionization efficiencies, due to enhanced confinement of hot (superthermal) electrons that are magnetically trapped between the two mirror (high-field) positions.

Because the gas pressure in these discharges can be as low as 0.1–0.01 mTorr and the field strengths can be large (large kicks), phase randomization due to nonlinear dynamical effects can be very important. Dynamical effects such as the influence of phase correlations in slowing the quasilinear heating rate, and the existance of adiabatic barriers to heating, have been observed experimentally in these discharges.

To determine the collisionless heating power from a Fermi acceleration model, the nonuniformity in the magnetic field profile $B(z)$ must be considered. For $\omega_{ce} \neq \omega$, an electron does not continuously gain energy, but rather its energy oscillates at the difference frequency $\omega_{ce} - \omega$. As an electron moving along z passes through resonance, its energy oscillates as shown in Fig. 21, leading to a transverse energy gained (or lost) in one pass.

For low power absorption, where the electric field at the resonance zone is known, the heating can be determined as follows. We expand the magnetic field near resonance as

$$\omega_{ce}(z') = \omega(1 + \alpha z'), \tag{5.1}$$

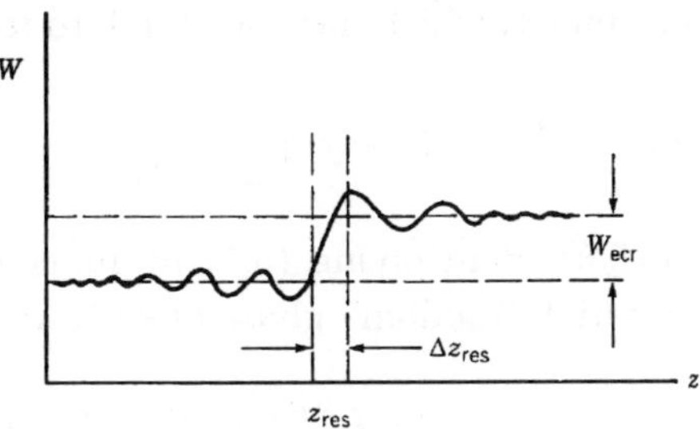

Fig. 21   Energy change in one pass through an ECR resonance zone (after [3]).

where $z' = z - z_R$ is the distance from exact resonance, $\alpha = \partial\omega_{ce}/\partial z'$ is proportional to the gradient in $B(z)$ near the resonant zone, and we approximate $z'(t) \approx v_R t$, where v_R is the parallel speed at resonance.

The complex force equation for the right hand component of the transverse velocity, $v_r = v_x + jv_y$, can be written in the form

$$\frac{dv_r}{dt} - j\omega_{ce}(z)v_r = -\frac{e}{m}E_r e^{j\omega t}, \tag{5.2}$$

where E_r is the amplitude of the RHP wave with

$$\mathbf{E} = \mathrm{Re}\left[(\hat{x} - j\hat{y})E_r e^{j\omega t}\right] \tag{5.3}$$

and $\hat{x}$ and $\hat{y}$ unit vectors along the x and y directions. Using (5.1) and substituting $v_r = \tilde{v}_r \exp(j\omega t)$ into (5.2), we obtain

$$\frac{d\tilde{v}_r}{dt} - j\omega\alpha v_R t\tilde{v}_r = -\frac{e}{m}E_r. \tag{5.4}$$

Multiplying by the integrating factor $e^{-j\theta(t)}$ and integrating (5.4) from $t = -T$ to $t = T$, we obtain

$$\tilde{v}_r(T)e^{-j\theta(T)} = \tilde{v}_r(-T)e^{-j\theta(-T)} - \frac{eE_r}{m}\int_{-T}^{T} dt'\, e^{-j\theta(t')}, \tag{5.5}$$

where

$$\theta(t) = \omega\alpha v_R t^2/2. \tag{5.6}$$

In the limit $T \gg (2\pi/\omega|\alpha|v_R)^{1/2}$, the integral in (5.5) is the integral of a

Gaussian of complex argument, which has the standard form

$$\int_{-T}^{T} dt'\, e^{-j\theta(t')} = (1 - j) \left(\frac{\pi}{\omega|\alpha|v_{\mathsf{R}}} \right)^{1/2}. \tag{5.7}$$

Substituting (5.7) into (5.5), multiplying (5.5) by its complex conjugate, and averaging over the initial "random" phase $\theta(-T)$, we obtain

$$|\tilde{v}_r(T)|^2 = |\tilde{v}_r(-T)|^2 + \left(\frac{eE_r}{m} \right)^2 \left(\frac{2\pi}{\omega|\alpha|v_{\mathsf{R}}} \right). \tag{5.8}$$

The average energy gain per pass is thus

$$W_{\mathrm{ecr}} = \frac{\pi e^2 E_r^2}{m\omega|\alpha|v_{\mathsf{R}}}. \tag{5.9}$$

This can also be written as

$$W_{\mathrm{ecr}} = \frac{1}{2} m (\Delta v)^2, \tag{5.10}$$

where $\Delta v = (eE_r/m)\Delta t_{\mathsf{R}}$, and

$$\Delta t_{\mathsf{R}} = \left(\frac{2\pi}{\omega|\alpha|v_{\mathsf{R}}} \right)^{1/2} \tag{5.11}$$

is the effective time in resonance. The effective resonance zone width (see Fig. 21) is

$$\Delta z_{\mathsf{R}} \equiv v_{\mathsf{R}}\Delta t_{\mathsf{R}} = \left(\frac{2\pi v_{\mathsf{R}}}{\omega|\alpha|} \right)^{1/2}, \tag{5.12}$$

which, for typical ECR parameters, gives $\Delta z_{\mathsf{R}} \sim 0.5$ cm.

The absorbed power per unit area is found by integrating (5.9) over the flux nv_{R} of electrons incident on the zone, yielding

$$S_{\mathrm{ecr}} = \frac{\pi n e^2 E_r^2}{m\omega|\alpha|}. \tag{5.13}$$

We can understand the form of Δt_{R} as follows: an electron passing through the zone coherently gains energy for a time Δt_{R} such that

$$\left[\omega - \omega_{ce}(v_{\mathsf{R}}\Delta t_{\mathsf{R}}) \right] \Delta t_{\mathsf{R}} \approx 2\pi. \tag{5.14}$$

Inserting (5.1) into (5.14) and solving for Δt_{R}, we obtain (5.11).

At high power absorption, the electric field is not known but must be determined self-consistently with the energy absorption, in the same manner as for inductive discharges. The propagation and absorption of microwave power in ECR sources is an active area of research and is not fully understood. However, the essence of the wave coupling, and transformation and absorption at the resonance zone, can be seen by considering the one dimensional problem of a righ-hand polarized wave propagating strictly along the magnetic field in a plasma that varies only along the axial direction z. This problem was originally studied in connection with wave propagation in the ionosphere, and the solution was obtained analytically by Budden [15] for the approximation of constant density and linear magnetic field variation. For a wave travelling into a decreasing magnetic field, he obtained the solution

$$
\begin{aligned}
S_{\text{abs}}/S_{\text{inc}} &= 1 - e^{-\pi\eta}, & (5.15)\\
S_{\text{trans}}/S_{\text{inc}} &= e^{-\pi\eta}, & (5.16)\\
S_{\text{refl}}/S_{\text{inc}} &= 0, & (5.17)
\end{aligned}
$$

where $\eta = \omega_{pe}^2/(\omega c|\alpha|)$. Hence some of the incident wave power is absorbed at the resonance while some tunnels through to the other side, but no power is reflected. Taking a typical case for which $\alpha = 0.1$ cm^{-1} and $k_0 = 0.5$ cm^{-1}, we find that $\eta > 1$ corresponds to $\omega_{pe}^2/\omega^2 > 0.2$. Thus at 2450 MHz we expect most of the incident power will be absorbed for a density $n_0 \gtrsim 1.5 \times 10^{10}$ cm^{-3}.

For high electric field strengths and low pressures, the energy gain per passage through resonance is large, and the nonlinear dynamical aspects of the problem come into play. For electrons heated to high energies by repeated interaction with the resonance zones, strong phase correlations can reduce the heating rate below that obtained from a random phase interaction, and an adiabatic barrier to heating can exist. Such phenomena were observed experimentally in a high field (50 kG) magnetic mirror compression experiment [16], in which a short pulse (0.25 μs) of high power (maximum 250 kW) microwaves was used to heat plasma electrons early during the compression, which were subsequently further heated by the increasing magnetic field. Figure 22 shows the results. The solid dots and open circles are from x-ray energy measurements. The adiabatic and stochastic limits to heating, and the Fokker-Planck heating rate are shown as the solid lines. The heating rate calculations determined by directly iterating the ECR

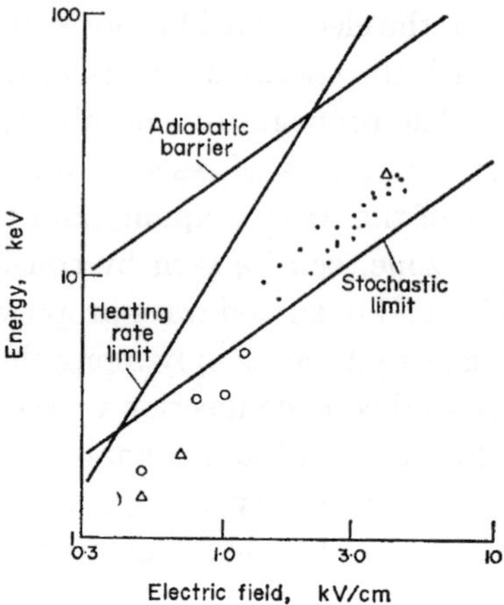

Fig. 22   Average electron energy as a function of microwave electric field.

mapping are shown as open triangles; these include the effects of the phase space correlations (see Section 2E). With increasing microwave power (field strength), a transition from a high to a lower heating rate was found that agreed well with the expected transition from a regime of random phase interaction to a regime of high phase correlations.

Another experiment was performed by Shoyama et al [17] in a magnetic mirror field configuration, ECR discharge at low pressures (3–8×10^{-5} Torr) using up to 4 kW of 2.45 GHz continuous wave power. A steady state plasma having a typical electron temperature of 6 V and density of 7×10^{11} cm^{-3} was formed, having a high energy electron tail of order 10 kV as determined by x-ray bremsstrahlung measurements. From the theory of Fermi acceleration and as confirmed by numerical iteration of the appropriate Fermi mapping for this system, the maximum possible energy $\mathcal{E}_{\mathrm{max}}$ of the heated electrons was found to scale with the input microwave power P_{in} as $\mathcal{E}_{\mathrm{max}} \propto P_{\mathrm{in}}^{0.5-0.6}$, due to the existance of an adiabatic barrier to heating. The experimental data confirmed this scaling.

6   Concluding discussion

Collisionless (stochastic) heating of electrons by time-varying fields is fundamental to the operation of radio frequency (rf) and microwave discharges.

Such heating is due to spatial variation of the fields, which lead to randomization of the electron phase during its thermal motion, even in the absence of collisions. Generally, electrons are heated collisionlessly by repeated interaction with fields that are localized within a sheath or skin depth layer inside the discharge. Consequently, the Fermi acceleration model of a ball bouncing elastically back and forth between a fixed and an oscillating wall is a paradigm to describe collisionless heating and phase randomization in capacitive, inductive, and ECR discharges. We introduced several mapping models for Fermi acceleration and showed how to use the Fokker-Planck formalism to determine the heating rate and the effects of phase correlations. We reviewed the role of collisionless heating in capacitive and inductive discharges, using simple Fermi models to determine the heating rates and comparing these with self-consistent (kinetic) calculations where available. We reviewed some experimental measurements and computer simulations and compared these to theoretical calculations of the heating. We described recent measurements and calculations of nonlocal heating effects, such as negative electron power absorption. The effects of partial phase randomization in reducing the heating rates were most clearly seen in low pressure ECR discharges. We described the use of Fermi acceleration models to determine the collisionless heating rates for these discharges, and showed that incomplete phase randomization could reduce the heating rate and lead to the existence of adiabatic barriers to heating.

The Fermi acceleration model has been shown to be an effective tool for describing collisionless heating in gas discharge plasmas having strong rf fields on the boundary. Amazingly, collisionless electron heating, which is usually associated with high temperature space and fusion plasmas, appears to be a fundamental process in the warm plasmas of low pressure discharges that are used in today's technology.

Acknowledgment

The author gratefully acknowledges the collaborations with V.A. Godyak and A.J. Lichtenberg, which form the basis of this work. This work was partially supported by NSF Grant ECS-9820836, California industries, the State of California UC-SMART Program under Contract SM99-10051, and by the Space Plasma and Plasma Processing Group at The Australian National University, Canberra.

References

L.D. Landau Oscillations of an electron plasma 1946 J. Phys. (USSR) 10 25

M.A. Lieberman and V.A. Godyak From Fermi acceleration to collisionless discharge heating 1998 IEEE Trans. Plasma Sci. 26 955

M.A. Lieberman and A.J. Lichtenberg 1994 Principles of Plasma Discharges and Materials Processing New York: J. Wiley

E. Fermi On the origin of the cosmic radiation 1949 Phys. Rev. 75 1169

A.J. Lichtenberg and M.A. Lieberman 1992 Regular and Chaotic Dynamics New York: Springer-Verlag

H. Goldstein 1951 Classical Mechanics New York: Addison-Wesley

V.A. Godyak The statistical heating of electrons by oscillating boundaries of the plasma 1972 Sov. Phys.—Tech. Phys. 16 1073

V.A. Godyak, O.A. Popov, and A.H. Khanna Effective collision frequency of the electrons in RF discharge 1976 Sov. J. Plasma Phys. 2 560

A.B. Pippard The high frequency skin resistance of metals at low temperatures 1949 Physica 15 45

G.E.H. Reuter and E.H. Sondheimer The theory of the anomalous skin effect in metals 1949 Proc. Roy. Soc. A195 336

E.S. Weibel Anomalous skin effect in a plasma 1967 Phys. Fluids 10 741

V. Vahedi, M.A. Lieberman, G. DiPeso, T.D. Rognlien, and D. Hewett Analytic model of power deposition in inductively coupled plasma sources 1995 J. Appl. Phys. 78 1446

V.A. Godyak, V.A. 1998 "ICP kinetics and electromagnetic fields in stochastically heated regime," in *Electron Kinetics in Glow Discharges*, U. Kortshagen and L.D. Tsendin, Eds., New York, Plenum Press

V.I. Kolobov 1998 "Anomalous skin effect in bounded plasmas," in *Electron Kinetics in Glow Discharges*, U. Kortshagen and L.D. Tsendin, Eds., New York, Plenum Press

K.G. Budden 1966 Radio Waves in the Ionosphere Cambridge, UK: Cambridge University Press

N.C. Wyeth, A.J. Lichtenberg, and M.A. Lieberman Electron cyclotron resonance heating in a pulsed mirror experiment 1975 Plasma Phys. 17 679

H. Shoyama, M. Tanaka, S. Higashi, Y. Kawai, and M. Kono Stochastic electron acceleration by an electron cyclotron wave propagating in an inhomogeneous magnetic field 1996 J. Phys. Soc. Japan 65 2860

Chapter 4

Large Resonances in Hamiltonian Systems, with Applications

Cathy Holmes

Department of Mathematics, University of Queensland,
Brisbane 4072 Australia

Abstract. Resonances are a familiar signature of nonintegrable Hamiltonian systems. We will look at how they appear and their importance for two applications: Cold atoms in an amplitude modulated optical periodic potential. Coupled Bose-Einstein condensates of alkali gases.

1 Some Experiments in Atom Optics

Hamiltonian Systems are conservative systems. We often obtain approximately Hamitonian systems by leaving out the dissipation. But, at the level of atoms, and well into the quantum regime, the world is conservative. While the dynamics is quantum mechanical under certain conditions it appears that understanding the corresponding classical dynamics can help us understand its quantum counterpart.

In this first section we will look at the classical dynamics of some experiments on cooled atoms and compare our results with the actual behaviour of the atoms.

As of about 10 years ago experimentalists have sucessfully cooled atoms using a Magneto-optic trap (MOT) [11], [3], where lasers and magnets work together to cool and trap the atoms. In our experiment a titanium saphire laser is then used to produce an optical standing wave, which induces a sinusoidal optical-dipole potential.

After some scaling the Hamiltonian is given by

$$H_0(q,\,p) = \frac{p^2}{2} + 2\kappa \sin^2\left(\frac{q}{2}\right)$$

where the variables q and p are scaled position and momentum [6] and κ

165

is proportional to the intensity of the saphire laser and can be manipulated at will. This Hamiltonian is a constant of the motion and the phase curves, curves of constant H_0, resemble a set of eyes.

For $\kappa > 0$ the solutions are either periodic, when $H < 2\kappa$ (inside a separatrix) or librational if $H > 2\kappa$. Actually on the separatrix $H = 2\kappa$.

The equations of motion then follow from Hamilton's canonical equations:

$$\frac{dp}{dt} = \frac{\partial H}{\partial q}$$

$$\frac{dq}{dt} = -\frac{\partial H}{\partial p}$$

(Note the simplicity of the second equation in this case. $\frac{dq}{dt} = p$.)

Using the fact that $\frac{dq}{dt} = p$, we can solve for $q(t)$ from

$$H(q, \frac{dq}{dt}) = constant.$$

Everwhere except on the separatrix and at the fixed points the solutions to the nonlinear problem are Elliptic functions. Inside the separatrix, where the motion is periodic the nonlinear frequency is given by

$$\omega(H) = \frac{\pi\sqrt{\kappa}F(\frac{\pi}{2}, \sqrt{\frac{H}{2\kappa}})}{2},$$

where $F(\frac{\pi}{2}, k)$ is the Elliptic integral of the first kind. It tends to zero as you approach the separatrix. In the experiment the atoms are loaded on the position axis. The trap is turned off, the potential turned on and the momentum distribution of the atoms measured a moment later. If they evolved classically the cloud would eventually become smeared around an anular shaped region, because the frequency is lower the further out you are from the origin. So the momentum distribution is also smeared around the origin, which is what you would see experimentally.

What happens if we now modulate the laser intensity in time so that the Hamiltonian becomes time dependent.

$$H(q, p, t) = \frac{p^2}{2} + 2\kappa(1 - 2\epsilon \sin t)\sin^2(\frac{q}{2})$$

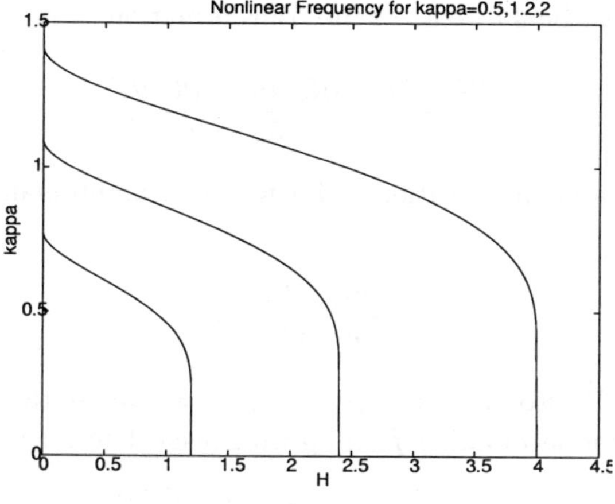

Fig. 1

where t is a scaled time variable [6]. Since any time-*in*dependent one degree of freedom system is integrable, if we can transform this, time *de*pendent system, to a time-*in*dependent Hamiltonian system the problem is solved.

1.1 *Canonical Transformations*

A system of equations may in certain cases be significantly simplified by an appropriate change of variables. However the choice of variables will be limited if at the same time we wish to preserve Hamilton's equations of motion.

The Hamiltonian generates rather special phase flows which preserve phase space volume. This is known as *Liouville's theorem*, and for 1 degree of freedom systems implies area preservation. So suppose we wish to change from $(p, q) \to (P, Q)$ then we require

$$A_R = \int \int_R dqdp = \int \int_S dQdP$$

where the area R in (q, p) space is transformed to S in (Q, P) space. But since $P(q, p)$ and $Q(q, p)$

$$\int \int_S dQdP = \int \int_R \frac{\partial(Q, P)}{\partial(q, p)} dqdp$$

where $\frac{\partial(Q,P)}{\partial(q,p)}$ is the Jacobian of the transformation

$$\frac{\partial(Q,\,P)}{\partial(q,\,p)} = \frac{\partial Q}{\partial q}\frac{\partial P}{\partial p} - \frac{\partial Q}{\partial p}\frac{\partial P}{\partial q}.$$

So to preserve area preservation and hence the Hamiltonian formulation we require

$$\frac{\partial(Q,\,P)}{\partial(q,\,p)} = 1$$

Or in terms of Poisson Brackets $\{Q,\,P\}_{\{q,p\}} = 1$ where the Poisson bracket of f and g is the jacobian of f and g with respect to p and q.

$$\{f,\,g\}_{\{q,p\}} = \frac{\partial f}{\partial q}\frac{\partial g}{\partial p} - \frac{\partial f}{\partial p}\frac{\partial g}{\partial q}$$

The Poisson bracket enables us to put Hamiltons equations in a symmetrical form

$$\frac{dq}{dt} = \{q,\,H\}$$

$$\frac{dp}{dt} = \{p,\,H\}$$

and in general can be used to describe the evolution of any function $f(q(t),\,p(t),\,t)$

$$\frac{df}{dt} = \{f,\,H\} + \frac{\partial f}{\partial t}$$

This in turn makes it simple to verify that Hamiltons equations are preserved if
$\{Q,\,P\}_{\{q,p\}} = 1$:

$$\frac{dQ}{dt} = \frac{\partial Q}{\partial q}\frac{dq}{dt} + \frac{\partial Q}{\partial p}\frac{dp}{dt} = \frac{\partial Q}{\partial q}\frac{\partial H}{\partial p} - \frac{\partial Q}{\partial p}\frac{\partial H}{\partial q} = \frac{\partial H}{\partial P}\{Q,\,P\} + \frac{\partial H}{\partial Q}\{Q,\,Q\} = \frac{\partial H}{\partial P}.$$

Coordinate Transformations which preserve Hamiltons equations of motion, i.e. such that $\{Q,\,P\}_{\{q,p\}} = 1$, are called *Canonical Transformations*. Although they can be found by trial and error, it is generally easier to find these new variables $P(q,\,p)$ and $Q(q,\,p)$ is via *Generating functions*.

1.2 *Generating Functions*

Hamilton's Principal of least action states that the motion is so as to minimise it's action.

$$\delta Action = \delta \int (p\frac{dq}{dt} - H(q,\,p,\,t))dt = 0$$

This principal implies Hamilton's equations of motion [1]. So that between any two points in $(q,\,p,\,t)$ space the path taken is such as to minimise the action. Now if we go to a new set of coordinates $(Q,\,P,\,t)$ and take the same two points in the new coordinates the same result must hold. This implies that the integrands differ by atmost a complete differential.

$$(p\frac{dq}{dt} - H(q,\,p,\,t)) = (P\frac{dQ}{dt} - K(Q,\,P,\,t)) + \frac{dF}{dt}$$

where Q and P depend on $(q,\,p,\,t)$. However the equation could also be thought of as a function of the independent variables $(q,\,Q,\,t)$. Then $F = F_1(q,\,Q,\,t)$ and

$$\frac{dF_1}{dt} = \frac{\partial F_1}{\partial q}\frac{dq}{dt} + \frac{\partial F_1}{\partial Q}\frac{dQ}{dt} + \frac{\partial F_1}{\partial t},$$

which imeadiately implies that

$$p = \frac{\partial F_1}{\partial q},\ P = -\frac{\partial F_1}{\partial Q}$$

and

$$K(Q,\,P,\,t) = H(q,\,p,\,t) + \frac{\partial F_1}{\partial t}$$

The are 4 possible functional dependencies for generating functions

$$F_1(q,\,Q,\,t),\ F_2(q,\,P,\,t),\ F_3(p,\,Q,\,t),\ F_4(p,\,P,\,t).$$

F_2 is the most useful. Both Action Angle variables and near identity transformations use F_2 generating functions. To find the relations between the old and new coordinates for F_2 first note that

$$\frac{d(PQ)}{dt} = P\frac{dQ}{dt} + \frac{dP}{dt}Q$$

then take $F_2 = F_1 + PQ$ with $F_2(q, P, t)$ so that

$$\frac{dF_2}{dt} = \frac{\partial F_2}{\partial q}\frac{dq}{dt} + \frac{\partial F_2}{\partial P}\frac{dP}{dt} + \frac{\partial F_2}{\partial t}$$

Now the equation for the integrands is satisfied if

$$p = \frac{\partial F_2}{\partial q}, \quad Q = \frac{\partial F_2}{\partial P}$$

and

$$K(Q, P, t) = H(q, p, t) + \frac{\partial F_2}{\partial t}$$

Throughout the calculation above time has been included explicitly, so that we can consider forced 1 degree of freedom systems ($1\frac{1}{2}$ dof systems), such as the model here involving forced cooled atoms, which are some of the simplest systems that can exhibit chaotic behaviour. However they are not always chaotic. In the following worked example we can use an F_2 generating function to show that the $1\frac{1}{2}$ dof system is integrable.

Example:
Suppose that

$$H(p, q, t) = \frac{p^2}{2} + \omega^2 \cos(q + t)$$

Now any time-*independent* 1 degree of freedom Hamiltonian system is integrable, H is the integral. Solutions can then be found by solving for $q(t)$ from

$$H(\frac{dq}{dt}, q) = constant.$$

So our aim is to transform to a time-*independent* Hamiltonian.
So let $Q = q + t$, then since $Q = \frac{\partial F_2}{\partial P}$

$$F_2(q, P, t) = P(q + t) + g(q, t)$$

for some function g. Now since $p = \frac{\partial F_2}{\partial q}$

$$p = P + \frac{\partial g}{\partial q}.$$

Taking $g = 0$ gives $p = P$ and the new Hamiltonian is

$$K(Q, P, t) = \frac{P^2}{2} + \omega^2 \cos Q + P$$

which is integrable. (Infact this is simply the nonlinear pendulum with a shifted momentum P.)

1.3   *Action Angle Variables*

An Integrable system is one that has as many independent integrals as degrees of freedom such that a canonical transformation can then be made where the integrals, which are constants of the motion, are functions of the new momenta. Each phase curve can then be labelled uniquely by the integrals, or equivalently the new momenta. Each point on these phase curves can also be labelled uniquely by the new position variable θ. For periodic motion, where θ is on S, it is typically taken as mod 2π.

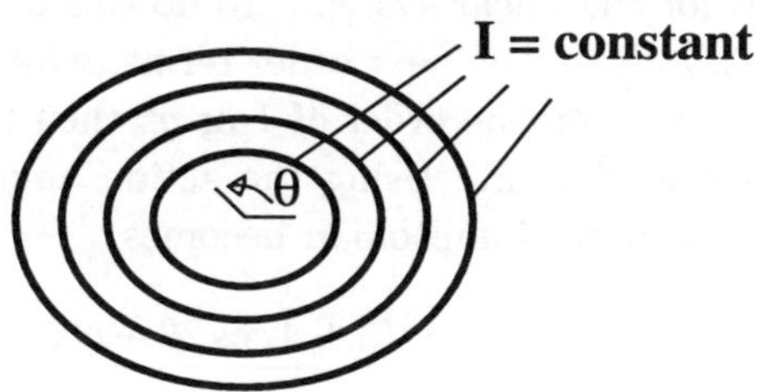

As an example let us consider the Hamiltonian above near the origin.

$$H(p, q, t) = \frac{p^2}{2} + 2\kappa(1 - 2\epsilon \sin t)\frac{q^2}{4} + \text{hot} \ \ \text{in} \ \ q,$$

where hot is short for higher order terms in q.
If ϵ is zero

$$H_0(p, q) = \frac{p^2}{2} + 2\kappa\frac{q^2}{4} + hot$$

and the phase curves are elliptical. Since the system is linear the period of the motion is constant $\frac{2\pi}{\sqrt{\kappa}}$. Now change to action angle variables (θ, I). The action, I as a constant of the motion, must be a function of H. While the angle coordinate must be periodic in time with period $\frac{2\pi}{\sqrt{\kappa}}$. However we know from Hamilton's equations of motion that the new frequency $\frac{d\theta}{dt} = \frac{\partial H}{\partial I}$

so that $\sqrt{\kappa} = \frac{\partial H}{\partial I}$ and then $H = \sqrt{\kappa}I$. Now we can use F_2 generating functions although they are usually denoted as $S(q, I) = F_2(q, I)$. As above we have $p = \frac{\partial S}{\partial q}$, $\theta = \frac{\partial S}{\partial I}$ and $K(\theta, I) = H(q(\theta, I), p(\theta, I))$ Substituting $p = \frac{\partial S}{\partial q}$ into the Hamiltonian above gives

$$\frac{\partial S}{\partial q} = \sqrt{(2\sqrt{\kappa}I - \kappa q^2)}.$$

From this we can work out $\frac{\partial S}{\partial I}$ and hence $\theta(q, I)$ to finally give

$$q = \left(\frac{4}{\kappa}\right)^{\frac{1}{4}} \sqrt{I} \sin\theta \ , \ p = (4\kappa)^{\frac{1}{4}} \sqrt{I} \cos\theta.$$

Exercise:
Use trig substitution to find S and then solve for q and p.

Before investigating the fully nonlinear system it is interesting to look at the the weekly nonlinear system near the origin in terms of the action angle variables for the linear system. To do this expand $\sin(\frac{q}{2})$ about zero as before, but now keep the next order terms, which are fourth order in q. If ϵ is assumed to be on the order of I or q^2 then the order ϵI term is the same order as the I^2 term. Using the action angle variables worked out above the approximate Hamiltonian becomes

$$H(I, \theta, t) \ = \ \sqrt{\kappa}I - \frac{I^2(3 + 4\cos 2\theta + \cos 4\theta)}{48}$$
$$+ \epsilon\sqrt{\kappa}I(\cos t + \frac{\cos(2\theta + t) + \cos(2\theta - t)}{2}) + \text{hot}.$$

To see if this is integrable we will try to change to Action Angle coordinates (ϕ, J) for the nonlinear system. That is we will try to change to new coordinates in which the Hamiltonian is only a function of the new momenta. Since the change of variables is only a perturbation of the original we can use a near identity generating function.

$$F_2(\theta, J, t) = \theta J + G(\theta, J, t),$$

where $G(\theta, J, t)$, is considered small, on the order of ϵJ or J^2. Then

$$I = \frac{\partial F_2}{\partial \theta} = J + \frac{\partial G}{\partial \theta}$$

and the new Hamiltonian

$$K(\phi(J,\theta),J) \;=\; H\!\left(\theta, J + \frac{\partial G}{\partial \theta}, t\right) + \frac{\partial F_2}{\partial t}$$

$$=\; H(\theta, J, t) + \frac{\partial H}{\partial J}\frac{\partial G}{\partial \theta} + \frac{\partial G}{\partial t} + \text{hot}$$

using Taylor series, where to this order $\frac{\partial H}{\partial J} = \sqrt{\kappa}$. Now *try to* choose $G(\theta, J, t)$ to eliminate all the harmonic terms. Say

$$G(\theta, J, t) \;=\; g_1(J)\sin 2\theta + g_2(J)\sin 4\theta$$

$$+g_3(J)\sin(2\theta + t) + g_4(J)\sin(2\theta - t) + g_5\sin t \,,$$

where $g_i(J)$ are order ϵJ or J^2. In fact we have a high success rate. $g_1(J) = \frac{J^2}{24\sqrt{\kappa}}$ eliminates the $\cos 2\theta$ term, provided κ is nonzero, g_2 can be chosen to eliminate the $\cos 4\theta$ term, g_3 the $\cos(2\theta + t)$, and g_5 the $\cos t$ term. But the coefficients of the final term

$$g_4(J) = -\frac{\epsilon\sqrt{\kappa}J}{2(2\sqrt{\kappa} - 1)}$$

may become very large if the denominator is small, i.e. if $\sqrt{\kappa} = \frac{1}{2}$. Away from this value, at least to this approximation, the system looks integrable. (To really prove integrability of course we need to tackle all the higher order terms as well!) At $\sqrt{\kappa} = \frac{1}{2}$ there is a resonance at the origin and for $\sqrt{\kappa} \approx \frac{1}{2}$ action angle variables appear not to exist. But what does this really mean? To investigate lets "remove" all the other harmonic terms, leaving only the resonant term and let $\sqrt{\kappa} = \frac{1}{2} + \kappa_1$, with κ_1 small.

$$K(\phi, J, t) = \frac{J}{2} + \kappa_1 J - \frac{J^2}{16} + \epsilon\frac{J}{4}\cos(2\phi - t) + \text{hot}$$

Since ϕ and t only appear as $(2\phi - t)$ we can transform to a rotating frame. An F_2 generation function can be used, with new variables (ϕ, L) where

$$\psi = \phi - \frac{t}{2} = \frac{\partial F_2}{\partial L}$$

Then $F_2(\phi, L) = L(\phi - \frac{t}{2})$ and $J = \frac{\partial F_2}{\partial \phi} = L$. So that

$$\bar{K}(\psi, L) = \kappa_1 L - \frac{L^2}{16} + \epsilon\frac{L}{4}\cos(2\psi) + \text{hot}$$

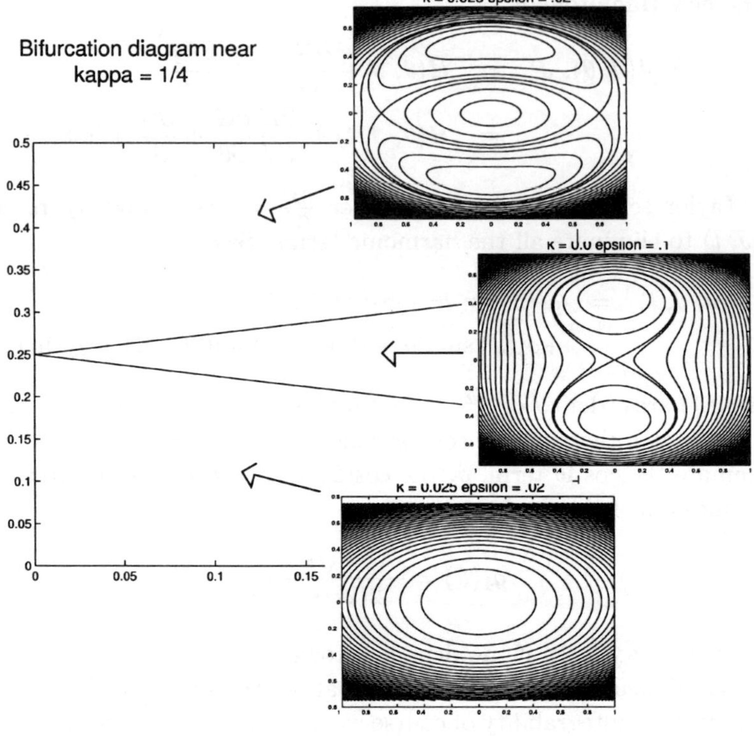

Fig. 2

Once in the rotating frame it is useful return to $(\bar{p}, \bar{q})$ type coordinates.

$$\bar{q} = 2\sqrt{L}\sin\theta, \quad \bar{p} = \sqrt{L}\cos\theta,$$

where

$$\bar{K}(\bar{q}, \bar{p}) = \kappa_1\left(\bar{p}^2 + \frac{1}{4}\bar{q}^2\right) - \frac{(\bar{p}^2 + \frac{1}{4}\bar{q}^2)^2}{16} + \epsilon\frac{1}{4}\left(\bar{p}^2 - \frac{1}{4}\bar{q}^2\right) + \text{hot}$$

and obtain the approximate equations of motion in the rotating frame.

$$\frac{d\bar{q}}{dt} = \bar{p}\left(2\kappa_1 + \frac{\epsilon}{2} - \frac{\bar{p}^2 + \frac{1}{4}\bar{q}^2}{4}\right)$$

$$\frac{d\bar{p}}{dt} = \frac{\bar{q}}{4}\left(2\kappa_1 - \frac{\epsilon}{2} - \frac{\bar{p}^2 + \frac{1}{4}\bar{q}^2}{4}\right)$$

There are two sets of fixed points. The first set on the momentum axis, existing for $4\kappa_1 + \epsilon > 0$ at $\bar{p} = \pm\sqrt{2(4\kappa_1 + \epsilon)}$ can be shown to be centers. While the second set, existing for $4\kappa_1 - \epsilon > 0$ at $\bar{q} = \pm 2\sqrt{2(4\kappa_1 - \epsilon)}$ can be shown to be saddles. So we can draw a local bifurcation diagram in (ϵ, κ) space for $\kappa \approx \frac{1}{4}$, Fig. 2.

As time evolves these pictures rotate. Take the figure eight and think of time on a circle, Fig. 3. The behaviour we see in the rotating frame is

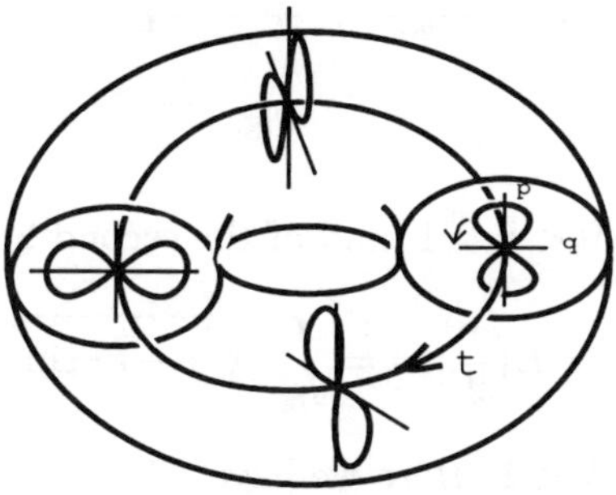

Fig. 3

local and only appoximate. Terms cubic in I, ϵ and κ_1 have been ignored. We can see that as κ is increased the nonzero fixed points of the rotating frame, which correspond to periodic orbits in the original system, move further out. To follow them we need to construct action angle variables for the fully nonlinear system with $\epsilon = 0$ and perturb off that.

Deriving action angle variables for the nonlinear system, $\epsilon = 0$, involves more complicated integrals, however the method is the similar. Qualitatively, the interesting difference lies in the fact that $H(I)$ is nonlinear, so that $\frac{\partial H}{\partial I}$ is nolonger a constant. In fact the nonlinear frequency is a monotonically decreasing function, from $\sqrt{\kappa}$ at the origin to zero on the separatrix. First we need to know about Elliptic functions.

The complete elliptic integral of the first kind is

$$F(\frac{\pi}{2}, k) = \int_0^{\frac{\pi}{2}} \frac{d\phi}{\sqrt{1 - k^2 \sin^2 \phi}}$$

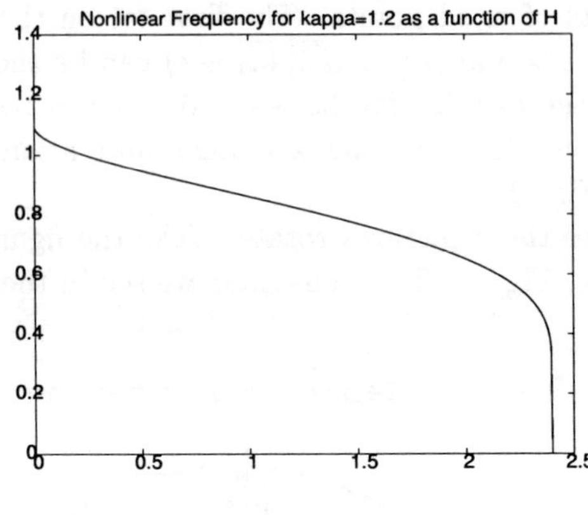

Fig. 4

which is valid for $0 < k < 1$, and of the second kind is

$$E(\frac{\pi}{2}, k) = \int_0^{\frac{\pi}{2}} \sqrt{1 - k^2 \sin^2 \phi} \, d\phi$$

also valid for $0 < k < 1$. Here we have

$$\frac{dq}{dt} = \sqrt{2(H - 2\kappa \sin^2 \frac{q}{2})}$$

from which you can solve for $q(t)$ and then work out the nonlinear frequency inside the separatrix

$$\omega(H) = \frac{dH}{dI} = \frac{\pi\sqrt{\kappa}}{2} F(\frac{\pi}{2}, k).$$

Note that as $k \to 0$ the frequency $\omega(H) \to \sqrt{\kappa}$. The new momentum $I(H)$ can then be found from

$$\frac{dH}{dI} = \omega(H).$$

Figure 4 shows a plot of the nonlinear frequency as a function of H for $\kappa = 1.2$. The frequency tends to zero as the separatrix is approached. So that the system has periodic motion for all periods above $\frac{2\pi}{\sqrt{\kappa}}$. Now if we perturb this system with a forcing term with period say $\frac{2\pi}{\omega}$ then the integrals commensurate with $\frac{2\pi}{\omega}$ may be resonant, depending on the actual form of the perturbation.

So lets look at our perturbation term; $-4\kappa\epsilon\sin\tau\sin^2(q/2)$. Inside the separatrix

$$\frac{\partial S}{\partial q} = \sqrt{2(H - 2\kappa\sin^2\frac{q}{2})}$$

from which you can show that

$$\theta = \frac{\partial S}{\partial I} = \frac{\pi}{2}\frac{F(q/2,k)}{F(\frac{\pi}{2},k)}$$

So that $\sin(q/2) = k\,sn(\frac{2F(\frac{\pi}{2},k)\theta}{\pi}, k)$. Then $\sin^2(q/2)$ can be expanded as a Fourier series in θ, with I dependent coefficients.

$$k\,sn(\frac{2F(\frac{\pi}{2},k)\theta}{\pi}, k) = \frac{2\pi}{F(\frac{\pi}{2},k)}\Sigma_{m=0}\frac{f^{m+1/2}}{1-f^{2m+1}}\sin(2m+1)\theta$$

where $f = e^{-\frac{\pi F(\frac{\pi}{2},\sqrt{1-k^2})}{F(\frac{\pi}{2},k)}}$. This means that

$$H(I,\theta,t) = H_0(I) + H_1(I,\theta,t)$$

where

$$H_1 = \sum_{m=0,1,\ldots} V_m(I)(\cos(\tau - 2m\theta) + \cos(\tau + 2m\theta)).$$

To see if any of these terms are resonant we try to remove them via a near identity generating function. In fact it is not hard to show that any harmonic term of the form $\cos(n\tau + l\theta)$ is resonant if $(n + l\frac{\partial H_0(I)}{\partial I})$ is zero. These are the so called first order resonance conditions. Here we have first order resonances when $\omega(I) = \frac{\partial H_0(I)}{\partial I} = \pm\frac{1}{2m}$. These will emerge from the origin when $\omega(0) = \sqrt{\kappa} = \frac{1}{2m}$. Our weekly nonlinear analysis about the origin, was able to pick up the resonance when $\kappa = \frac{1}{4}$. If you include the next order terms in the weekly nonlinear analysis the resonance at $\kappa = \frac{1}{16}$ is becomes apparent and the next order again brings in the $\kappa = \frac{1}{36}$ resonance, etc. However the resonances that bifurcate out of the origin at $\kappa = \frac{1}{(2m)^2}$ for $m > 1$ are degenerate. Suddenly all $2m$ eyes pop out of the origin at once.

We know from the local analysis of the period-2 resonance at $\kappa = \frac{1}{4}$ that they move out as κ is increased . But the analysis is only local. To extend this we need to use canonical transformations on the perturbed nonlinear problem.

$$H = H_0(I) + \sum_{m=0,1\dots} V_m(I)(\cos(t - 2m\theta) + \cos(t + 2m\theta))$$

(Where I is the nonlinear action.) First use a near identity transformation to remove all the nonresonant terms, assuming I is close to the resonance in question. That is $I \approx I_{0m}$, where I_{0m} is defined through $\omega(I_{0m}) = \frac{\partial H_0}{\partial I} = \frac{1}{2m}$. Then

$$H = H_0(\bar{I}) + V_m(\bar{I}) \cos(t - 2\bar{\theta})$$

where

$$\bar{I} = I + \sum_{l=0,\pm1,\pm2,\pm3\dots,l\neq m} V_l(I)\frac{\sin(t + 2l\theta)}{1 + 2l\omega(I)}$$

and $\omega(I) = \frac{\partial H_0}{\partial I}$. Then, as before, transform to a rotating frame:

$$J = \bar{I} - \bar{I}_{0m}, \ \phi = \theta + \frac{\tau}{2m}$$

$$H = H_0(\bar{I}_{0m}) + \frac{\partial^2 H_{0m}(\bar{I}_{0m})}{\partial I^2}\frac{I^2}{2} + V_1(\bar{I}_{0m})\cos(\phi) + \text{hot}$$

The final system is just the nonlinear pendulum wrapped around the integral $I = I_{0m}$. There are m eyes. This picture rotates one complete revolution each $\Delta t = 2m\pi$. So if you take a stroboscopic map $\Delta t = 2\pi$ the center of each eye is one iterate of a stable period $2m$ point. Between each of these is an unstable period $2m$ point. (See Birchoff Fixed point theorem). ϵ controls the width of the resonances (width $\approx o(\sqrt{\epsilon})$),and as κ increases the resonances move out.

If we plot the position of the resonances versus κ we see that they accumulate on the separatrix. Even for ϵ small the region close to the separatrix contains an infinite number of resonances. In fact the motion near the separatrix is chaotic. If this were the full story though, the phase space away from resonances would be regular. But the neglected higher order terms can be resonant too! Suppose we remove all the first order resonances. This assumes that $I \neq \frac{1}{2m}$ for any m. Then the new hamiltonian is

$$K = H_0(\bar{I}) + H_2(\bar{I}, \bar{\theta})$$

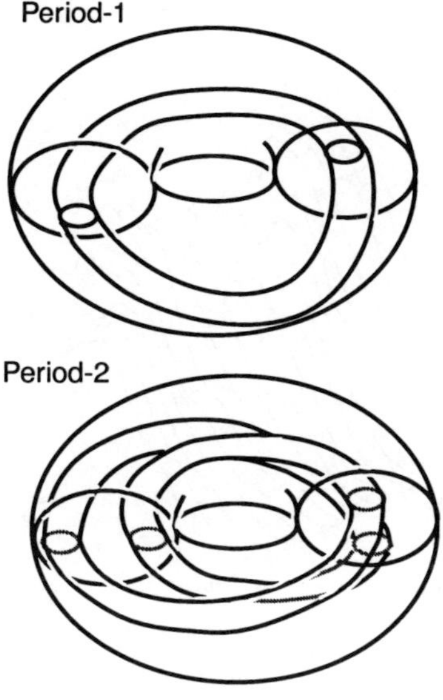

Fig. 5

Now the perturbing Hamiltonian has new resonance terms; second order resonances, when $\omega(I) = \frac{2}{2m}$. For small ϵ they are harder to see as their width is $o(\epsilon)$, as opposed to $o(\sqrt{\epsilon})$ for first order resonances. They emerge from the origin when $\kappa = \frac{1}{m^2}$ and appear as m eyes. But they rotate twice as fast as the first order resonances. The sequence of Poincare Maps near to the resonance at $\kappa = 1$ is similar to that near $\kappa = \frac{1}{4}$, but the bifurcation curves are order ϵ^2 and the resonances are period-1. In fact the topology in the full 3 dimensional space is quite different from that of the first order period-2 we looked at earlier. See Fig. 5. Just as we removed the first order resonances we could remove the second order resonances and look at the third order resonance terms. Infact this can be done to all orders and at order n there are new resonances at $\omega(I) = \frac{n}{2m}$. Most of these are very small for ϵ small. But it does mean that for $\epsilon \neq 0$ a substantial amount of phase space is taken up with resonances, although there may still be some invariant tori left as proved in the KAM Theorem [1].

As ϵ is increased the chaotic separatricies widen and eventually the phase space resembles a sea of chaoic behaviour relieved by islands of approximately regular motion. For the atoms in the trap ϵ is actually quite large, on the order of $0.2 - 0.3$. Also decifering what is actually happening is

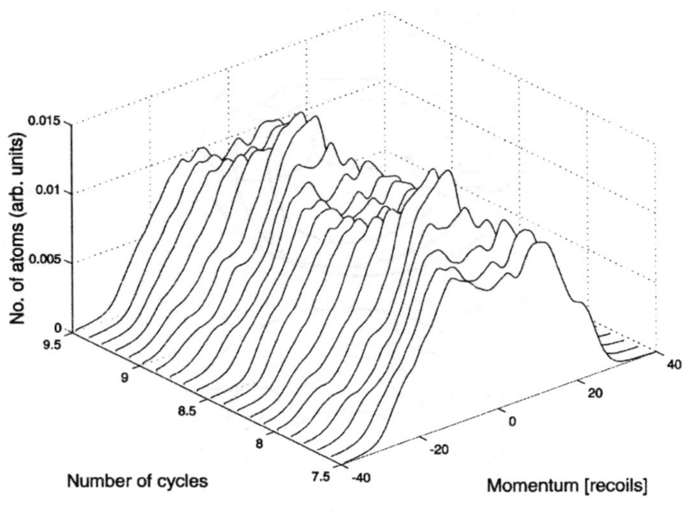

Fig. 6

not quite straight forward because experimentally we can only measure the momentum distribution. To get some idea of what we might see imagine starting a gaussian set of initial data (atoms) with some positive momentum. Now as the system is allowed to evolve the points closer in rotate faster than those further out and eventually the momentum distribution becomes smeared around the origin.

Next suppose we start the same gaussian initial condition in a, say, period-2 resonance. The gaussian becomes smeared around the resonance island, but is contained within the island so that after two forcing periods the distribution will still have approximately returned to its initial position, Fig. 6. Or take the period-1 resonance, after one forcing period the distribution will approximately return to its initial position, Fig. 7.

Because ϵ is quite large in the experiments the chaotic sea is fairly extensive and the points initially in the chaotic region become spread uniformly in momentum. But the points in the resonance islands remain contained and the resonance islands can be observed experimentally as peaks in momentum. The resonances with only two symmetrically placed resonance islands, i.e. those with $m = 1$, can be clearly identified. For lower values of κ that is the period-2, which emerges from the origin at $\kappa = \frac{1}{4}$, the period-1, emerging at $\kappa = 1$ and the period $\frac{3}{2}$, emerging at $\kappa = \frac{9}{4}$. (In general a

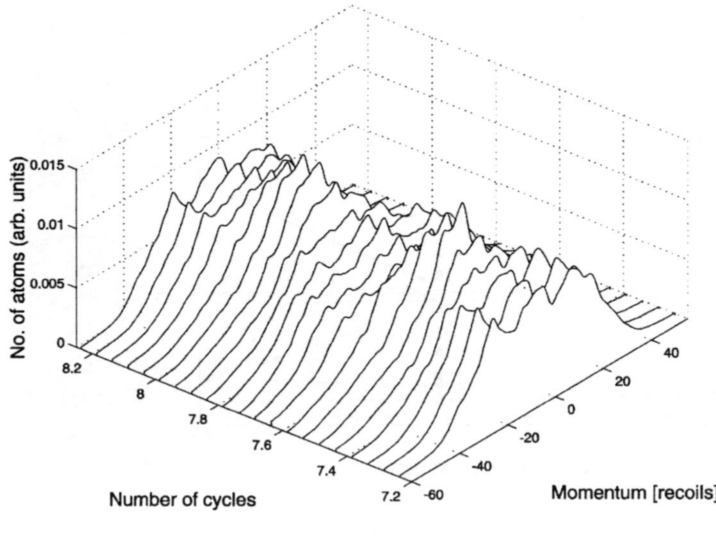

Fig. 7

period-$\frac{n}{m}$ emerges from the origin at $\kappa = \frac{m^2}{n^2}$.) For each of these resonances a weakly nonlinear analysis near the origin yields a similar bifurcation sequence. Qualitatively this classical bifurcation sequence is reflected in the experiments.

In fact the atoms are well into the quantum regime and quantum mechanics is fundamentally different from classical mechanics. One manifestation of this is that quantum mechaics allows for tunneling between symmetrical states. To see if this can happen we took the two period-1 resonance islands. Classically there are two sets of independent, but symmetrically placed tori corresponding to each of the islands in the stroboscopic map. In 1993 we showed that, theorectically, on a much longer time scale, Quantum Mechanical tunneling can occur between these period-1 solutions [12]. Very recently [7] it was also shown to occur experimentally as well!

2 Two and Three mode Bose-Einstein Condensation

Bose-Einstein Condessation (BEC) was first proposed by Einstein in 1925. Bosons, which are particles with integer spin for example phonons, may be made up of an even number of fermions, which have $\frac{1}{2}$ integer spin, such

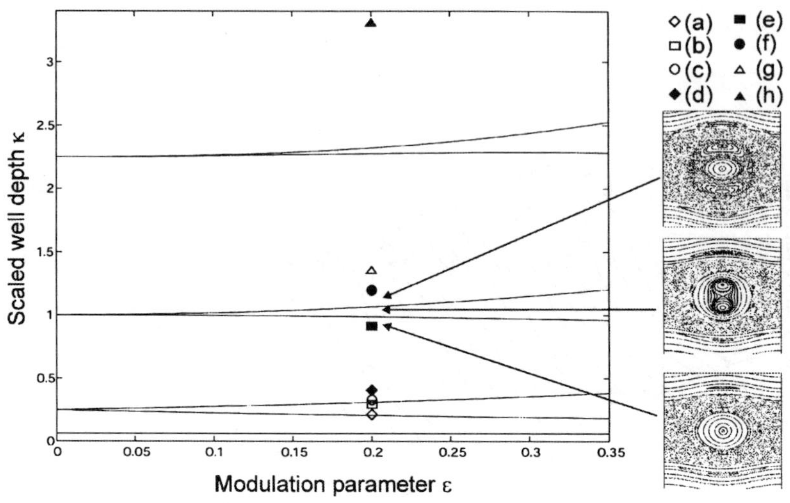

Fig. 8

as Rubidium-87 atoms (^{87}Rb), used by Anderson et al or Sodium-25 atoms (^{25}Na), used by Davis et al, in the first experiments to produce a condensate in 1995 [2] [10]. To produce a condensate the atoms need to be cooled down to the ground state, while remaining as a gas. The temperatures you need are actually much colder than that needed for liquification and down at this energy level the wavelength of the atoms is comparable to the distance between the atoms in the gas and the bosons behave more like a wave. Hence a BEC is called a quantum fluid.

The atoms are cooled down by using an magneto-optical trap (MOT) followed by evapourative cooling. A very good account of the proceedure is given in *Very Cold Indeed....* Eric Cornell J. Res. Natl. Inst. Stand. Technol. **101**, 419 (1996). Here we consider two and three mode approximations to a Bose Eistein Condensate.

Rather than talk about the Quantum Mechanics, which is explained in [5], [9], [8], lets start with the real Gross-Pitaevskii equation, which provides a mean field approach to describing the interacting Bose gases.

$$\frac{i\partial \psi}{\partial t} = -\frac{\hbar}{2m}\nabla^2\psi + V(\mathbf{x})\psi + \frac{U_0}{2}|\psi|^2\psi$$

where $V(\mathbf{x})$ is the confining potential and the last term is due to the non-

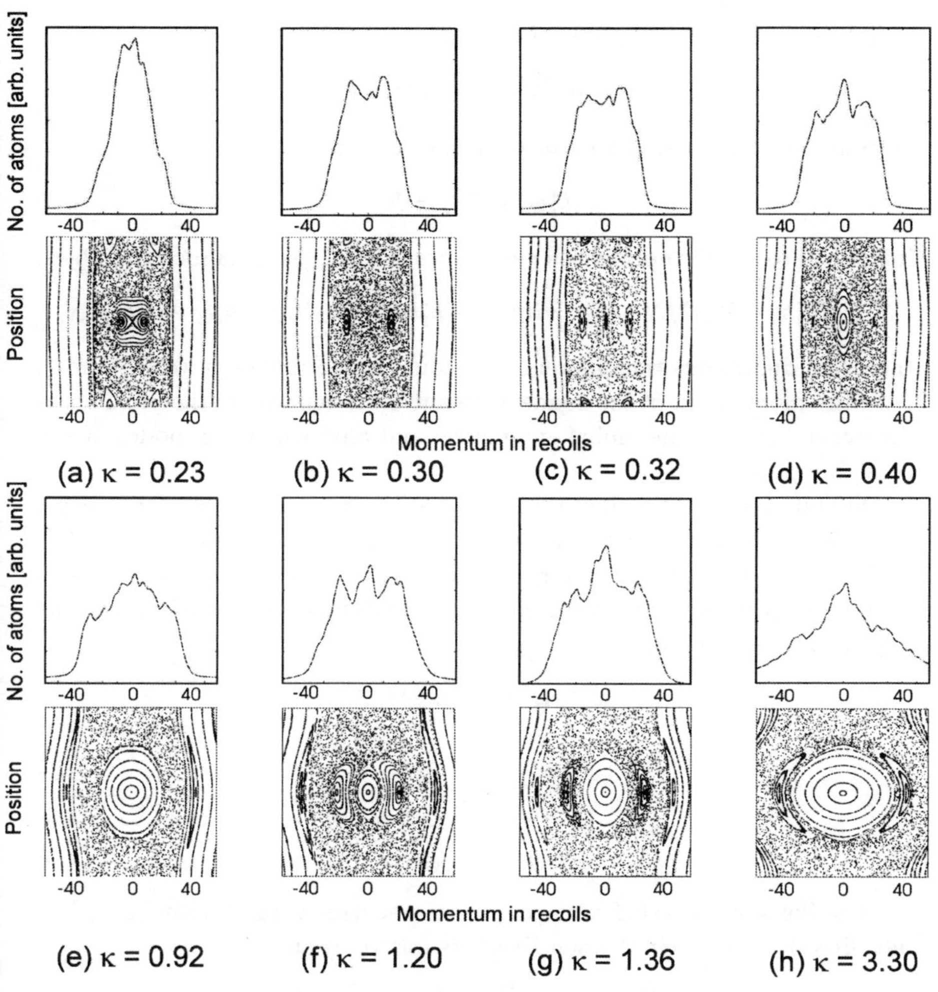

(a) $\kappa = 0.23$ (b) $\kappa = 0.30$ (c) $\kappa = 0.32$ (d) $\kappa = 0.40$

(e) $\kappa = 0.92$ (f) $\kappa = 1.20$ (g) $\kappa = 1.36$ (h) $\kappa = 3.30$

Fig. 9

linear collisions of bosons. For a two mode approximation the potential is
a quartic where

$$V(\mathbf{x}) \approx \frac{1}{2} m \omega_0^2 (\mathbf{x} - \mathbf{x_i})^2,$$

for $\mathbf{x} \approx \mathbf{x_i}$. We then expand in terms of the local modes u_i, where

$-\frac{\hbar}{2m}\nabla^2 u_i + V_i u_i = E_0 u_i$ and E_0 is the ground state energy;

$$\psi = e^{-\frac{iE_0 t}{\hbar}} \Sigma_{i=1,2} b_i(t) u_i(x).$$

Since the total number of atoms is a constant

$$b_1^* b_1 + b_2^* b_2 = N$$

it is useful to use the normalised generators of SU(2) as dynamical variables.

$$S_x = (b_2^* b_2 - b_1^* b_1)/N, \quad S_y = i(b_2^* b_1 - b_1^* b_2)/N, \quad S_z = (b_1^* b_2 + b_2^* b_1)/N,$$

then NS_x represents the difference in atom number between wells 2 and 1, NS_y the momentum of the condensate and NS_z represents the population difference between the global symmetric and antisymmetric modes of the confining potential [5].

Number conservation implies that $S_x^2 + S_y^2 + S_z^2 = 1 + \frac{2}{N} \approx 1$ for N large.

This means that the geometry of this Hamiltonian system is not flat. In fact it has the same geometry as the angular momentum variables of a spinning top where the total angular momentum is conserved. Since the angular momentum is

$$\mathbf{L} = \mathbf{q} \times \mathbf{p} \rightarrow$$

$$L_x = q_y p_z - q_z p_y \tag{2.1}$$
$$L_y = q_z p_x - q_x p_z \tag{2.2}$$
$$L_z = q_x p_y - q_y p_x. \tag{2.3}$$

The Poisson brackets for the angular momentum variables can be worked out directly from their Poisson Brackets for (p, q) and are;

$$\{L_x, L_y\} = \frac{\partial L_x}{\partial q_x}\frac{\partial L_y}{\partial p_x} - \frac{\partial L_y}{\partial q_x}\frac{\partial L_x}{\partial p_x} + \frac{\partial L_x}{\partial q_y}\frac{\partial L_y}{\partial p_y} - \,.... = L_z$$

$$\{L_y, L_z\} = L_x$$

$$\{L_z, L_x\} = L_y$$

So that the equations of motion are then simply given by

$$\frac{dL_\alpha}{dt} = \{L_\alpha, H\}$$

Here the Hamiltonian for the 2mode BEC is

$$H = \Omega S_z + \frac{\chi N}{2} S_x^2$$

and the equations of motion of the corresponding semiclassical system are

$$\dot{S}_x = -\Omega S_y, \tag{2.4}$$
$$\dot{S}_y = \Omega S_x - \chi N S_x S_z, \tag{2.5}$$
$$\dot{S}_z = \chi N S_x S_y, \tag{2.6}$$

where Ω is the tunnel frequency and χ, which is proportional to U_0, measures the strength of the nonlinear collision term and N is the number of atoms.   In these normalised corordinates the two modes are positioned on the equator of the unit sphere at $S_x = \pm 1$, $S_y = 0$, $S_z = 0$. The resulting system is integrable. If $\chi = 0$ or $\Omega = 0$ solutions are planar oscillations in the $S_z = constant$ or $S_x = constant$ planes respectively. Between these two values and for $-1 < \frac{\chi N}{\Omega_0} < 1$ the two poles are still centers, but at $\frac{\chi N}{\Omega_0} = 1$ the north pole undergoes a supercritical pitchfork bifurcation and two stable fixed points are created at $S_x = \pm\sqrt{1 - \frac{\Omega_0}{\chi N}^2}$, $S_y = 0$, $S_z = \frac{\Omega_0}{\chi N}$.

If we now imagine starting in one mode, i.e. at $(-1, 0, 0)$ for different strengths of the particle-particle interaction, χ, a qualitative change in the solutions occurs as the separatrix crosses the equator. To actually see this occuring it is probably easier to use variables in the stereoprojected plane via the coordinate change

$$a = \frac{S_x}{1 + S_z},$$

$$b = \frac{S_y}{1 + S_z}.$$

This maps the the points on the unit sphere to points on the plane $S_z = 1$ so that the equator is mapped to the unit circle, the northern hemisphere is mapped inside and the southern hemisphere outside.

In these variables the Hamiltonian becomes

$$H = \Omega \frac{2}{(a^2 + b^2 + 1)} + 2\chi N \frac{a^2}{(a^2 + b^2 + 1)^2} - \Omega,$$

with the separatrix is given by $H = \Omega$.

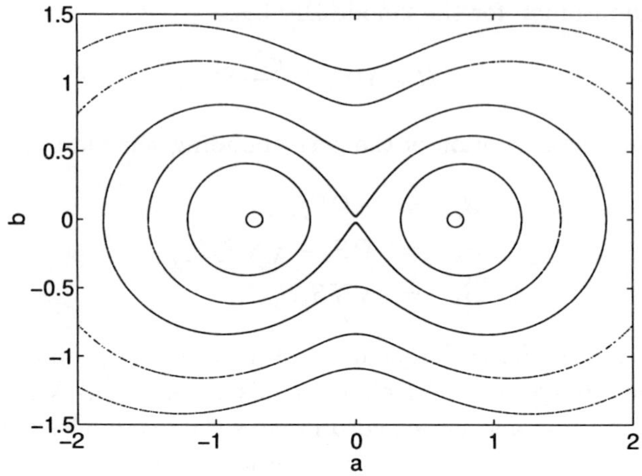

Fig. 10   Phase space plot of the semiclassical dynamics in the stereographicaly projected variables, showing the period-1 fixed points and separatrix. $\frac{\kappa N}{\omega_0} = 3.2$.

The equations of motion are

$$\dot{a} = -\Omega b - 2\chi N \frac{a^2 b}{(1 + a^2 + b^2)}$$

$$\dot{b} = \Omega a - \chi N a \frac{(1 - a^2 + b^2)}{(1 + a^2 + b^2)}.$$

The stereographical projection is not a canonical transformation, so Hamilton's equations of motion are not valid for the new $(a,\ b)$ variables. But the Hamiltonian is still a constant of the motion.The separatrix first crosses the equator on the a axis, at $(b = 0,\ a = 1)$. Setting $H = \Omega$ and $b = 0$ gives

$$a = \pm \sqrt{\frac{(\chi N - \Omega)}{\Omega}}$$

Setting this equal to 1 gives

$$\frac{\chi N}{\Omega} = 2.$$

For $\frac{\chi N}{\Omega} < 2$ solutions started in one of the modes lie outside the separatrix and therefore execute large scale oscillations, between both modes.

But for $\frac{\chi N}{\Omega} > 2$ solutions started in one of the modes lie inside the separatrix and therefore stay on one side of the sphere, remaining localised in one mode. If you look at the corresponding initial conditions for the full quantum system the results are similar. Small localisations occur for $\frac{\chi N}{\Omega} > 2$ and tunneling between the two modes occurs for $\frac{\chi N}{\Omega} > 2$. But what happens if the tunnel coupling is modulated in time?

A periodic modulation of the tunnel coupling between the two modes results in a parametrically excited system. We let

$$\Omega(t) = \Omega_0(1 + \epsilon \cos \omega_D t),$$

where ϵ is considered small. Integrals with a period commensurate with the period of the forcer, $\frac{2\pi}{\omega_D}$, may be resonant. So we need to know the actual period of the motion of the unperturbed system. Since the system can be integrated in terms of Elliptic integrals it is enough to know the period of the Elliptic integrals as a function of the parameters and H. This is a little messy, but straight forward.

$$T_{sys}(c, \lambda) = \frac{4K(k)}{\sqrt{w_+ - w_-}},$$

where $K(k)$ is a complete elliptic integral of the first kind, $k^2 = \frac{-2\lambda w_-}{w_+ - w_-}$, where $\lambda = \frac{\chi N}{\Omega_0}$ and w_+, w_- are given by

$$w_\pm = \frac{-(\lambda - 1) \pm \sqrt{(\lambda - 1)^2 - 4\lambda H/\Omega}}{2}.$$

In particular if the period of the unperturbed system equals that of the perturbation we expect period-1 resonances and that is exactly what we find. The integral curve given by $T_{sys}(c, \lambda) = \frac{2\pi\Omega_0}{\omega_D}$ is resonant, breaking up under perturbation to give two stable period-1 resonances which for $t = 0$ lie on the momentum axis. These can be seen in the Poincaré section, $t = 0 \ mod(\frac{2\pi\Omega_0}{\omega_D})$, which here is simply a stroboscopic map taken at $t_n = n\frac{2\pi\Omega_0}{\omega_D}$. In such a stroboscopic map, the resonances appear as fixed points. They lie outside the separatrix, which has been replaced by a 'chaotic sea', and are distinct from the two period-1 fixed points nearer the origin. Here $\frac{\omega_D}{\Omega_0}$ was taken as 1.37 and the period-1 resonances are fairly large.

There are other resonances, large period-2s etc, but as before the period-1's are of particular interest from the point of comparing the quantum de-

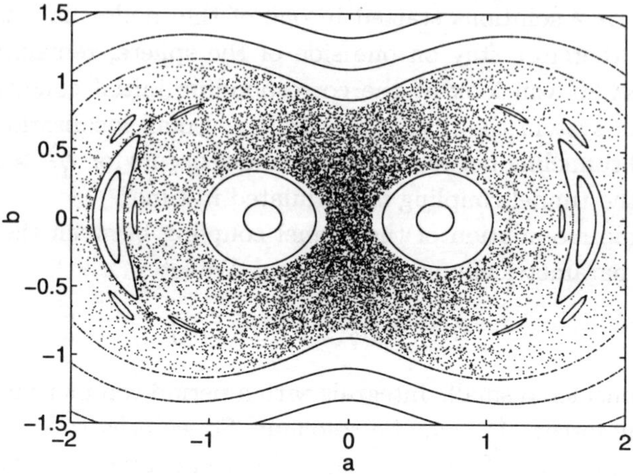

Fig. 11   Stroboscopic map of the semiclassical driven dynamics showing period-1 resonances at $a = \pm 1.72, 0.0$. $\epsilon = 0.3$, $\lambda = 3.2$, $\omega_D = 1.37$.

scription with our semi-classical model. Because, while the tori surrounding the two period-1 resonances are identical but for a rotation in phase space, classically they are totally separate. However this symmetry appears to allow for quantum mechanical tunneling between the two states. The tunneling period is about 303 times the period of modulation [8].

2.1   *Three mode Bose-Einstein Condensate*

The semiclassical model for the three mode Bose-Einstein condensate

$$\psi = e^{-\frac{iE_0 t}{\hbar}} \Sigma_{i=1,2,3} b_i(t) u_i(x).$$

also has number conservation and this leads to a representation in terms of the generators of SU(3).

$$x_1 \;\; = \;\; \frac{\sqrt{3}}{2}(b_1^* b_1 - b_2^* b_2)/N, \tag{2.7}$$

$$x_2 \;\; = \;\; \frac{1}{2}(b_1^* b_1 + b_2^* b_2 - 2b_3^* b_3)/N, \tag{2.8}$$

$$y_i \;\; = \;\; i(b_i^* b_j - b_j^* b_i)/N, \tag{2.9}$$

$$z_i \;\; = \;\; (b_i^* b_j + b_j^* b_i)/N, \tag{2.10}$$

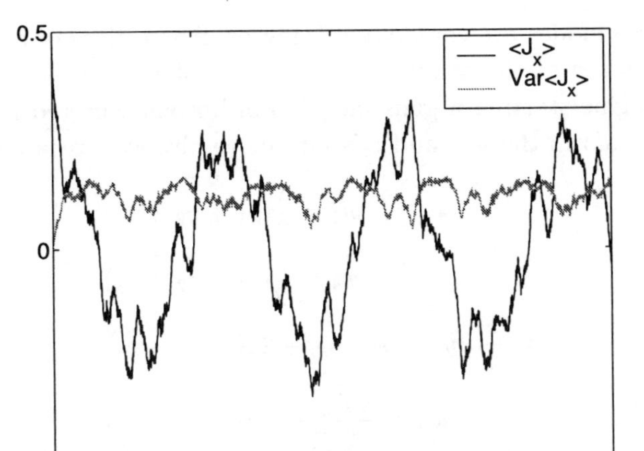

Fig. 12   Plot of quantum model population difference $< J_x >$ and variance $Var < J_x >$, with initial atomic coherent state centered on the resonance $a = 1.72, b = 0.0$. $\epsilon = 0.3$, $\lambda = 3.2$, $N = 100$, $\omega_D = 1.37$, showing tunneling between the two period-1 resonances.

where $i = 1, 2, 3$ and $j = (i + 1)\,mod\,3$.

The three mode Hamiltonian is

$$H_3 = \Omega(z_1 + z_2 + z_3) + \frac{2\chi}{3}(x_1{}^2 + x_2{}^2)$$

and as a consequence of number conservation

$$x_1{}^2 + x_2{}^2 + \frac{3}{4}(y_1{}^2 + y_2{}^2 + y_3{}^2 + z_1{}^2 + z_2{}^2 + z_3{}^2) = 1 + \frac{3}{N} \approx 1.$$

If only pure states are considered there are a further three integrals of the motion

$$y_1^2 + z_1^2 \;=\; \frac{4}{9}((x_2 + N)^2 - 3x_1^2), \tag{2.11}$$

$$y_2^2 + z_2^2 \;=\; \frac{4}{9}((x_2 - \sqrt{3}x_1 + N)(N - 2x_2), \tag{2.12}$$

$$\frac{y_1}{z_1} + \frac{y_2}{z_2} + \frac{y_3}{z_3} \;=\; \frac{y_1 y_2 y_3}{z_1 z_2 z_3}, \tag{2.13}$$

which restricts behaviours that can be observed [14]. The behaviour considered here, however, can be observed with or without these constraints.

The three modes are equidistant lying on the vertices of an equilateral triangle centered at the origin of the (x_1, x_2) plane. As in the two mode case we imagine starting a gaussian blob of bosons centered in one mode. First we transform the y_i's and z_i's into more physical variables:

$$w_3 = \frac{1}{2}(y_1 + y_2 + y_3) \tag{2.14}$$

$$w_4 = \frac{1}{2}(z_1 + z_2 + z_3) \tag{2.15}$$

$$w_5 = \frac{1}{4}(2y_1 - y_2 - y_3 + 2z_1 - z_2 - z_3) \tag{2.16}$$

$$w_6 = \frac{\sqrt{3}}{4}(y_2 - y_3 + z_2 - z_3) \tag{2.17}$$

$$w_7 = \frac{1}{4}(2y_1 - y_2 - y_3 - (2z_1 - z_2 - z_3)) \tag{2.18}$$

$$w_8 = \frac{\sqrt{3}}{4}(y_2 - y_3 - (z_2 - z_3)). \tag{2.19}$$

$$\tag{2.20}$$

then w_3 represents the total angular momentum of the condensate.

In the new variables the equations of motion are

$$\dot{x}_1 = -\Omega(w_5 + w_7), \tag{2.21}$$

$$\dot{x}_2 = -\Omega(w_6 + w_8), \tag{2.22}$$

$$\dot{w}_3 = \frac{2}{3}\chi N(x_1(w_7 - w_5) + x_2(w_8 - w_6)), \tag{2.23}$$

$$\dot{w}_4 = \frac{2}{3}\chi N(x_1(w_7 + w_5) + x_2(w_8 + w_6)), \tag{2.24}$$

$$\dot{w}_5 = -\Omega(w_6 - x_1)) + \frac{2}{3}\chi N(x_1(w_3 - w_4 + w_7) - x_2 w_8), \tag{2.25}$$

$$\dot{w}_6 = \Omega(w_5 + x_2) + \frac{2}{3}\chi N(x_2(w_3 - w_4 - w_7) - x_1 w_8), \tag{2.26}$$

$$\dot{w}_7 = \Omega(w_8 + x_1) + \frac{2}{3}\chi N(-x_1(w_3 + w_4 + w_5) + x_2 w_6), \tag{2.27}$$

$$\dot{w}_8 = -\Omega(w_7 - x_2) + \frac{2}{3}\chi N(-x_2(w_3 + w_4 - w_5) + x_1 w_6), \tag{2.28}$$

where time has been scaled $t-> \sqrt{3}t$ and the variables are now confined, for N large, to the unit sphere

$$x_1^2 + x_2^2 + w_3^2 + w_4^2 + w_5^2 + w_6^2 + w_7^2 + w_8^2 = (1 + \frac{3}{N}) \approx 1,$$

and the Hamiltonian is

$$H = 2\Omega w_3 + \frac{2\chi}{3}(x_1{}^2 + x_2{}^2).$$

Now imagine starting in just one mode, say the mode on the x_2 axis. Then solutions lie on an integrable subspace of the full system; $x_1 = 0$, $w_3 = 0$, $w_5 = -w_7$ and $w_6 = w_8$.

$$\dot{x}_2 = -\Omega 2 w_6, \tag{2.29}$$

$$\dot{w}_4 = \frac{4}{3}\chi N x_2 w_6, \tag{2.30}$$

$$\dot{w}_5 = -\Omega w_6 + \frac{2}{3}\chi N(-x_2 w_6), \tag{2.31}$$

$$\dot{w}_6 = \Omega(w_5 + x_2) + \frac{4}{3}\chi N(x_2(-w_4 + w_5)). \tag{2.32}$$

Number conservation implies that solutions lie on the ellipsoid $x_2^2 + w_4^2 + 2w_5^2 + 2w_6^2 = 1$, but from the equations we can see that they lie on planes $w_4 + 2w_5 - x_2 = P = \text{constant}$. Further the Hamiltonian is a constant of the motion.

$$H = 2\Omega w_3 + \frac{2}{3}\chi N x_1^2.$$

Once again the analysis is simpler in the stereoprojected space, although the equations are rather messy even for this sub case. Take $a_i = \frac{x_i}{1 + w_4}$, for i=1,2, and $a_i = \frac{w_i}{1 + w_4}$, for i=3,5-8. This reduces the dimension of the system by 1. So in the subcase we have

$$\dot{a}_2 = -\Omega 2 a_6 - \frac{4}{3}\chi N \frac{2a_2^2 a_6}{(1 + a_2^2 + 2a_5^2 + 2a_6^2)}, \tag{2.33}$$

$$\dot{a}_5 = -\Omega a_6 + \frac{4}{3}\chi N \frac{(-a_2 a_6)}{(1 + a_2^2 + 2a_5^2 + 2a_6^2)}$$
$$\qquad - \frac{4}{3}\chi N \frac{2a_5 a_2 a_6}{(1 + a_2^2 + 2a_5^2 + 2a_6^2)}, \tag{2.34}$$

$$\dot{a}_6 = \Omega(a_5 + a_2) - \frac{4}{3}\chi N a_2 + \frac{4}{3}\chi N \frac{a_2(a_2^2 + 2a_5^2 + 2a_6^2 + a_5)}{(1 + a_2^2 + 2a_5^2 + 2a_6^2)}$$
$$\qquad - \frac{4}{3}\chi N \frac{2a_2 a_6^2}{(1 + a_2^2 + 2a_5^2 + 2a_6^2)}, \tag{2.35}$$

The invariant planes now become more complicated surfaces

$$\frac{1 - (a_2^2 + 2a_5^2 + 2a_6^2) + 2(2a_5 - a_2)}{(1 + a_2^2 + 2a_5^2 + 2a_6^2)} = P$$

on which we have a two dimensional flow. Rather than isolated fixed points there is a fixed line in the $a_6 = 0$ plane, which intersects the surfaces $P = P_0$ in 1 or 3 points. On a given surface the number of fixed points can change as χ changes via a saddle-node bifurcation which creats two new fixed points and an unsymmetrical separatrix.

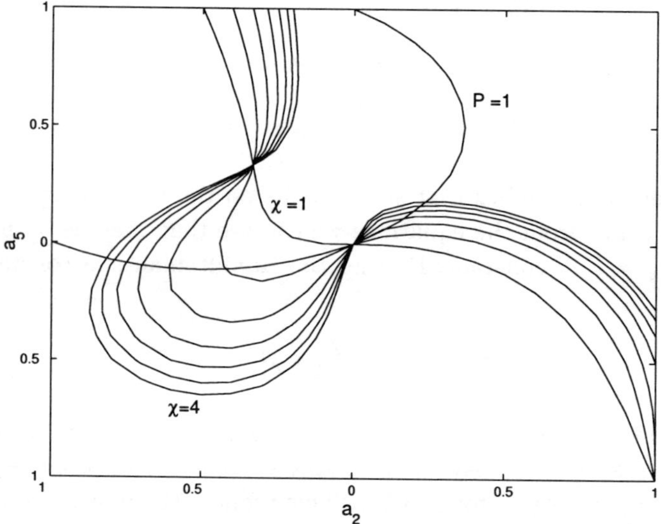

The intersection of the surface P =1 with the position
of the fixed points for   χ = 1, 1.5,2,2.5,3,4.

Fig. 13

If we now start at the mode $x_2 = -1$, $x_5 = x_6 = 0$ then the surface $P(-1,0,0)=1$, on which the solutions lie also passes through the origin and in fact the origin is always a fixed point. For $\frac{\chi N}{\Omega}$ small it is the only fixed point and remains a center. At $\frac{\chi N}{\Omega} \approx 1.9$ a saddle node bifurcation takes place and as $\frac{\chi N}{\Omega}$ is increased the two new fixed points move apart. At $\frac{\chi N}{\Omega} = \frac{9}{4}$ the origin undergoes a transcritical bifurcation becoming a saddle

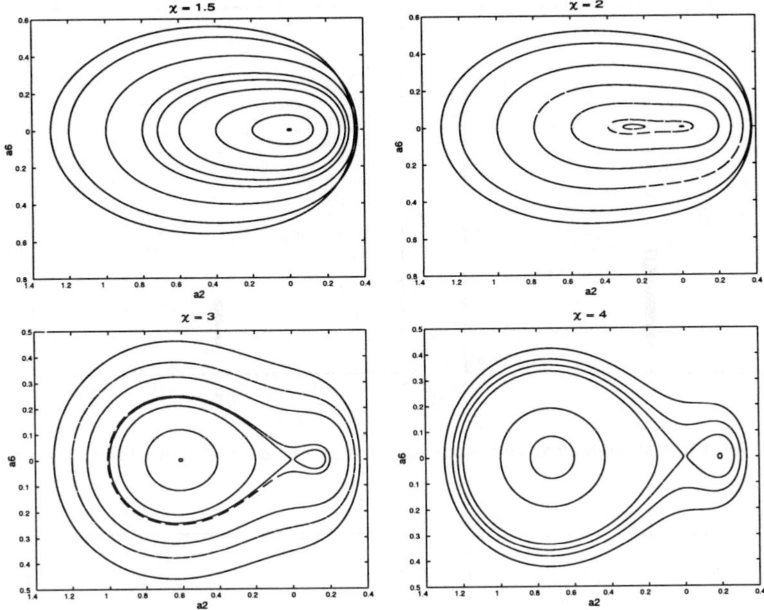

Phase Portraits on the integral subspace P=1.

Fig. 14

in the process. Now the separatrix with the saddle at the origin plays a similar role to the separatrix in the two mode case. For $\frac{\chi N}{\Omega} < 3$ solutions started at the mode execute large scale oscillations, but for $\frac{\chi N}{\Omega} > 3$ localised behaviour takes over as the initial point $(-1, 0, 0)$ lies inside the separatrix.

Symmetry between the modes implies that there are two other invariant planes, associated with starting in the each of the other two modes, which have exactly the same dynamics. But just off these invariant planes the dynamics is a little more complicated, however we can still make some general statements. If χ is large, which corresponds to the case inside the separatrix, solutions still remain localised and appear to be fairly regular, although the extra dimensions make deciphering any actual surfaces difficult.

If χ is small solutions also look almost regular and tunnel between the modes, some times only between two modes. But inbetween, say at $\chi = 2$, solutions are fairly complicated and can certainly be chaotic.

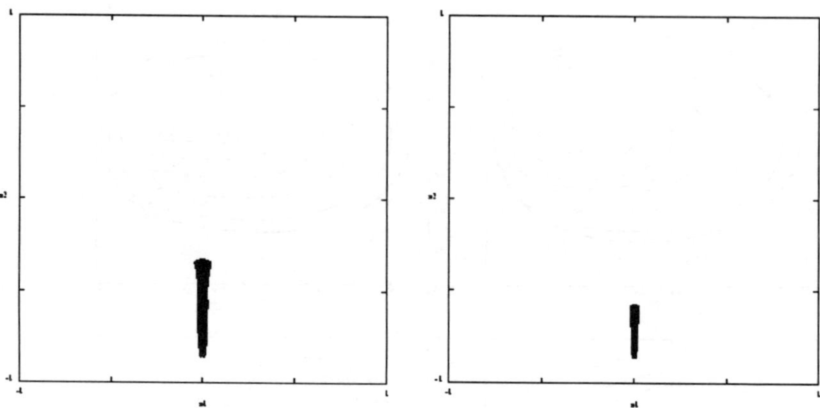

Fig. 15 Solutions started near the mode at $x_2 = -1$ in (a_1, a_2) space for $\chi = 4$ and 3.

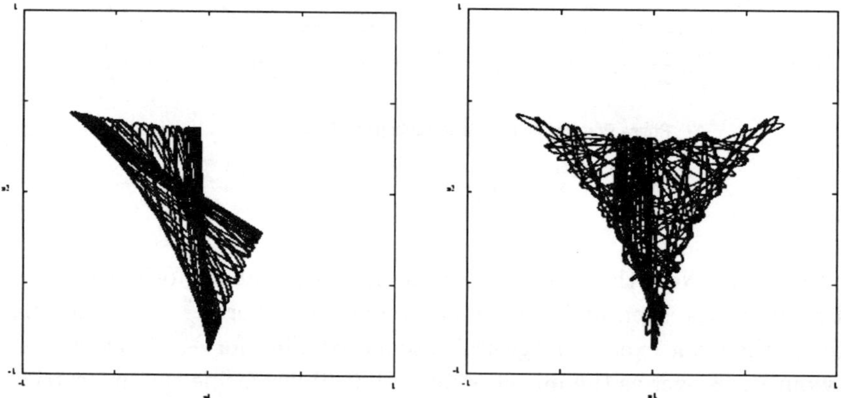

Fig. 16 Solutions started near the mode at $x_2 = -1$ for $\chi = 1$ and 2.

For $\chi \approx 3$ and starting close to the origin, but not actually on an invariant plane, the motion remains close to one of the invariant planes but when near the origin it may randomly switch to one of the other two planes. While close to the invariant planes the solutions almost follow the homoclinic orbit, which passes close to the ground state, at the origin where the atoms are equally distributed between the wells, and then out to a single well, where all atoms lie in one well. However the actual well chosen is unpredictable.

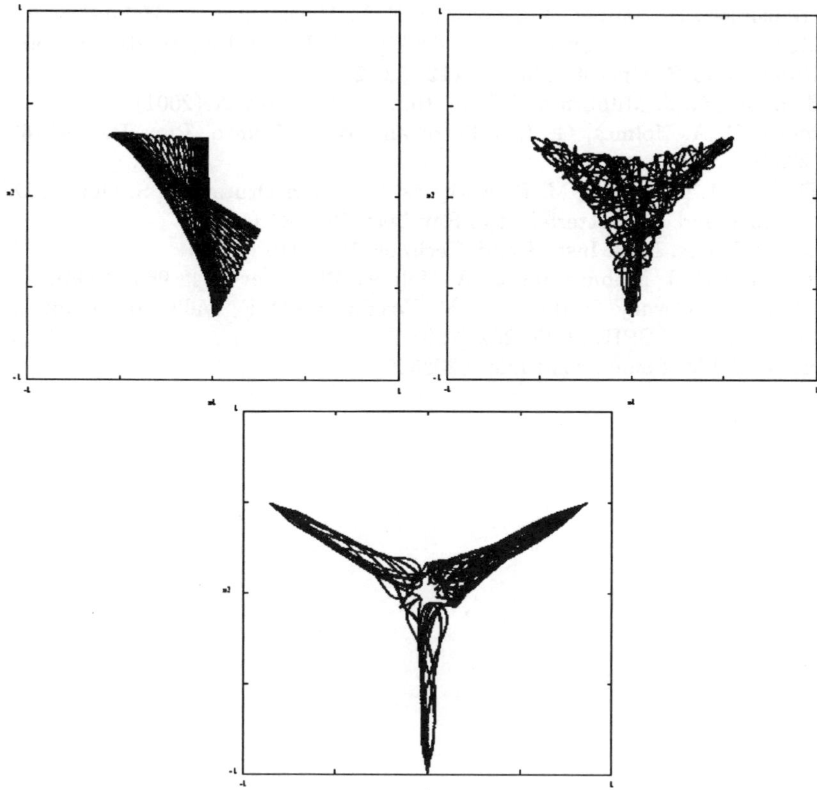

Fig. 17   Solutions started near the origin for $\chi = 1$, 2 and 3.

References

A. J. Lichtenberg and M. A. Lieberman, *Regular and Chaotic Dynamics* Springer Verlag, Berlin (1992).

M. H. Anderson, J. R. Ensher, M. R. Matthews, C. E. Wieman and E. A. Cornell, Science **269**, 198 (1995).

G.J.Milburn. Quantum Technology. Allen and Unwin, Sydney,(1996)

H. -J. Stöckmann, *Quantum Chaos - An Introduction*, Cambridge University Press (1999).

G. J. Milburn, J. Corney, E. M. Wright and D. F. Walls, Phys. Rev. A **55**, 4318 (1997).

W.K. Hensinger, B. Upcroft, C.A. Holmes, N. R. Heckenberg, G. J. Milburn and H. Rubinsztein-Dunlop, Phys Rev **A** 64 ,063408 (2001).

W. K. Hensinger, H. Häffner, A. Browaeys, N. R. Heckenberg, K. Helmerson, C. McKenzie, G. J. Milburn, W. D. Phillips, S. L. Rolston, H. Rubinsztein-Dunlop and B. Upcroft, Nature **412**, 52 (2001).

G. L. Salmond, G. J. Milburn and C. A. Holmes Phys Rev **A** (2001)

K. Nemoto, C. A. Holmes, G. J. Milburn and W. J. Munro, Phys Rev **A** 63 (2000).

K. B. Davies, M. O. Mewes, M. R. Andrews, N. J. van Druten, D. S. Durfee, D. M. Kurn and W. Ketterle, Phys Rev Lett **75** 1687 (1995)

Eric Cornell J. Res. Natl. Inst. Stand. Technol. **101**, 419 (1996)

S. Dryting and G. J. Milburn and C. A. Holmes, Phys Rev **E** 48 969 (1993).

G. J. Milburn, J. Corney, D. Harris, E. M. Wright and D. F. Walls, Atom Optics: Proceedings of SPIE **2995**, 232 (1997).

R. Franzosi and V. Penna, cond-mat/0203509

Chapter 5

Structure Functions, Cumulants and Breakdown Criteria for Wave Turbulence

L.J. Biven, C. Connaughton, and A.C. Newell[1]

Mathematics Institute, University of Warwick, Coventry, CV4 7AL, UK

Abstract.   We study the structure functions and cumulants of wave turbulence at small and large separations, respectively. We show that the criteria for breakdown obtained previously by examining the uniform validity of the asymptotic closure govern how close wave turbulence stays to joint Gaussianity. The results for large separations are new and include the remarkable result that, in certain circumstances, correlations of surface elevation decay very slowly. There are also new results in the case of small separations in that the system behavior is organized by a special point in the (α, β) plane where α and β are the homogeneity parameters for the linear and nonlinear coupling coefficients. Finally, we explore how modifications of the breakdown criteria are necessary because of the strengths of non-local long wave- short wave interactions.

1   Introduction

In [1; 9], we developed simple criteria for the breakdown of the wave turbulence approximation for situations where the energy transfer is sufficiently local. Here we both extend these results and discover some new behaviors of the structure functions

$$S_N(r) = < (u(\underline{x} + \underline{r}) - u(\underline{x}))^N >$$
(1.1)

when $r = |\underline{r}|$ is small and the two-point, N-correlators

$$C_N(r) = \{u^{N-1}(\underline{x})u(\underline{x} + \underline{r})\}.$$
(1.2)

[1] also Department of Mathematics, University of Arizona, Tucson, AZ, 85721, USA

197

when r is large. The brackets $\{\}$ denote the cumulant which is the combination of averages which decay as $r \to \infty$. The variable $u(\underline{x}, t)$ is a spatially homogeneous, random, bounded field and we assume that, initially, its cumulants decay strongly as separations become large. It represents a wave field which, for example, one may think of as the surface elevation of the sea. To leading order, propagation of deformations of the surface obey linear dispersion relations but, over long times, energy is exchanged between waves via r-wave resonances ($r = 3, 4, ...$). This exchange is described by a kinetic equation for the wave action spectral density whose most relevant stationary states are the Kolmogorov-Zakharov (KZ) finite flux spectra for both energy and wave action (particles). Using these solutions, we can directly calculate the corresponding physical space averages. To examine small scale behavior, the structure functions $S_N(r)$ are most useful. The main reason is that their dependence on r for small r indicates the degree of non-smoothness of the variable $u(x)$ in realizations of the field. For example, we will find that $S_N \sim r^{\zeta_N}$, $\zeta_N < N$, which indicates that the variable (e.g. the sea surface) has lost differentiability and the values ζ_N will tell us something of how scalloped it looks. In section 2 we examine their behavior in considerable detail. For the large scales, we examine the two-point N-cumulants $C_N(r)$. In section 3, we do this using gravity waves as a concrete example. Finally, we briefly consider how the breakdown criteria change as non-local (long wave- short wave) interactions compete with the local interactions which give rise to the KZ spectra.

The new results of this paper are:

1. Previously, we have shown that the expansions for the structure functions exhibit a departure from joint Gaussian statistics at small values $r < r_{NL}$. The scale r_{NL} exactly coincides with the wave number k_{NL}^{-1} we obtained for wave turbulence breakdown by examining the uniform validity of the kinetic equation's asymptotic expansion. What is new here is that we examine universality of the structure functions by calculating whether the dominant contribution to their behaviors arise from the universal Kolmogorov-Zakharov (KZ) spectra, or from the forcing terms associated with large scales. It turns out that their universality can be conveniently characterized in the (α, β) plane where α and β are numbers capturing the relative strengths of the linear and nonlinear coupling coefficients. We provide colorful road-maps in the cases of three and four wave interactions. We argue why there is a special point in the (α, β) plane about

which all the different behaviors are organized. Further, we show that there is an open set of (α, β) values such that one can always find an appropriate variable for which all the structure functions are universal.

2. The r-wave resonances induce strong correlations over large distances. Indeed, ratios such as $C_N(r)/C_N(0)$, which initially may decay exponentially, decay only weakly in r when calculated on the stationary KZ spectra corresponding to the constant flux of wave action (particle flux). In particular, in section 3, we show that $C_2(r)/C_2(0)$, with $u(\underline{x})$ taken as the surface elevation, decays as $J_0(k_*(t)r)$ for large $k_*(t)r$. Here, k_* is the position of the traveling front in wave-number space as the wave action propagates from a wide range of intermediate forcing scales to longer and longer waves. The reasons for the weak decay are the high degree of correlation induced by the four wave resonances and the fact that the front which carries waveaction to longer and longer waves is sharp.

3. The analysis of wave turbulence breakdown to date has assumed that the nonlinear coupling coefficients are sufficiently local so that all the integrals multiplying the large k behaviors converge. This may not always be the case. In section 4 we ask whether certain divergences which, depending on coupling coefficients, can emerge due to strong non-local long wave- short wave interactions, modify the simple breakdown criteria and find that a modification can indeed occur. We discuss this modification and, as a concrete example, look at the case for deep water gravity waves.

The wave turbulence approximation proceeds as follows. We begin with the equation

$$\frac{\partial}{\partial t} A^s(\underline{k}, t) - is\omega(\underline{k})A^s(\underline{k}, t)$$

$$= \varepsilon^{m-1} \sum_{m=2}^{\infty} \sum_{s_1 \cdots s_m} \int L_{\underline{k}\underline{k}_1\underline{k}_2\cdots\underline{k}_m}^{ss_1s_2\cdots s_m} A^{s_1}(\underline{k}_1, t) A^{s_2}(\underline{k}_2, t) \cdots A^{s_m}(\underline{k}_m)$$

$$\times \delta(\underline{k}_1 + \underline{k}_2 + \cdots + \underline{k}_m - \underline{k})d\underline{k}_1 d\underline{k}_2 \cdots d\underline{k}_m \tag{1.3}$$

for the generalized Fourier transform $\varepsilon A_{\underline{k}}^s$ of the canonical physical field $u^s(\underline{x})$ where $0 < \varepsilon \ll 1$. In (1.3), $\omega^s(\underline{k})$ is the dispersion relation and $L_{\underline{k}\underline{k}_1\cdots}^{ss_1\cdots}$ is the nonlinear coupling coefficient. The letter s is an integer which counts the degeneracy of the dispersion relation; here we take it to be plus or minus one so that $\omega^s(\underline{k}) = s\omega(\underline{k})$. The functions $\omega(\underline{k})$ and $L_{\underline{k}\underline{k}_1\cdots\underline{k}_m}^{ss_1\cdots s_m}$ are taken to be homogeneous ($\omega(\lambda\underline{k}) = \lambda^\alpha\omega(\underline{k})$; $L_{\lambda\underline{k}\ \lambda\underline{k}_1\ \lambda\underline{k}_2}^{ss_1s_2} = \lambda^\beta L_{\underline{k}\ \underline{k}_1\ \underline{k}_2}^{ss_1s_2}$; $L_{\lambda\underline{k}\ \lambda\underline{k}_1\ \lambda\underline{k}_2\ \lambda\underline{k}_3}^{ss_1s_2s_3} = \lambda^\gamma L_{\underline{k}\ \underline{k}_1\ \underline{k}_2\ \underline{k}_3}^{ss_1s_2s_3}$) of degrees α, β and γ respectively. From (1.3) one can build the BBGKY hierarchy of equations for the Fourier cumulants, namely the Fourier transforms, $Q^{ss'\cdots s^{(N-1)}}(\underline{k},\underline{k}',\cdots,\underline{k}^{(N-1)})\delta(\underline{k}+\cdots+\underline{k}^{(N-1)})$, of physical space cumulants $R^{(N)ss'\cdots s^{(N-1)}}(\underline{r},\underline{r}',\cdots,\underline{r}^{(N-2)})$ which are defined as N^{th} order moments, $< u^s(\underline{x})u^{s'}(\underline{x}+\underline{r})\cdots u^{s^{(N-1)}}(\underline{x}+\underline{r}^{(N-2)}) >$, from which the appropriate combinations of products of lower order moments are subtracted so that $R^{(N)} \rightarrow 0$ as any of the separations $|\underline{r}|,|\underline{r}'|, \ldots, |\underline{r}^{(N-2)}| \rightarrow \infty$. We solve the hierarchy iteratively in powers of amplitude

$$Q^{(N)ss'\cdots s^{(N-1)}}(\underline{k},\underline{k}',\cdots,\underline{k}^{(N-1)}) =$$
$$q_0^{(N)ss'\cdots s^{(N-1)}}(\underline{k},\underline{k}',\cdots,\underline{k}^{(N-1)})e^{i(s\omega_k+s'\omega_{k'}+\cdots)t}$$
$$+\varepsilon Q_1^{(N)ss'\cdots}(\underline{k},\underline{k}',\cdots)+\cdots \qquad (1.4)$$

and demand that (1.4) is a uniformly valid asymptotic expansion in time. To achieve this goal, we choose the slow time behaviors of the $q_0^{(N)ss'\cdots s^{(N-1)}}(\underline{k},\cdots)$ to remove any secular growth. The resulting hierarchy of equations for the $q_0^{(N)}$ is closed. This is what we call *asymptotic closure*. Because the equations for $q_0^{(N)}$, $N > 2$, appear as expressions for $\frac{d}{dt}lnq_0^{(N)},\frac{d}{dt}lnq_0^{ss}$ which only depend on $n_k^s = q_0^{-ss}(\underline{k})$, the asymptotic closure reduces to a kinetic equation

$$\frac{d}{dt}n_k^s = \varepsilon^2 T_2[n_k^s] + \varepsilon^4 T_4[n_k^s] + \cdots + \varepsilon^{2p}T_{2p}[n_k^s] + \cdots, \qquad (1.5)$$

and a frequency renormalization

$$s\omega_k \rightarrow s\omega_k + \varepsilon^2\Omega_2^s[n_k^s] + \cdots. \qquad (1.6)$$

For time scales $\varepsilon^{2p}t = O(1)$, $\varepsilon^{2p+2}t \ll 1$, $p \geq 1$, we seek solutions of the kinetic equation (1.5) truncated at order $2p$. The collision term $T_2[n_k]$ is built from three wave resonances; the terms $T_4[n_k]$ from four wave resonances and spectral gradients of the three wave resonances; and so on. Here we give the expression for $T_2[n_k]$ and $\Omega_{2k}^s[n_k]$ where we have used

$$n_k \equiv n_k^+ = n_k^- = q_0^{(2)s-s}.$$

$$T_2[n_k] = 4\pi \sum_{s_1 s_2} \int L_{\underline{k}\underline{k}_1\underline{k}_2}^{ss_1 s_2} (n_k n_{k_1} n_{k_2})$$

$$\times \left(\frac{L_{-\underline{k}-\underline{k}_1-\underline{k}_2}^{-s-s_1-s_2}}{n_k} + \frac{L_{\underline{k}_1-\underline{k}_2\underline{k}}^{s_1-s_2 s}}{n_{k_1}} + \frac{L_{\underline{k}_2\underline{k}-\underline{k}_1}^{s_2 s-s_1}}{n_{k_2}} \right)$$

$$\times \delta(s_1\omega(\underline{k}_1) + s_2\omega(\underline{k}_2) - s\omega(\underline{k}))\delta(\underline{k}_1 + \underline{k}_2 - \underline{k})d\underline{k}_1 d\underline{k}_2 \qquad (1.7)$$

$$= 4\pi \int |L_{\underline{k}\underline{k}_1\underline{k}_2}^{+++}|^2 n_k n_{k_1} n_{k_2}$$

$$\times \left(\left(\frac{1}{n_{k_1}} + \frac{1}{n_{k_2}} - \frac{1}{n_k} \right) \delta(\omega_{\underline{k}_1} + \omega_{\underline{k}_2} - \omega_{\underline{k}}) \right.$$

$$+ \left(\frac{1}{n_{k_1}} - \frac{1}{n_{k_2}} - \frac{1}{n_k} \right) \delta(\omega_{\underline{k}_1} - \omega_{\underline{k}_2} - \omega_{\underline{k}})$$

$$+ \left. \left(\frac{1}{n_{k_2}} - \frac{1}{n_{k_1}} - \frac{1}{n_k} \right) \delta(\omega_{\underline{k}_2} - \omega_{\underline{k}_1} - \omega_{\underline{k}}) \right) \delta(\underline{k}_1 + \underline{k}_2 - \underline{k})d\underline{k}_1 d\underline{k}_2$$

$$(1.8)$$

$$\Omega_{2k}^s[n_k] = 4 \sum_{s_1 s_2} \int L_{\underline{k}\underline{k}_1\underline{k}_2}^{ss_1 s_2} L_{\underline{k}_1\underline{k}-\underline{k}_2}^{s_1 s-s_2} n_{k_2} \left(PV\left(\frac{1}{(s_1\omega_{\underline{k}_1} + s_2\omega_{\underline{k}_2} - s\omega_{\underline{k}})} \right) \right.$$

$$\left. - i\pi\delta(s_1\omega_{\underline{k}_1} + s_2\omega_{\underline{k}_2} - s\omega_{\underline{k}}) \right) \delta(\underline{k}_1 + \underline{k}_2 - \underline{k})d\underline{k}_1 d\underline{k}_2. \qquad (1.9)$$

$\delta(\cdot)$ is the Dirac delta function; PV denotes principal value and we will often write Ω_{2k} for $\Omega_2^s[n_k]$. In equation (1.8) we have made use of the relation $L_{\underline{k}\underline{k}_1\underline{k}_2}^{+++} = L_{\underline{k}\underline{k}_1-\underline{k}_2}^{++-}$ which is an essential symmetry of three wave interactions. The truncated equations can be solved [11] for their finite flux Kolmogorov-Zakharov solutions. For three wave resonances,

$$n_k = C_1 P^{\frac{1}{2}} k^{-(\beta+d)}. \qquad (1.10)$$

For cases where $T_2 = 0$, either by virtue of the resonant manifold ($\underline{k}_1 + \underline{k}_2 = \underline{k}$; $\pm\omega_{\underline{k}_1} \pm\omega_{\underline{k}_2} = \omega_{\underline{k}}$) being empty, or by $L_{\underline{k}\underline{k}_1\underline{k}_2}^{ss_1 s_2}$ being zero, the kinetic equation truncated at $p = 2$ yields the finite (energy and particle) flux solutions

$$n_k = C_2 P^{\frac{1}{3}} k^{-(\frac{2\gamma}{3}+d)} \qquad (1.11)$$

$$n_k = C_3 Q^{\frac{1}{3}} k^{-(\frac{2\gamma}{3}+d)+\frac{\alpha}{3}}. \qquad (1.12)$$

Breakdown occurs when these solutions no longer keep the asymptotic expansions (1.5) and (1.6) uniformly valid in k and where, equivalently, the ratios of linear to nonlinear time scales (which should be small) approach unity. For example

$$\frac{t_L}{t_{NL}} = \frac{1}{\omega_k n_k}\frac{\partial n_k}{\partial t} \sim \frac{\varepsilon^2 T_2}{\omega_k n_k}$$
$$= \varepsilon^2 P^{\frac{1}{2}} k^{\beta - 2\alpha} I_2 \tag{1.13}$$

where I_2 is the integral

$$I_2 = 4\pi \int L^{s s_1 s_2}_{\hat{k}\hat{k}_1 \hat{k}_2}\left(n_{\hat{k}} n_{\hat{k}_1} n_{\hat{k}_2}\right)$$
$$\times \left(\frac{L^{-s-s_1-s_2}_{-\hat{k}-\hat{k}_1-\hat{k}_2}}{n_{\hat{k}}} + \frac{L^{s_1-s_2 s}_{\hat{k}_1-\hat{k}_2\hat{k}}}{n_{\hat{k}_1}} + \frac{L^{s_2 s-s_1}_{-\hat{k}_2\hat{k}-\hat{k}_1}}{n_{\hat{k}_2}}\right)$$
$$\times \delta(s_1\omega_{\hat{k}_1} + s_2\omega_{\hat{k}_2} + s\omega_{\hat{k}})(\hat{k}_1)^{d-1} d\hat{k}_1 d\Omega_1 \tag{1.14}$$

where $\hat{\underline{k}}_i = (\underline{k}_i)/(|\underline{k}|)$ has modulus $\hat{k}_i$; and $d\Omega_1$ denotes the integration over angles when $\hat{\underline{k}}_1$ is expressed in d-dimensional polar coordinates. We have taken n_k to be the homogeneous Kolmogorov-Zakharov solution, (1.10).

If the interactions are sufficiently local (namely the coefficients $L_{\underline{k}\underline{k}_1 k_2}$ go to zero sufficiently fast as their arguments go to zero or infinity), the integral I_2 is finite. In that case the breakdown criterion is simple because it only depends on α, β and P. Furthermore, t_L/t_{NL} approaches unity at a scale $k_{NL} = P^{-1/(2(\beta-2\alpha))}$ where we have incorporated ε into P ($\varepsilon^4 P \to P$). An analysis of T_4/T_2, T_6/T_4, $\Omega_{2k}/\omega_k,...$ yields similar ratios. Therefore, if the coefficient integrals, I_N, are finite then, for $\beta > 2\alpha$, k_{NL} is large and wave turbulence breaks down for $k > k_{NL}$; and for $\beta < 2\alpha$, k_{NL} is small and breakdown occurs for $k < k_{NL}$.

For four wave interactions (when three wave resonances are not present), the corresponding ratios of linear to nonlinear time scales; $T_4/\omega_k n_k$ and the ratios of T_6/T_4, T_8/T_6, Ω_2^s/ω_k are given by $P^{2/3}k^{(2\gamma/3)-2\alpha}I_4$ and $P^{1/3}k^{(\gamma/3)-\alpha}(I_6/I_4, T_8/T_6, ...)$ where the I_N are integrals like in (1.14). In particular, I_4 is given by

$$I_4 = \int |G_{\hat{k}\hat{k}_1,\hat{k}_2\hat{k}_3}|^2 n_{\hat{k}} n_{\hat{k}_1} n_{\hat{k}_2} n_{\hat{k}_3}$$
$$\times \left(\frac{1}{n_{\hat{k}}} + \frac{1}{n_{\hat{k}_1}} - \frac{1}{n_{\hat{k}_2}} - \frac{1}{n_{\hat{k}_3}} \right)$$
$$\times \delta(\omega_{\hat{k}} + \omega_{\hat{k}_1} - \omega_{\hat{k}_2} - \omega_{\hat{k}_3})(\hat{k}_1 \hat{k}_2)^{d-1} d\hat{k}_1 d\hat{k}_2 d\Omega_1 d\Omega_2 \qquad (1.15)$$

where $k_i = (\underline{k}_i)/(|\underline{k}|)$, $i = 1, 2$ and $\underline{k}_3 = \underline{k} + \underline{k}_1 - \underline{k}_2$. $d\Omega_i$ denotes integration over angles when $\hat{\underline{k}}_i$ is expressed in d-dimensional polar coordinates. (T_4 is given is section 4 along with the definition of G.)

If the interactions are sufficiently local so that all integrals converge, then the criteria for breakdown are simple and depends only on α, γ and P. Breakdown occurs for large (small) k if $\gamma > 3\alpha$ ($\gamma < 3\alpha$) at $k_{NL} = P^{-1/(\gamma-3\alpha)}$. Therefore, in both cases, for sufficiently local interactions, the criterion for breakdown are very simple.

In many practical situations, however, the coefficients $L_{\underline{k}\underline{k}_1\underline{k}_2}$ and $L_{\underline{k}\underline{k}_1\underline{k}_2\underline{k}_3}$ for the non-local interactions between long and short waves, (for three wave interactions: $\underline{k}_1$ is small while $\underline{k}$, $\underline{k}_2$ are large or $\underline{k}$ is small while $\underline{k}_1$, $\underline{k}_2$ are large; for four wave interactions: $\underline{k}_1$, $\underline{k}_3$ are small and $\underline{k}$, $\underline{k}_2$ are large) are not sufficiently small to ensure convergence. Formally this would necessitate the introduction of a large scale cut-off k_I and then the breakdown wave-number k_{NL} will depend on k_I as well as the universal parameters P, α and β. We discuss this in section 4. For now, we will assume that the coupling coefficients are such that interactions are local enough so that all coefficient integrals converge.

2 Structure Functions

We now turn our attention to the structure functions based upon the KZ solution and the long time surviving parts of each of the cumulants. We note that, because $Im\Omega^s_{2k} > 0$ and indeed because the zeroth order cumulant is multiplied by the exponential of a fast phase, all the initial information on cumulants of order higher than two (and of order two when $s' = s$) is lost. In what follows, we calculate only those terms which survive asymptotically and which are given by the integrals over products of n_k. We begin with the case of three wave resonances.

We intend to plot the behavior of the various structure functions in the

(α, β) plane, showing where the functions behave universally and where they are dominated by the forcing. We show that, along the line $\beta = 2\alpha$, where the KZ solution (1.10) inherits the scaling symmetries of the governing equation (1.3), the behavior of structure functions is either universal for all N, or dominated by the forcing for all N. The boundary between these two regimes is the point $\alpha = 2$, $\beta = 4$. This point organizes all of the intermediate cases which occur off the $\beta = 2\alpha$ line where some structure functions are universal while others are not.

We give explicit calculations for $S_2(r)$, $S_3(r)$ and $S_4(r)$ and then state the results for the higher orders. These calculations will define what we mean by the universal strip for S_N and describe the behavior of S_N in and outside of the strip. For convenience, we omit the superscript s denoting signs. Because we are dealing with KZ solutions which carry a finite amount of energy, we use as our physical variable that quantity $v(\underline{x})$ whose power spectrum is the spectral energy $\omega_k n_k$. This means that the generalized Fourier transform of v is $\omega_k^{1/2} A_k$. Formally,

$$S_2(r) = <(v(\underline{x} + \underline{r}) - v(\underline{x}))^2>$$
$$= 2 \int \omega_k n_k (1 - cos(\underline{k} \cdot \underline{r})) d\underline{k} \tag{2.1}$$

We split the integral into three regions in $\underline{k}$ space. The first, $k < k_I$, is the region in which the non-universal forcing is applied. The second, $k_I < k < k_U$, is the region in which the universal KZ spectrum is obtained. The third, $k > k_U$, is the non-universal dissipation region.

We say that the result is universal if, as $r \to 0$, the dominant contribution comes from the inertial range, $k_I < k < k_U$, and we can take the limits $k_I \to 0$, $k_U \to \infty$. In the region $k < k_I$, $\omega_k n_k = F(k)$, some non-universal forcing which we can take to be analytic in k. Therefore, as $r \to 0$, this part of the integral scales as r^2. In the region $k_U < k$, we assume that the solution in the dissipation region decays fast enough as a function of wave-number such that the contribution from this integral is always less than the contribution from the other regions. In the inertial region $k_I < k < k_U$, a simple change of variables will show that this contribution goes as $P^{\frac{1}{2}} r^{\beta - \alpha} I$. The integral factor I converges at small k (i.e. near k_I where $1 - cos(\underline{k} \cdot \underline{r}) = O(r^2)$) if $2 > \beta - \alpha$ and converges at large k (i.e. near k_U where $1 - cos(\underline{k} \cdot \underline{r}) = O(1)$) if $\beta - \alpha > 0$. The former condition also ensures that the inertial range, universal contribution to the scaling

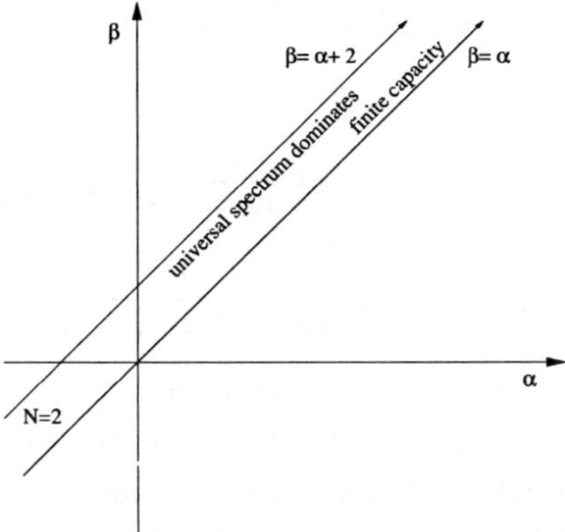

Fig. 1   Strip in which S_2 is universal

dominates that of the forcing interval. The latter condition is simply the condition that the KZ spectrum has finite energy capacity. Therefore, if α and β lie in the strip $0 < \beta - \alpha < 2$ in the (α, β)-plane, we can say, for small r $(k_U^{-1} \ll r \ll k_I^{-1})$,

$$S_2(r) \sim C_1 P^{\frac{1}{2}} r^{\beta - \alpha} \tag{2.2}$$

is universal. For $2 < \beta - \alpha$, $S_2 \sim F^2 r^2$ is non-universal. The parameter F is a measure of the amplitude of the forcing and the width of the region over which it is applied. For example, $F^2 r^2$ will be proportional to k_I times some value of $F(k)(1 - cos(kr))$ in $0 < k < k_I$. In what follows, we use $F^N r^N$ to indicate the size of the contribution to $S_N(r)$ coming from the forcing region. Although we will not write all dimensional factors in our calculations, we follow the dependence of the structure functions on F since this will show when the breakdown scale is non-universal. We follow the dependence on the energy flux P since this is the universal parameter which we take to be small. On the line $\beta = \alpha + 2$, S_2 has universal scaling with logarithmic corrections. The organizing point $\alpha = 2$, $\beta = 4$ can already be seen as the intersection between the top border of the S_2 strip and the breakdown line $\beta = 2\alpha$. Above the line $\beta = \alpha + 2$, non-universal effects contaminate the second order structure function and its scaling. The line $\beta = 2\alpha$, for small r, separates the region of wave turbulence breakdown from wave turbulence validity.

We can calculate $S_3(r)$ from the first surviving term in the asymptotic expansion for $Q^{(3)ss's''}(\underline{k},\underline{k}',\underline{k}'')$ which is given by

$$q^{(3)ss's''}\delta(\underline{k}+\underline{k}'+\underline{k}'') =$$

$$\cdots + \varepsilon^2 \left(L^{s-s'-s''}_{\underline{k}-\underline{k}'-\underline{k}''} n_{k'}n_{k''} + L^{s'-s''-s}_{\underline{k}'-\underline{k}''-\underline{k}} n_{k''}n_k + L^{s''-s-s'}_{\underline{k}''-\underline{k}-\underline{k}'} n_k n_{k'} \right)$$

$$\times \left(\pi\delta(s\omega_k+s'\omega_{k'}+s''\omega_{k''})+iPV\left(\frac{1}{s\omega_k+s'\omega_{k'}+s''\omega_{k''}}\right) \right)\delta(\underline{k}+\underline{k}'+\underline{k}'')+\cdots$$

where PV denotes the principal value. This is, of course, only the first term in an asymptotic series. The ratios of successive terms will be of the same order as we will find when we calculate the deviation from joint Gaussianity. The exception will occur when non-local, long wave- short wave interactions are important. We return to this point in section 4.

The third order structure function is given by

$$S_3(r) = \int (\omega_{\underline{k}}\omega_{\underline{k}'}\omega_{\underline{k}''})^{\frac{1}{2}}q^{(3)}(\underline{k},\underline{k}',\underline{k}'')\delta(\underline{k}+\underline{k}'+\underline{k}'')$$

$$\times 2i\left(sin(\underline{r}\cdot(\underline{k}+\underline{k}')) - sin(\underline{r}\cdot\underline{k}) - sin(\underline{r}\cdot\underline{k}')\right)d\underline{k}d\underline{k}'d\underline{k}'' \quad (2.3)$$

which is identically zero due to the fact that the *sin* functions are odd functions of the angles between the $\underline{k}$ vectors and $\underline{r}$ and because we have chosen the variable v to preserve the isotropy of the system. In these cases, the dispersion relation depends on $|\underline{k}|$ and the coefficients, $L_{\underline{k}\underline{k}_1\underline{k}_2}$, are even functions of the wave vector angles. The same holds for all odd structure functions. The main thrust of the breakdown result, however, does not depend on this property.

The fourth order structure function

$$S_4(r) =< (v(\underline{x}+\underline{r}) - v(\underline{x}))^4 >$$

$$= \int \sqrt{\omega_k\omega_{k_1}\omega_{k_2}\omega_{k_3}} < A_{\underline{k}}A_{\underline{k}_1}A_{\underline{k}_2}A_{\underline{k}_3} >$$

$$[1 - e^{i\underline{k}\cdot\underline{r}}][1 - e^{i\underline{k}_1\cdot\underline{r}}][1 - e^{i\underline{k}_2\cdot\underline{r}}][1 - e^{i\underline{k}_3\cdot\underline{r}}]d\underline{k}d\underline{k}_1 d\underline{k}_2 d\underline{k}_3$$

splits into two parts just as the fourth order moment splits into its fourth order part plus products of second order cumulants.

$$S_4(r) = 3S_2^2(r) + \mathfrak{S}_4(r) \quad (2.4)$$

where $\mathfrak{S}_4$ is calculated from the first term of the surviving fourth order cumulant which is proportional to a product of three n_k's. We find that

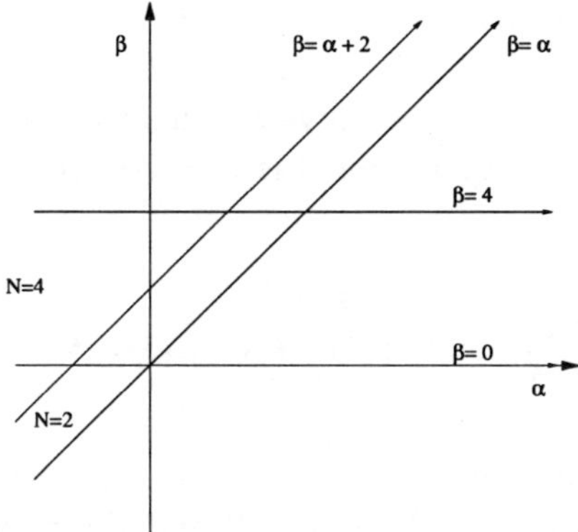

Fig. 2   Strip in which $\mathfrak{S}_4$ is universal. S_4 is universal in the intersection of the two strips.

$\mathfrak{S}_4$ is universal if α, β lie in the strip $0 < \beta < 4$. Outside of this strip, $\mathfrak{S}_4$ is nonuniversal. On the line $\beta = 4$ which defines the top of the strip, $\mathfrak{S}_4$ will have the universal scaling with logarithmic corrections. Furthermore, this line intersects the analogous line for the S_2 strip ($\beta = \alpha + 2$) and the breakdown line ($\beta = 2\alpha$) at the special point ($\alpha = 2, \beta = 4$). This pattern will continue for all strips.

For α,β inside the $\mathfrak{S}_4$ strip and for small r,

$$S_4(r) = C_1^2 P r^{2\beta - 2\alpha} + C_1^3 P^{\frac{3}{2}} r^{\beta}. \tag{2.5}$$

For a joint Gaussian field, S_4 would be simply $3S_2^2(r)$ and thus the remainder, $\mathfrak{S}_4$, measures deviations from joint Gaussianity. Observe that

$$\frac{\mathfrak{S}_4}{S_2^2} \sim P^{\frac{1}{2}} r^{2\alpha - \beta}. \tag{2.6}$$

Wave turbulence requires that the fields remain close to joint Gaussian at all scales. The right hand side of (2.6) fails to do this at small scales if $\beta > 2\alpha$ and indeed the length scale, r_{NL}, for which $P^{\frac{1}{2}} r^{2\alpha - \beta}$ becomes of order unity is precisely k_{NL}^{-1} where k_{NL} was defined in the introduction.

The strips, together with the breakdown line determine the behavior of the structure functions and the presence of breakdown. For instance, we find that the Nth order structure function can be written as a series

$$S_N(r) = \mathfrak{S}_N + \cdots + C_0 (S_2)^{\frac{N}{2}}.$$

$\mathfrak{S}_N$ is universal in the strip for $0 < \beta + (N/2 - 2)\alpha < N$. The intersection $\bigcap(0 < \beta + (N/2 - 2)\alpha < N)$ of all the strips is the black and white parallelogram, which we call the universal rhombus, shown in figure 3 ($\alpha < \beta < \alpha + 2$, $0 < \alpha \le 2$). Its diagonal is $\beta = 2\alpha$. Inside this parallelogram we have

$$\frac{S_N}{(S_2)^{\frac{N}{2}}} = C_0 + \sum_{1}^{\frac{N}{2}-1} C_{N-2s}(P^{\frac{1}{2}}r^{2\alpha-\beta})^s. \tag{2.7}$$

Depending on the value of the exponent, $2\alpha - \beta$, there are three possible scenarios at small scales: i) $2\alpha - \beta > 0$ and the ratio of structure functions can be approximated by the Gaussian value of C_0; $S_N/(S_2)^{N/2} \sim C_0$. Furthermore, this approximation gets better as $r \to 0$ so wave turbulence looks more and more Gaussian at small scales. ii) $2\alpha - \beta = 0$. In this case, the ratio of non-Gaussian to Gaussian parts of the structure functions is independent of r and universally small. Wave turbulence has complete self-similarity with respect to k_{NL}. The KZ spectrum exactly inherits the scaling properties of the original equation (1.3) and wave turbulence is uniformly valid over all scales. iii) $2\alpha - \beta < 0$ in which case the r-dependent corrections to the Gaussian constant grow as $r \to 0$. At small scales, structures appear which are outside of the remit of weak turbulence theory. In this case, we identify a breakdown region at small scales as we have discussed. Looking at figure 3, the universal rhombus contains two triangles. In the triangle $\beta > 2\alpha$, we still have breakdown for $k > k_{NL}$. In the triangle $\beta < 2\alpha$, the wave turbulence approximation is applicable for all small r (large k) and is universal.

Throughout the (α, β) plane, the occurrence of breakdown is governed by the line $\beta = 2\alpha$. For $\beta > 2\alpha$, the ratio of time scales becomes of the order of unity at $k \sim k_{NL}$ and asymptotic closures of the weak turbulence theory fail. Furthermore, when universal, the real space structure function can illustrate the breakdown by measuring strong corrections to Gaussianity. Inside the universal rhombus where all S_N are universal, corrections get proportionally larger at smaller scales

$$\frac{\mathfrak{S}_N}{(S_2)^{\frac{N}{2}}} \sim (P^{\frac{1}{2}}r^{-\beta+2\alpha})^{\frac{N}{2}-1} \tag{2.8}$$

and we define r_{NL} as k_{NL}^{-1} as before.

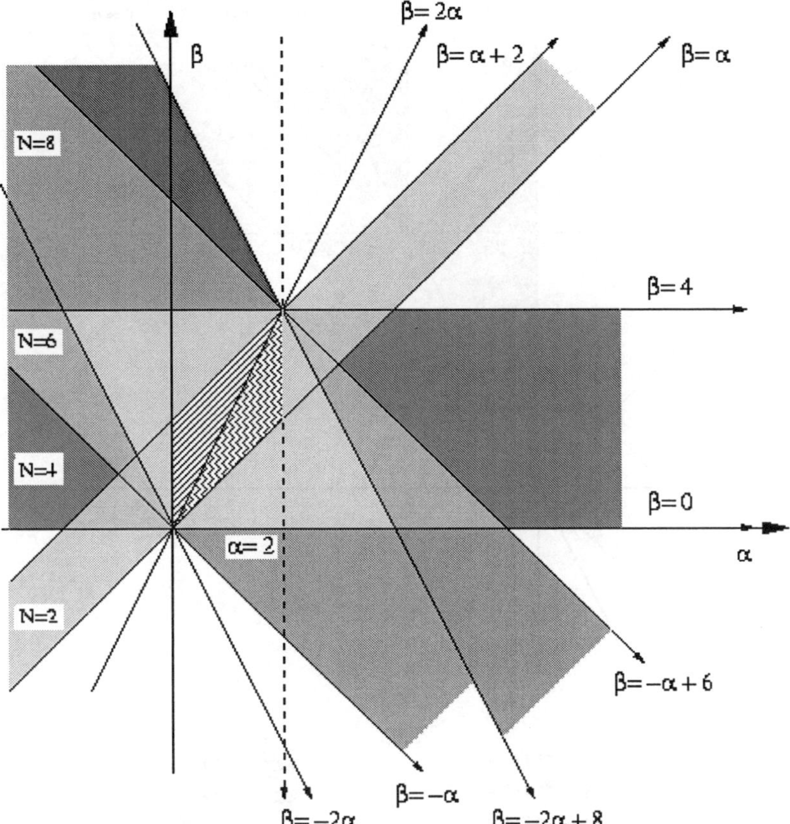

Fig. 3 Strips for the three wave energy spectrum: Each strip is the condition for some $\mathfrak{S}_N$ to be universal. The black and white parallelogram shows the region in which all S_N are universal. The breakdown line is $\beta = 2\alpha$ and breakdown at small scales occurs for $\beta > 2\alpha$.

Outside this rhombus, the general features of breakdown remain. However the breakdown scale may have a dependence on non-universal effects. As examples, we consider the regions 1,5 and 7 of figure 4 where we have singled out the area in the (α, β) plane corresponding systems which have capacity to absorb only a finite amount of energy at small scales. In section 3, we study an infinite capacity case when we look at the large r behavior of the cumulants C_N.

In region 1, only S_2 is universal and scales as $P^{\frac{1}{2}} r^{\beta - \alpha}$ while all S_N,

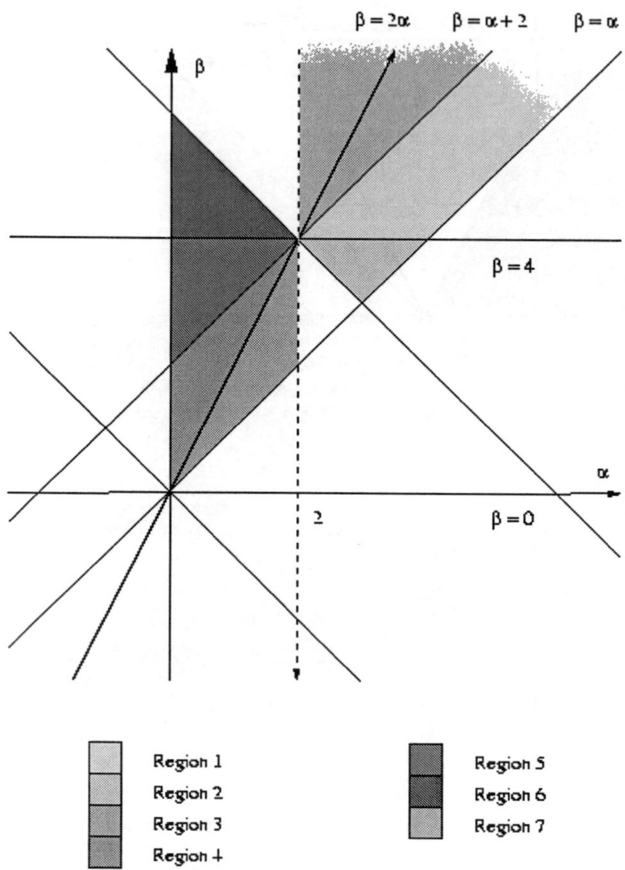

Fig. 4   Sectors for three wave energy case: This figure shows all the regions for which we can take $k_U \to \infty$. The forcing plays a role in all regions except 3 and 4. For example, both S_2 and S_4 are universal in region 2; only S_2 is universal in region 1.

$N > 2$ scale as $F^N r^N$ (For convenience, we can assume that there is only one universal dimensional parameter in the system which we take to be P [3].) The normalized corrections to Gaussianity are

$$\frac{\mathfrak{S}_N}{(S_2)^{\frac{N}{2}}} = F^N \left(P^{-\frac{1}{2}} r^{2-\beta+\alpha} \right)^{\frac{N}{2}}. \tag{2.9}$$

Since $2 + \alpha - \beta > 0$ for the entire region, there is no breakdown at small scales. Similar expressions are obtained for the ratio of $\mathfrak{S}_N/(S_2)^{N/2}$ in regions 2,... for N sufficiently high such that $\mathfrak{S}_N$ feels the forcing.

Now consider region 5 for which $\alpha + 2 < \beta < 4$ and $\alpha < 2$. In this region, $S_2 \sim F^2 r^2$ is non-universal while $\mathfrak{S}_N \sim P^{\frac{1}{2}(N-1)} r^{\beta+(N/2-2)\alpha}$ are universal for all $N > 2$. In this case, the ratio of structure functions gives

$$\frac{\mathfrak{S}_N}{(S_2)^{\frac{N}{2}}} = (P^{-\frac{1}{2}} r^{\beta-2\alpha})(PF^{-2} r^{\alpha-2})^{\frac{N}{2}} \qquad (2.10)$$

Since $\beta > 2\alpha$ in this region, the first bracket suppresses universal breakdown at small scales . In its place, we see a non-universal breakdown for $r < (P^{-1} F^2)^{1/(\alpha-2)}$ because $\alpha < 2$. We call the breakdown *non-universal* because the breakdown scale depends on the non-universal parameters collected in F.

In region 7, all the structure functions are dominated by the forcing at small scales. $\mathfrak{S}_N \sim r^N$ for all N. Weak turbulence is irrelevant for the small scale picture.

Before briefly discussing how similar calculations can be done for four wave interactions, we comment on the usefulness of the strip diagram. On the $\beta = 2\alpha$ line, all normalized corrections to Gaussianity are scale independent; a property which we call quasi-Gaussianity. The behavior on this line of the S_N is either universal for all N or non-universal for all N. The border between these two regions is the point $\alpha = 2$, $\beta = 4$ which is common to all (top) strip borders.

The position of the universal rhombus and all strips depends on the field v used to form the structure functions. We chose for our calculations that quantity whose second order moment has as its Fourier transform the spectral energy– a natural choice for trying to capture the behavior due to the finite energy flux KZ spectrum. If we had chosen a different, perhaps more easily measurable, variable such as the surface velocities for capillary waves $(< u_{\underline{x}+\underline{r}} u_{\underline{x}} > \sim \int k^{5/2} n_k e^{i\underline{k}\underline{r}})$, we would have found the stripes shown in figure 5.5(a).

We note that, for α, β inside the strip defined by $2\alpha - 2 < \beta < 2\alpha + 2$ which is centered about the breakdown line, one can always find an appropriate v for which all the S_N on the KZ spectrum are universal. We call this region the *Universal High Way* for the (α, β)- plane road map.

We now briefly describe how a very similar analysis can be done for the case of four-wave resonances.

The second order structure function for the case of four wave resonances can be defined just as in equation (2.1). In this case, the Kolmogorov-

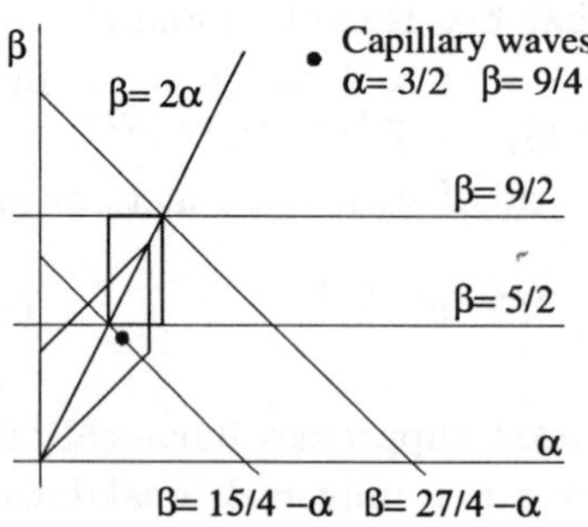

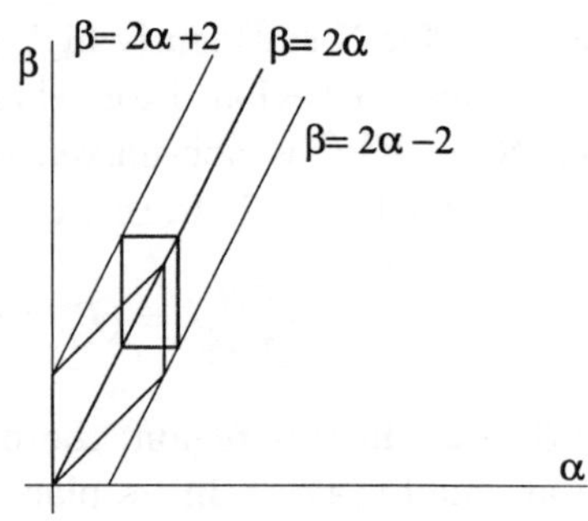

(a) Universal rhombus for surface velocity

(b) Universal Highway

Fig. 5   Figure (a) shows the strip diagram using surface velocities as the field variable. The red parallelogram is the universal rhombus in this case; the blue parallelogram is the universal rhombus from our previous calculations. Figure (b) shows the universal High Way.

Zakharov finite energy flux solution is given by $P^{\frac{1}{3}}k^{-(2/3)\gamma-d}$ where we again allow P to absorb the small parameter $\varepsilon^{12}P \to P$. The arguments following (2.1) have a direct analog in the four wave case. We find that $S_2(r)$ scales universally as

$$S_2(r) \sim P^{\frac{1}{3}}r^{\frac{2}{3}\gamma-\alpha} \tag{2.11}$$

provided α and γ lie within the strip defined by

$$0 < \frac{2}{3}\gamma - \alpha < 2. \tag{2.12}$$

In general, we will find that the term in S_N due to the Nth order moment has universal scaling

$$\mathfrak{S}_N(r) \sim P^{\frac{1}{3}(N-1)}r^{\frac{1}{3}(\frac{N}{2}+1)\gamma-\alpha} \tag{2.13}$$

provided

$$0 < \frac{1}{3}(\frac{N}{2}+1)\gamma - \alpha < N. \tag{2.14}$$

Again, the lower bound ensures that the Fourier integral in the definition of $\mathfrak{S}_N$ is independent of the high wave-number cut-off while the lower bound

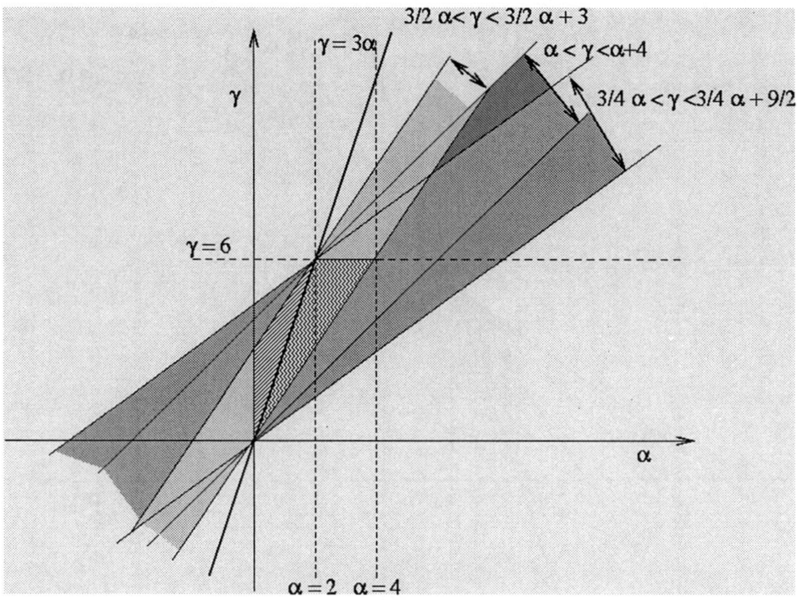

Fig. 6   Strips for the four wave energy spectrum: Each strip is the condition for some $\mathfrak{S}_N$ to scale universally. All S_N are universal in the black and white region.

ensures independence with respect to the lower cut-off.

Figure 6 shows the first three strips for $N = 2, 4, 6$ and their inter-sections. All structure functions will scale universally at small scales in $\bigcap(0 < \frac{1}{3}(N/2 + 1)\gamma - \alpha < N)$ (with the added condition that $0 < \alpha$). This region is shown in black and white in figure 6 $((3/2)\alpha + 3 < \gamma < (3/2)\alpha;$ $0 < \gamma < 6$ and $0 < \alpha < 4)$, and as regions 3 and 4 in figure 7.)

Within this (truncated) parallelogram, the ratio of structure functions gives the following corrections to Gaussianity

$$\frac{\mathfrak{S}_N}{(S_2)^{\frac{N}{2}}} = (P^{\frac{1}{3}}r^{\alpha - \frac{1}{3}\gamma})^{(\frac{N}{2} - 1)}. \tag{2.15}$$

Breakdown at small scales, $r < r_{4NL} = P^{-1/(3\alpha - \gamma)}$, occurs for $\gamma > 3\alpha$. The breakdown line, $\gamma = 3\alpha$ bisects the rhombus of universal scaling.

The point $(\alpha = 2, \gamma = 6)$ is the four wave analogue of $(\alpha = 2, \beta = 4)$ from the three-wave case. It's the unique point where all structure functions are self similar with respect to k_I and k_{NL}. Along the line $\gamma = 3\alpha$, this point separates the regions where all structure functions are universal, and

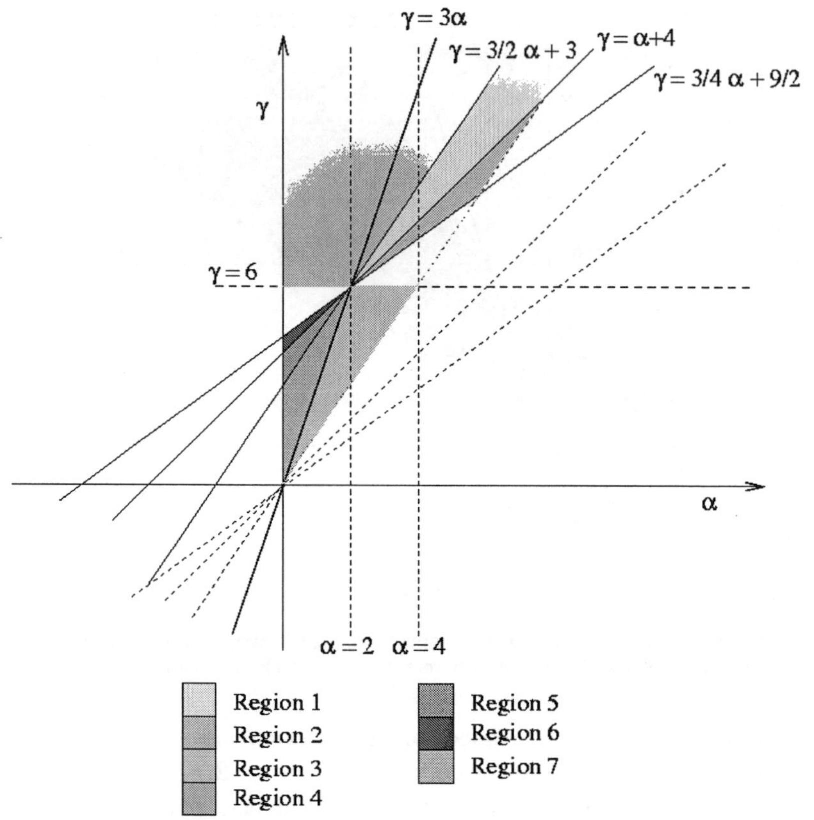

Fig. 7   Sectors for the four wave energy case. This figure shows the sections around $(2, 6)$ for which we can take $k_U \to \infty$. Different regions display different scaling behavior.

where they are all dominated by the forcing. Just as in the three wave case, we can consider the sectors surrounding this point for which only some $\mathfrak{S}_N$ are universal. We consider the regions identified in figure 7, all of which correspond to finite energy capacity.

Region 1, defined by $\alpha + 4 < \gamma < (3/2)\alpha + 3$, is characterized by $S_2 \sim P^{\frac{1}{3}} r^{(2/3)\gamma - \alpha}$ universal at small scales while $\mathfrak{S}_N \sim F^N r^N$, $N > 2$ non-universal where again, F^N represents a collection of non-universal parameters. For $N > 2$, the ratio of structure functions gives

$$\frac{\mathfrak{S}_N}{(S_2)^{\frac{N}{2}}} = F^N (P^{-\frac{1}{2}} r^{3 + \frac{3}{2}\alpha - \gamma})^{\frac{N}{3}}. \tag{2.16}$$

Since $\gamma < (3/2)\alpha + 3$, these corrections get smaller at small scales and there is no breakdown. Similar results hold for regions 2,3,...4.

Consider region 5 given by $(3/2)\alpha + 3 < \gamma < \alpha + 4$, $0 < \alpha < 2$, which lies above the breakdown region. The second order structure function scales with the non-universal forcing while those of order $N > 2$ scale universally at small scales. We have $S_2 \sim F^2 r^2$ non-universal and $\mathfrak{S}_{N>2} \sim P^{\frac{1}{3}(N-1)} r^{\frac{1}{3}(N/2+1)\gamma - \alpha}$ universal. The ratio of structure functions gives

$$\frac{\mathfrak{S}_N}{(S_2)^{\frac{N}{2}}} = (P^{-\frac{1}{3}} r^{\frac{1}{3}\gamma - \alpha})(P^2 F^{-6} r^{\gamma - 6})^{\frac{N}{6}}, \tag{2.17}$$

Again, universal breakdown in the first bracket is suppressed. A non-universal breakdown occurs for $r < (P^2 F^{-6})^{-1/(\gamma-6)}$ since $\gamma < 6$.

In region 7, the small scale behavior of all structure functions is totally determined by the forcing. $\mathfrak{S}_N \sim r^N$ for all N.

We have so far examined how the structure functions behave on the finite flux energy spectra of weak turbulence where we have taken structure functions to mean moments of Δv_r where the power spectrum of v is the spectral energy. If we had chosen v in a different way (still evaluating it on the KZ energy spectrum), the universal rhombus and all the strips would be shifted in the (α, γ) plane. We note that it is always possible to find a v such that the universal rhombus includes the point (α, β) provided this point lies in the universal high way defined by $3\alpha - 6 < \gamma < 3\alpha + 6$.

So far, we have considered the breakdown of weak turbulence on the energy spectra in the case of three and four wave interactions and we have examined the consequences of a breakdown at small scales. In the next section, we will consider the behavior of the two-point N-correlators on the wave-action spectrum (1.12) to assess deviations from the weak turbulence picture at large scales. We will consider the case of gravity waves on deep water which has the added property of infinite capacity for wave action at small wave-numbers.

3 Large r behavior of C_N

We now examine the behavior of $C_N(r)$ for large r calculating these quantities on the universal wave action /particle flux (Q) spectrum

$$n_k = c_2 Q^{\frac{1}{3}} k^{\frac{1}{3}(\alpha - 2\gamma - 3d)} \tag{3.1}$$

which obtains at large scales. We assume that long scales have infinite capacity ($\alpha < 2\gamma$) to absorb waves and for definiteness, we will focus on the case of gravity waves on deep water for which $\alpha = 1/2$ and $\gamma = 3$ and

$$n_k = c g^{\frac{1}{6}} Q^{\frac{1}{3}} k^{-\frac{23}{6}}. \tag{3.2}$$

The surface elevation $\eta(\underline{x})$ is related to $n_{\underline{k}}$ via

$$\eta(\underline{x}) = \int \sqrt{\frac{\omega_k}{2g}} \, (A^+_{\underline{k}'} + A^-_{\underline{k}'}) \, e^{i\underline{k}' \cdot \underline{x}} \, d\underline{k}' \tag{3.3}$$

where the normal canonical variables have the property, $< A^+_{\underline{k}'} A^-_{\underline{k}} > = < A^-_{\underline{k}'} A^+_{\underline{k}} > = \delta(\underline{k} + \underline{k}') n_{\underline{k}}$.

For spectra with infinite capacity, a non-stationary solution of the kinetic equation, which approaches equation (3.1) as $t \to \infty$ has the following form.

$$n_k(t) = c g^{\frac{1}{6}} Q^{\frac{1}{3}} (k_*)^{-x_0} n_0 \left(\frac{k}{k_*} \right) \tag{3.4}$$

$$= c g^{\frac{1}{6}} Q^{\frac{1}{3}} k^{-x_0} f \left(\frac{k}{k_*} \right) \tag{3.5}$$

where

$$k_*(t) = g^{\frac{1}{11}} Q^{-\frac{4}{11}} t^{-\frac{6}{11}} \tag{3.6}$$

and x_0 is the KZ exponent 23/6. The function $n_0(k/k_*)$ behaves as $(k/k_*)^{-x_0}$ for $k > k_*$ and rapidly decays to zero for $k < k_*$. The function $f(k/k_*)$ behaves as 1 for $k > k_*$ and rapidly decays to zero for $k < k_*$. The traveling front leaves in its wake the KZ spectrum (3.2) [4]. In our Fourier integral calculations of the C_N, k_* will be the low integral limit. Physically, $k_*/2\pi$ is the inverse of the largest wavelength, which at time t, has been generated nonlinearly by interactions with short waves on the stormy sea. It grows as longer and longer waves are fed by the cascade process to small wave-numbers.

We now calculate C_2 and C_4. The calculations of the large $k_* r$ behavior will assume that the front is sharp. We will thus find a weak decay for $C_2(r)$ proportional to $J_0(k_* r)$. The reason for the weak decay is that the four wave interactions produce a waveaction profile which is dominated by the longest wave $2\pi/k_*(t)$ at any given time t. If, on the other hand, the front is broad and contains many wavenumbers about k_*, the interference of these waves will lead to a steeper decay of $C_2(r)$. It is our, as yet unproven, conjecture that the front is indeed sharp and the results below are valid.

$$C_2(r) = \langle \eta(x)\eta(x+r) \rangle \tag{3.7}$$

$$= \frac{1}{g} c \int_{-\infty}^{\infty} \omega_k n_k e^{i\underline{k}\cdot\underline{r}} d\underline{k} \tag{3.8}$$

$$= c \left(\frac{Q}{g}\right)^{\frac{1}{3}} \int_{k_*}^{\infty} \int_0^{2\pi} k^{-\frac{7}{3}} e^{ikr\cos(\theta)} dk d\theta \tag{3.9}$$

$$= 2\pi c \left(\frac{Q}{g}\right)^{\frac{1}{3}} \int_{k_*}^{\infty} k^{-\frac{7}{3}} J_0(kr) dk. \tag{3.10}$$

To take account of the front width, we would replace the last integral by

$$2\pi c \left(\frac{Q}{g}\right)^{\frac{1}{3}} \int_0^{\infty} k^{-\frac{7}{3}} f\left(\frac{k}{k_*}\right) J_0(kr) dk \tag{3.11}$$

where $f(u)$ describes the transition from zero at $u < 1$ to unity for $u > 1$. A little analysis will show that if the inverse width μ of the front is greater than or equal to $k_* r$, itself large, then $f(k/k_*)$ can be replaced by the Heaviside function $H((k/k_*) - 1)$. Strictly the upper limit of integration in (3.10) and (3.11) should be at k_u, the dissipation cut-off; but because the capacity of (3.2) is finite, we can take $k_u = \infty$.

We proceed to evaluate equation (3.10) using integration by parts and find

$$C_2 = \frac{3}{2}\pi c \left(\frac{Q}{g}\right)^{\frac{1}{3}} \left(k_*^{-\frac{4}{3}} J_0(k_* r) + r^{\frac{4}{3}} \int_{k_* r}^{\infty} u^{-\frac{4}{3}} J_0'(u) du\right) \tag{3.12}$$

where the final integral remains convergent as $k_* r \to 0$. Taking the ratio with $< \eta^2(\underline{x}) >= (3/2)\pi c Q^{1/3} g^{-1/3} k_*^{-4/3}$, we have

$$\frac{< \eta(\underline{x}+\underline{r})\eta(\underline{x}) >}{< \eta(\underline{x})\eta(\underline{x}) >} = J_0(k_* r) + (k_* r)^{\frac{4}{3}} \int_{k_* r}^{\infty} u^{-\frac{4}{3}} J_0'(u) du. \tag{3.13}$$

Looking at correlations over distances much larger than the longest wave, we see that, after a little calculation,

$$\frac{< \eta(\underline{x}+\underline{r})\eta(\underline{x}) >}{< \eta(\underline{x})\eta(\underline{x}) >} \to J_0(k_* r); \tag{3.14}$$

as $k_* r \to \infty$. Correlations decay very slowly.

We now calculate a higher order cumulant to determine if this slow decay is accompanied by a strong deviation from Gaussianity. We find

$$C_4 = \frac{1}{(2g)^2} \int_{k_*} \sqrt{\omega_k \omega_{k'} \omega_{k''} \omega_{-k-k'-k''}} \; q^{(4)}(k, k', k'') e^{i\underline{k}\cdot\underline{r}} d\underline{k} d\underline{k}' d\underline{k}'' \quad (3.15)$$

where C_4 is the part of $< \eta(\underline{x} + \underline{r})\eta^3(\underline{x}) >$ which comes from the fourth order Fourier space cumulant. We find

$$C_4 = \frac{1}{2}k_*^{-2}\Big(J_0(k_*r) + (k_*r)J_0'(k_*r)$$
$$+ (k_*r)^2 J_0''(k_*r)ln(k_*r) - (k_*r)^2 \int_{k_*r} ln(w) J_0'''(w)dw. \Big) \quad (3.16)$$

Dividing by $< \eta^4(\underline{x}) >= -\frac{1}{2}k_*^{-2}$ we find

$$\frac{C_4}{< \eta^4(\underline{x}) >} \to J_0(k_*r) \quad (3.17)$$

also decays very slowly. This slow decay does not correspond to any breakdown, however, since the correction to Gaussianity at large scales remains small because

$$< \eta^3(\underline{x})\eta(\underline{x} + \underline{r}) > \; = \; 3 < \eta(\underline{x})\eta(\underline{x} + \underline{r}) >< \eta^2(\underline{x}) > +C_4(\underline{r}) \quad (3.18)$$

and the ratio of C_4 to $C_2(0)C_2(r)$ is of the order $k_*^{2/3}$ which tends to zero as t tends to infinity.

4   Non-universal corrections to the breakdown criteria

Zakharov [12] has pointed out to us that in some situations, the coupling strengths of long wave- short wave interactions may not be small enough to guarantee convergence of all the integrals which arise. A particular example is the four-wave interaction of gravity waves. In this case, $T_2 = 0$ and

$$\Omega_{2k} = \int G_{\underline{kk_1},\underline{kk_1}} n_{k_1} d\underline{k}_1 \tag{4.1}$$

$$\Omega_{4k} = \int_\Delta |G_{\underline{kk_1},\underline{k}_2\underline{k}_3}|^2 n_{k_1} n_{k_2} n_{k_3}$$
$$\times \left(\frac{1}{n_{k_1}} - \frac{1}{n_{k_2}} - \frac{1}{n_{k_3}} \right)$$
$$\times \left(-i\pi\delta(\omega + \omega_{k_1} - \omega_{k_2} - \omega_{k_3}) + PV \frac{1}{\omega + \omega_{k_1} - \omega_{k_2} - \omega_{k_3}} \right)$$
$$\times \delta(\underline{k} + \underline{k}_1 - \underline{k}_2 - \underline{k}_3) d\underline{k}_1 d\underline{k}_2 d\underline{k}_3 \tag{4.2}$$

$$T_4[n_k] = \int_\Delta |G_{\underline{kk_1},\underline{k}_2\underline{k}_3}|^2 n_k n_{k_1} n_{k_2} n_{k_3}$$
$$\times \left(\frac{1}{n_k} + \frac{1}{n_{k_1}} - \frac{1}{n_{k_2}} - \frac{1}{n_{k_3}} \right)$$
$$\times \delta(\omega + \omega_1 - \omega_2 - \omega_3)\delta(\underline{k} + \underline{k}_1 - \underline{k}_2 - \underline{k}_3) d\underline{k}_1 d\underline{k}_2 d\underline{k}_3 \tag{4.3}$$

Note that T_4 has the form $F[n_{k_1}, n_{k_2}, n_{k_3}] - n_k Im\Omega_{4k}$ so that $(Im\Omega_{4k})^{-1}$ is a relaxation time scale for the redistribution of energy. In equations (4.1),(4.2) and (4.3), the coefficient $G_{\underline{kk_1},\underline{k}_2\underline{k}_3}$ is $L^{++--}_{\underline{kk_1},\underline{k}_2\underline{k}_3}$ from which quadratic products of $L^{+s_1s_2}_{\underline{kk_1k_2}}$ are subtracted. The integration region Δ in equations (4.2) and (4.3), after angle averaging, is that region of the ω_{k_1}, ω_{k_2} plane defined by $\omega_{k_2} > 0$, $\omega_{k_3} > 0$ and $\omega_{k_1} = \omega_{k_2} + \omega_{k_3} - \omega_k > 0$. The neighborhood of $\omega_{k_1} = \omega_{k_2} = \omega_{k_3} = \omega_k$ is that of ultra-local interactions while the neighborhood of $\omega_{k_1} = \omega_{k_3} = 0$, $\omega_{k_2} = \omega_k$ or $\omega_{k_1} = \omega_{k_2} = 0$, $\omega_{k_3} = \omega_k$ correspond to long wave- short wave interactions for which the group velocity of the short wave at k in the direction of $\underline{k}_1 - \underline{k}_3$ is equal to the phase velocity of the long wave second harmonic $exp(i(\underline{k}_1 - \underline{k}_3)\underline{x} - i(\omega_{\underline{k}_1} - \omega_{\underline{k}_3})t)$.

For gravity waves, $\omega_k \sim (gk)^{\frac{1}{2}}$, $\gamma = 3$ and $d = 2$ so that equation (1.11) becomes $n_k = C_2 P^{\frac{1}{3}} k^{-4}$. The coefficient $G_{\underline{kk_1}\underline{k}_2\underline{k}_3}$ behaves as kk_1^2 for k_2 close to k and $k_1, k_3 \ll k, k_2$. A little calculation shows that Ω_{2k} diverges logarithmically near $k_1 = 0$ and is given by $2\pi C_2 P^{\frac{1}{3}} k ln(k_I/k)$ where k_I is some infrared cutoff at the forcing scale. The frequency correction $Im\Omega_{4k}$ (due to the term containing $n_{k_1} n_{k_3}$) is proportional to $(P^{2/3}k^{3/2})/g^{\frac{1}{2}}(k_I/k)^{-\frac{1}{2}}$. The ratios Ω_{2k}/ω_k and $Im\Omega_{4k}/\omega_k$ are $(P^{\frac{1}{3}}k^{\frac{1}{2}}/g^{\frac{1}{2}})ln(k_I/k)$ and $(P^{2/3}k/g)(k_I/k)^{-\frac{1}{2}}$ in contrast to $(P^{\frac{1}{3}}k^{\frac{1}{2}}/g^{\frac{1}{2}})$ and $P^{2/3}k/g$ if there were no divergences. As a consequence of the extra factor, the revised breakdown wave number, k_{LS}, is still large but less than $k_{NL} = gP^{-2/3}$ calculated before. The collision integral, T_4, on the other hand, has no such divergence because of the cancellation between the terms $n_{k_1} n_{k_2} n_{k_3}$ and $-n_k n_{k_1} n_{k_3}$ in the integrand.

The open questions are: what is the consequence of the fact that $Im\Omega_{4k}/\omega_k$ becomes of order unity at $k_{LS} = g^{\frac{2}{3}} k_I^{\frac{1}{3}} P^{-\frac{4}{9}}$ rather than at $k_{NL} = gP^{-\frac{2}{3}}$, the point at which the ratio $T_4/\omega_k n_k$ becomes unity? Is the ratio T_8/T_4 of order $P^{2/3}k/g$ or $(P^{2/3}k/g)(k/k_I)^{1/2}$ reflecting the ratio Ω_{4k}/ω_k? Can there be worse divergences?

To attempt to answer these questions, we have carried out calculations in order to determine the long time behaviors of the next corrections $T_6[n_k]$ and $T_8[n_k]$ in the collision integral asymptotic expansion, equation (1.5). In $T_8[n_k]$, there arise two types of secular terms. The first involve interactions which are genuine compositions of four wave interactions (e.g. $\underline{k}, \underline{k}_1 \to \underline{k}_4, \underline{k}_5 \to \underline{k}_6, \underline{k}_7 \to \underline{k}_2, \underline{k}_3$ which comes from the two loop correction to the fourth order cumulant) that lead to nontrivial energy exchange. The ratio of these terms in T_8/T_4 is $P^{2/3}k/g$. The second involve modal interactions which are simply the next terms in the expansion of the Dirac delta function $\delta(\omega_k + \omega_{k_1} - \omega_{k_2} - \omega_{k_3})$ in T_4 with the frequency ω_k replaced by its renormalized value; namely

$$\omega_k \to \omega_k + \Omega_{2k} + \Omega_{4k} + \cdots \tag{4.4}$$

with Ω_{2k}, Ω_{4k} given by equations (4.1) and (4.2) respectively.

Accordingly, T_6 and T_8 contain (respectively) the combinations

$$(\Omega_{2k} + \Omega_{2k_1} - \Omega_{2k_2} - \Omega_{2k_3})\delta'(\omega_k + \omega_{k_1} - \omega_{k_2} - \omega_{k_3})$$

and

$$(\Omega_{4k} + \Omega_{4k_1} - \Omega_{4k_2} - \Omega_{4k_3})\delta'(\omega_k + \omega_{k_1} - \omega_{k_2} - \omega_{k_3})$$
$$\frac{1}{2}(\Omega_{2k} + \Omega_{2k_1} - \Omega_{2k_2} - \Omega_{2k_3})^2\delta''(\omega_k + \omega_{k_1} - \omega_{k_2} - \omega_{k_3}) \tag{4.5}$$

in their integrands in place of the $\delta(\omega_k + \omega_{k_1} - \omega_{k_2} - \omega_{k_3})$ appearing in T_4. There are two observations we can now make. The first is that the ratios T_6/T_4 and T_8/T_4 will be at least of orders $(P^{1/3}k^{1/2}/g^{1/2})\ln(k/k_I)$ and $(P^{2/3}k/g)(k/k_I)^{1/2}$ respectively as the combinations $\Omega_{jk} + \Omega_{jk_1} - \Omega_{jk_2} - \Omega_{jk_3}$, $j = 1, 2$ will not in general vanish on the resonant manifold. On the other hand, it will vanish on that part of the manifold where k_1, k_3 are small and k, k_2 are large. Therefore, there will be no worse divergences which might otherwise arise because the derivative delta function will act as $\delta(\omega_k + \omega_{k_1} - \omega_{k_2} - \omega_{k_3})\frac{\partial}{\partial k_3}$ (one can show that $\delta(\omega_k + \omega_{k_1} - \omega_{k_2} - \omega_{k_3}) \simeq \delta(\omega_{k_1} - \omega_{k_3} + O(\underline{k}_1 - \underline{k}_3)) \simeq \delta(g^{1/2}(k_1 - k_3)/(k_1^{1/2} + k_3^{1/2}))$. The potential divergence resulting from the term $n_1 n_3 \frac{\partial n_2}{\partial k_2}\frac{\partial k_2}{\partial k_3}$ is killed by the vanishing of $\Omega_{jk} + \Omega_{jk_1} - \Omega_{jk_2} - \Omega_{jk_3}$ near $k_1, k_3 = 0$.

Physically, the origin of these non-local effects is the interaction between the given wave vectors $\underline{k}$ and a nearby wave vector $\underline{k}_2 = \underline{k} + \underline{k}_1 - \underline{k}_3$ and a family of long waves which are the first superharmonic $exp(i(\underline{k}_1 - \underline{k}_3)\underline{x} - i(\omega_{k_1} - \omega_{k_3})t)$ of the long waves $\underline{k}_1, \underline{k}_3$. The resonance criterion is that $\omega_{k_1} - \omega_{k_3} = \omega(\underline{k} + \underline{k}_1 - \underline{k}_3) - \omega(\underline{k}) \simeq |\underline{k}_1 - \underline{k}_3|\omega'_k cos\theta$ where ω'_k is the group velocity of the short wave packet and θ is the angle between $\underline{k}$ and the long wave superharmonic wave vector $\underline{k}_1 - \underline{k}_3$. For short enough waves $k > k_{LS}$ ($> k_{NL}$), the short wave packets remain attached (phase locked) to the long wave for a sufficiently long time for energy exchange to take place. The resonance does not have to be exact. All wave vectors $\underline{k}, \underline{k}_1, \underline{k}_2 = \underline{k} + \underline{k}_1 - \underline{k}_2, \underline{k}_3$ in a finite window of width $Im\Omega_k$ around the resonant locus will participate.

Where are such effects manifested in the structure functions? We conjecture, but have not yet completely proved, that they arise when we calculate the relative magnitudes of successive surviving terms in each of the cumulants. For example, in $\mathfrak{S}_4$, the ratio of the next surviving term to the one we have calculated in (2.13) for $N = 4$ will exhibit the same divergence that appears in $Im\Omega_{4k}/\omega_k$ except that it will be manifested in physical space and lead to a revision of the breakdown scale from $r_{NL} = k_{NL}^{-1}$ to $r_{LS} = k_{LS}^{-1}$. It is important to emphasize, therefore, that the results of section 2 only obtain when coupling coefficients are sufficiently local.

Our conclusion is then that the wave-number range of validity of wave turbulence may be further shortened by the effects of non-local, long wave -short wave interactions. How should we interpret this behavior? The imaginary part of Ω_{4k} (and higher order corrections) can be interpreted as a broadening of the resonant manifold. Quartets of wave vectors $\underline{k}, \underline{k}_1\underline{k}_2\underline{k}_3$ in an $(Im\Omega_{4k})$ neighborhood of the resonant manifold (shifted from the one based on the linear dispersion relation because of real frequency corrections) play a role in the energy exchange to and from wave-numbers k greater than k_{LS} (where $(P^{2/3}k/g)(k/k_I)^{1/2}$ is unity) given by $(gP^{-2/3})^{2/3}k_I^{1/3}$ or $k_{NL}^{2/3}k_I^{1/3}$ for gravity water waves. For the MMT model [8], there is a stronger long wave-short wave coupling as $|G_{\underline{kk}_1\underline{k}_2\underline{k}_3}|^2 \sim k^3 k_1^{3/2} k_3^{3/2}$. This leads to a k_{LS} of $k_{NL}^{2/5} k_I^{3/5}$ and a much shorter wave-number range for which the wave turbulence approximation is valid.

The evidence from experiments, numerical and physical, suggest that the KZ spectrum $C_2 P^{\frac{1}{3}} k^{-4}$ remains valid all the way either to the surface tension scale, $K_0 = \sqrt{g/\sigma}$, where $\sigma = S/\rho_w$, S is surface tension and ρ_w is the density of water, or to $k_{NL} = g P^{-2/3}$, whichever is less [10; 6]. k_{NL}, as we have noted in [1], is precisely the point that the KZ spectrum intersects the Phillips spectrum, $n_k = C_3 g^{1/2} k^{-9/2}$, which corresponds to derivative discontinuities in the ocean surface, nearly whitecaps. The criterion for whitecaps is thus $k_{NL} < K_0$ which translates into $P > P_c = (\sigma g)^{3/4}$ or a wind speed of $V = (\rho_w/\rho_a)^{\frac{1}{2}} P^{\frac{1}{3}}$ (ρ_a is the density of air) greater than $6 m/s$ [10]. (One gets exactly the same criterion if one calculates k_{NL} for surface tension waves which exchange energy by three wave resonances and compares k_{NL} to K_0.) On the other hand, there is no evidence for the KZ spectrum in the MMT experiments.

Similar calculations apply to three wave interactions. However, in most situations (and definitely if the system contains only one additional dimensional parameter, e.g. surface tension waves [3]) $\beta < 2\alpha$ and so breakdown occurs at low values of k. In either case, the long wave- short wave, nonlocal interaction which might cause the multiplying integrals to diverge involve a small k and large k_1, k_2 close to one another. The resonant condition becomes $\underline{k} \cdot \nabla_{\underline{k}_2} \omega = \omega_{\underline{k}}$. However, for large k_2, the integrals would appear, in almost all cases, to converge.

To date we have left out the question of how the cut off wave-number k_I is chosen. For wind generated gravity waves, one might choose k_I to be the wave-number at which energy is injected. This is imprecise as energy is transferred over a large range of scales. Moreover, from four wave resonances with the collision integral (4.3), there is a secondary inverse cascade of wave-action (particles) from short waves to long waves. This has been discussed in section 3. If we choose k_I to be the front $k_*(t)$ of the long wave spectrum, namely the point to which wave action has been transferred after time t by wave-wave interactions, we also have to take into account that the particle flux spectrum is less steep and therefore the long wave-short wave wave-number k_{LS} is closer to k_{NL}. For water waves, the ratio $Im\Omega_{4k}/\omega_k$ is $(g^{1/3} Q^{2/3} k^{3/2}/g) k_I^{-1/6} \equiv (P^{2/3} k/g)(k/\bar{k})^{1/3}(k/k_I)^{1/6}$ where we have written the particle flux Q as $P/\bar{\omega}$ and $\bar{\omega} = \sqrt{g\bar{k}}$ where $\bar{k}$ is some injection wave-number which is not necessarily small. We now see that $k_{LS} = (gP^{-2/3})^{2/3}\bar{k}^{2/9}k_I^{1/9}$. It is not unreasonable to take $\bar{k} \sim k_{NL} << k_I$. If indeed we take $\bar{k}$ to be of the same order as (but numerically less than) k_{NL}, then $k_{LS} = k_{NL}(k_I/k_{NL})^{1/9}$. The dependence on the cutoff is very weak. This may well account for the fact that the KZ spectrum is observed over such a large range.

Definite conclusions concerning the validity of the wave turbulence approximations are further inhibited by the manner in which the KZ spectrum is realized. In finite capacity situations for which $\int \omega_k n_k d\underline{k} \sim \int k^{\alpha - 2\gamma/3 - 1} dk < \infty$ (for water waves and MMT $\alpha = 1/2$, $\gamma = 3$), a companion paper in this volume [4] points out that the stationary spectrum is achieved in a nontrivial manner. For a finitely supported initial power spectrum, the energy spreads to higher wave-numbers at an accelerating rate behind a moving front $k_* = (t_* - t)^b$, where $b = -(2(x - x_0) - (2\gamma - 3\alpha)/3)$ where $x_0 = 2\gamma/3 + d$. In the wake of this front lies a steeper than KZ spectrum $n_k \sim k^{-x}$ where $x = 2\gamma/3 + d + (2\gamma - 3\alpha)/12$ which, for water waves and MMT is $4\frac{3}{8}$. Only when $t > t_*$, after the spectrum reaches $k = k_d$, the small scale at which dissipation occurs, can a finite flux P of energy begin to set up the KZ spectrum. It does so as the wake of a backward traveling front which invades the region of steeper slope. One might ask how it manages to achieve the KZ spectrum if either k_{NL} or k_{LS} is smaller than k_d and the backward moving front starts out in the fully nonlinear region.

We know no answer for this at present. While the results presented here have added some new insights to the delicate questions concerning the applicability of wave turbulence, it is clear that the subject is very much alive and that there are still many open and intriguing challenges.

Acknowledgments

The authors thank Sergey Nazarenko for constant help throughout this work; Oleg Zaboronski for frequent and useful discussions and Volodja Zakharov for several crucial insights. The authors are also grateful to the Erwin Schrödinger Institute for hospitality during the latter stages of this work. This work was partially supported by NSF Grant DMS 0072803.

References

Biven, L., Newell, A.C., and Nazarenko, S.V. (2001). Breakdown of weak turbulence and the onset of intermittency. *Physics Letters A*, **280** (1–2), 28–32.

Cai, D. Majda, A., McLaughlin, D., and Tabak, E. (2001) Dispersive wave turbulence in one dimension. *Physica D*, **152–153**, 551–572

Connaughton, C., Nazarenko, S.V., and Newell, A.C. (2003) Dimensional Analysis and Weak Turbulence. *To appear in Physica D.*

Connaughton, C., Newell, A.C., and Pomeau, Y., (2003) Nonstationary spectra of local wave turbulence *To appear in Physica D.*

Dyachenko, A., Newell, A.C., Pushkarev, A., and Zakharov, V.E. (1992) Optical turbulence: weak turbulence, condensates and collapsing filaments in the nonlinear Schrödinger equation *Physica D*,**57**, 96–160.

Forristall, G.Z. (1981) *J. Geophys. Res.*, **86**, 8075–8084.

Frisch,U., (1995). Turbulence. *Cambridge University Press.*

Majda, A.J., McLaughlin, D.W., and Tabak, E.G. (1997) *J. Nonlinear Sci* **6** 9–44

Newell,A.C., Nazarenko, S.V., and Biven, L., (2001). Wave Turbulence and Intermittency. *Physica D*, **152/153**, 520–550.

Newell, A.C., and Zakharov, V.E., (1992). Rough Sea Foam *Phys. Rev. Lett.*, **69** (8), 1149.

Zakharov,V.E., L'vov, V.S., and Falkovich, G. (1992) Kolmogorov Spectra of Turbulence vol. 1. *Springer-Verlag.*

Zakharov, V.E. *private communication*

Chapter 6

Renormalized Closure Theory and Subgrid-scale Parameterizations for Two-Dimensional Turbulence

Jorgen S. Frederiksen[1]

CSIRO Atmospheric Research, Aspendale 3195, Australia

Abstract.   In this chapter, recent developments in renormalized closure theory and its application for developing subgrid-scale parameterizations are reviewed, focussing on two-dimensional turbulence. A derivation of the direct interaction approximation closure (DIA) based on the barotropic vorticity equation for two-dimensional turbulence is presented. Because the vorticity is a scalar, the derivation presented is particularly simple as it avoids the tensor notation needed if the closure is based on the velocity fields for Navier Stokes flows. The formulation applies perturbation theory followed by a heuristic renormalization of the propagators, the second-order cumulant and response function, with the vertex function being unrenormalized or bare and equal to the interaction coefficient. A variant of the DIA closure, termed the regularized DIA closure (RDIA) that includes an empirical vertex renormalization is also considered. The performance of the closures is compared with ensemble averaged direct numerical simulations (DNS) and their application for developing subgrid-scale parameterizations of renormalized viscosity and stochastic backscatter is discussed. Such parameterizations are needed for large-eddy simulations (LES) in which simulations are carried out at relatively coarse resolution to reduce computational costs and focus on the large-scale features of the flows. Kinetic energy spectra of LES with dynamical subgrid-scale parameterizations have been found to compare closely with those of DNS.

[1]e-mail: jorgen.frederiksen@csiro.au

225

1   Introduction

The thickness of the earth's troposphere is approximately one thousandth of the diameter of the earth and, in relative terms, may be compared with the skin of an apple. Large-scale motions in the atmosphere are quasi-geostrophic, with an approximate balance between Coriolis and pressure forces, and in many respects share properties with two-dimensional turbulence. The kinetic energy spectrum of atmospheric transients has a -3 inertial range power law, characteristic of the enstrophy cascading inertial range of two-dimensional turbulence, for total wavenumbers ranging from about 10 to 100. This corresponds to length scales from about 2000 to 200 km, with internal gravity wave and three-dimensional effects becoming important at smaller scales (Nastrom and Gage, 1985; Koshyk et al., 1999). The first numerical weather predictions 50 years ago were carried out with the barotropic vorticity equation, which describes two-dimensional turbulence, but also including the beta-effect due to the earth's differential rotation and a long-wave stabilization factor.

A major difficulty in developing prognostic equations describing the statistics of turbulence, represented for example by the Navier Stokes equations with quadratic nonlinearity, is that the resulting equation for the moment of a given order involves the next higher order moment. The problem is how to close the infinite hierachy of moment or cumulant equations in terms of the lower order terms. Kraichnan (1959a) made a major contribution to the development of renormalized closure theory with his formulation of the Eulerian DIA which he based on assumptions of "maximal randomness" and "weak dependence" of the field variables. The DIA equations were subsequently derived by more formal mathematical approaches that were previously developed for quantum field theories. Wyld (1961) used a Feynman diagram approach to derive the DIA through a partial summation of a perturbation series in the interaction coefficient. Martin et al. (1972) generalized the Schwinger-Dyson functional formalism to apply to classical field theories and showed that the DIA corresponds to a bare vertex approximation. Elegant and powerful path integral formalisms allowing generalization to non-Gaussian initial conditions, multiplicative random forces and nonlocal interactions were developed by Phythian (1977) and Jensen (1981).

The Eulerian DIA contains no arbitrary parameters and is in satisfactory agreement with simulations in the energy containing range of the large scales. However, at high Reynolds numbers, it leads to power laws which differ slightly from the $k^{-\frac{5}{3}}$ energy and k^{-3} enstrophy cascading inertial ranges (Kraichnan, 1964a; Herring et al., 1974). In some problems such as plasma transport calculations (Bowman et al., 1993) or large-scale atmospheric flows (Frederiksen and Davies, 1997) the interest is primarily with the energy containing eddies and inertial range deficiencies may not be terribly important. In this chapter, we shall also consider two other closure models, the regularized DIA (RDIA) (Kraichnan, 1964b; Frederiksen and Davies, 2002) and eddy-damped quasi-normal Markovian model (EDQNM) (Orszag, 1970) which do yield the correct inertial ranges. Both of these closures involve one specified parameter and may be established as modifications of the Eulerian DIA.

By far the majority of closure calculations have been formulated for continuous wavenumbers and compared with discrete wavenumber formulations of direct numerical simulations (DNS) (McComb, 1990 reviews the literature). This has the numerical advantage that it is possible to use logarithmic discretization of wavenumber space in the closures (Kraichnan, 1964b; Leith, 1971; Leith and Kraichnan, 1972) to perform high wavenumber calculations. However, it is not easy to determine whether differences between the closures and DNS are intrinsic or due to the different wavenumber formulations or discretizations. In addition, logarithmic discretization of wavenumber space may remove non-local interactions (Leith, 1971) that may be particularly important for two-dimensional turbulence. Recently, discrete wavenumber non-Markovian closures were implemented and compared with DNS for two-dimensional turbulence (Frederiksen and Davies, 2000, 2002). This results in a much larger computational task but ensures that any differences between the closures and DNS are intrinsic differences. Studies were made with the DIA and RDIA closures and as well with the related self-consistent field theory (SCFT) (Herring, 1965) and local energy-transfer theory (LET) (McComb, 1974; McComb et al., 1992). The SCFT and LET closures have separate historical developments (McComb, 1990) but in retrospect may be formally obtained from the DIA by invoking the fluctuation-dissipaton theorem (FDT) (Kraichnan, 1959b; Leith, 1975; Carnevale and Frederiksen, 1983a) out of strict canonical equilibrium. The discrete wavenumber closures of Frederiksen and Davies (2000) were also found to perform considerably better than the continuous wavenumber closures of Herring et al. (1974) in comparison with DNS for runs starting from very similar initial spectra.

In general, non-Markovian closures may result in a large computational task because they are integro-differential equations with potentially long time-history integrals. Frederiksen and Davies (2000, 2002) also implemented at high resolution an approximate method of calculating the long time-history integrals much more efficiently. This method follows the low resolution works of Rose (1985) and Frederiksen et al. (1994) who developed cumulant update closures in which the problem of the long time-history integrals is overcome by periodically stopping the time integrations, calculating the third-order cumulant and using this in the new non-Gaussian initial conditions for subsequent integrations.

Closure models may also be used for determining self-consistent subgrid-scale parameterizations needed in LES at relatively coarse resolution to successfully simulate the same large-scale flow structures as in higher resolution DNS (Frederiksen and Davies, 1997; Frederiksen, 1999, and references therein). Here we also review recent developments in the application of closure theory for deriving subgrid-scale parameterizations of renormalized viscosity and stochastic backscatter.

The plan of this chapter is as follows. In Section 2, we present the barotropic vorticity equation for forced dissipative two-dimensional turbulence in both physical and spectral space and for the case of flows on the doubly periodic domain. We also present the equation for the response function which measures the change in the vorticity due to an infinitessimal change in the force. In Section 3, we outline a simplified derivation of the DIA closure equations for homogeneous two-dimensional turbulence with zero mean field based on perturbation theory and a heuristic renormalization procedure following Kraichnan (1959a). The DIA and RDIA equations specialized to isotropic turbulence are presented in Section 4 in a suitable way for subsequent development of subgrid-scale parameterizations. Section 5 defines a number of diagnostic kinematic quantities which are used in Section 6 where the closure results are compared with DNS for a range of large-scale Reynolds numbers between ≈ 50 and ≈ 4000. In Section 7, we consider the barotropic vorticity equation in spherical geometry and note the relationship between the DIA closure for isotropic turbulence on the sphere and on the doubly periodic domain. Section 8 describes how the integro-differential equations for the DIA closure can be reduced to the simpler differential equations for the EDQNM closure by using the FDT and at the cost of introducing a parameter specifying the strength of the eddy damping in the response function. In Section 9, we consider the problem of establishing self-consistent subgrid-scale parameterizations based on DIA and EDQNM closures. Section 10 describes comparisons of LES, employing renormalized drain viscosity together with stochastic backscatter parameterizations or renormalized net viscosity, with the results of higher resolution DNS. The conclusions are summarized in Section 11.

2 Barotropic vorticity equation

The barotropic vorticity equation for two-dimensional turbulence takes the form

$$\frac{\partial \zeta}{\partial t} \;=\; -J(\psi, \zeta) - D_0 + f^0 \,, \tag{2.1a}$$

where ζ is the vorticity, ψ is the streamfunction, D_0 is the bare dissipation and f^0 is the bare forcing. The vorticity and stream function are related by

$$\zeta \;=\; \nabla^2 \psi \,. \tag{2.1b}$$

The Jacobian

$$J(\psi, \zeta) \;=\; \frac{\partial \psi}{\partial x}\frac{\partial \zeta}{\partial y} - \frac{\partial \psi}{\partial y}\frac{\partial \zeta}{\partial x} \,, \tag{2.1c}$$

and the winds in the x and y directions are

$$u \;=\; -\frac{\partial \psi}{\partial y} \,; \quad v \;=\; \frac{\partial \psi}{\partial x} \,. \tag{2.1d}$$

Also, the bare dissipation is related to ν_0 the bare viscosity, through

$$D_0 \;=\; -\nu_0 \nabla^2 \zeta \,. \tag{2.1e}$$

For simplicity, we consider flow on a doubly periodic domain. We can make the equations non-dimensional by introducing suitable length and time scales. In particular we consider flows on the domain $0 \le x \le 2\pi, 0 \le y \le 2\pi$. The equations are most conveniently analysed and solved in spectral space. We expand each of the functions in a Fourier series; for example

$$\zeta(\mathbf{x}, t) = \sum_{\mathbf{k}} \zeta_{\mathbf{k}}(t) exp(i\mathbf{k}.\mathbf{x}) \,, \tag{2.2a}$$

where

$$\zeta_{\mathbf{k}}(t) \;=\; (2\pi)^{-2} \int_0^{2\pi} d^2\mathbf{x}\,\zeta(\mathbf{x},t)exp(-i\mathbf{k}.\mathbf{x})\,, \qquad (2.2b)$$

and $\mathbf{x} = (x,y)$ and $\mathbf{k} = (k_x, k_y)$. Then multiplying Eq. (2.1a) by $exp(-i\mathbf{k}.\mathbf{x})$ and integrating over the (x,y) domain we find that

$$\left(\frac{\partial}{\partial t} + D_0(k)\right)\zeta_{\mathbf{k}}(t) \;=\; \lambda \sum_{\mathbf{p}} \sum_{\mathbf{q}} \delta(\mathbf{k}+\mathbf{p}+\mathbf{q})\, K(\mathbf{k},\mathbf{p},\mathbf{q})\, \zeta_{-\mathbf{p}}(t)\, \zeta_{-\mathbf{q}}(t)$$
$$+\, f_{\mathbf{k}}^0(t)\,, \qquad (2.3a)$$

where

$$D_0(k) \;=\; \nu_0 k^2\,. \qquad (2.3b)$$

In Eq. (2.3a) we have introduced the ordering parameter λ which is equal to 1, but which we shall use to order terms in perturbation expansions in Section 3. The interaction coefficient or bare vertex function is given by

$$K(\mathbf{k},\mathbf{p},\mathbf{q}) \;=\; \frac{1}{2}\left(\frac{p^2 - q^2}{p^2 q^2}\right)(p_x q_y - p_y q_x)\,, \qquad (2.4a)$$

$$k \;=\; (k_x^2 + k_y^2)^{\frac{1}{2}}\,, \qquad (2.4b)$$

$$\delta(\mathbf{k}+\mathbf{p}+\mathbf{q}) \;=\; \begin{cases} 1 & \text{if } \mathbf{k}+\mathbf{p}+\mathbf{q} = 0\,, \\ 0 & \text{otherwise}\,. \end{cases} \qquad (2.4c)$$

We note that the interaction coefficient satisfies the symmetry properties

$$K(\mathbf{k},\mathbf{p},\mathbf{q}) \;=\; K(\mathbf{k},\mathbf{q},\mathbf{p})\,, \qquad (2.5a)$$

$$K(\mathbf{k},\mathbf{p},\mathbf{q}) + K(\mathbf{p},\mathbf{q},\mathbf{k}) + K(\mathbf{q},\mathbf{k},\mathbf{p}) \;=\; 0\,. \qquad (2.5b)$$

In developing our perturbation theory, we shall also need to consider the response function

$$\hat{R}_{\mathbf{k}}(t,t') \;=\; \frac{\delta \zeta_{\mathbf{k}}(t)}{\delta f_{\mathbf{k}}^0(t')}\,, \qquad (2.6)$$

which measures the change in the vorticity due to an infinitessimal change in the force. Suppose in Eq. (2.3a)

$$f_{\mathbf{k}}^0(t) \;\longrightarrow\; f_{\mathbf{k}}^0(t) + \delta f_{\mathbf{k}}^0(t)\,, \qquad (2.7a)$$

and this effects a change in the vorticity

$$\zeta_{\mathbf{k}}(t) \quad \longrightarrow \quad \zeta_{\mathbf{k}}(t) + \delta\zeta_{\mathbf{k}}(t) \ . \tag{2.7b}$$

Then, for infinitessimal perturbations we have

$$\left(\frac{\partial}{\partial t} + D_0(k)\right)\delta\zeta_{\mathbf{k}}(t) =$$

$$2\lambda \sum_{\mathbf{p}} \sum_{\mathbf{q}} \delta(\mathbf{k}+\mathbf{p}+\mathbf{q}) \, K(\mathbf{k},\mathbf{p},\mathbf{q}) \, \delta\zeta_{-\mathbf{p}}(t) \, \zeta_{-\mathbf{q}}(t) + \ \delta f_{\mathbf{k}}^0(t) \ . \tag{2.8}$$

Thus, the response function satisfies the equation

$$\left(\frac{\partial}{\partial t} + D_0(k)\right)\hat{R}_{\mathbf{k}}(t,t') =$$

$$2\lambda \sum_{\mathbf{p}} \sum_{\mathbf{q}} \delta(\mathbf{k}+\mathbf{p}+\mathbf{q}) \, K(\mathbf{k},\mathbf{p},\mathbf{q}) \, \hat{R}_{-\mathbf{p}}(t,t') \, \zeta_{-\mathbf{q}}(t) + \ \delta(t-t') \ , \tag{2.9}$$

where $\delta(t-t')$ is the Dirac delta function. The solution to Eq. (2.8) is

$$\delta\zeta_{\mathbf{k}}(t) \ = \ \int_{t_0}^{t} ds \, \hat{R}_{\mathbf{k}}(t,s) \, \delta f_{\mathbf{k}}^0(s) \ . \tag{2.10}$$

3 Perturbation theory and the closure problem

In developing statistical prognostic equations from dynamical equations with quadratic nonlinearity one faces the problem of closure. The equation for the statistical mean field involves the second-order moment, the second-order moment equation involves the third-order moment etc., resulting in an infinite hierarchy of moment equations. Millionshtchikov (1941) proposed expressing the fourth-order moment in terms of the fourth-order cumulant and lower order moments, setting the fourth-order cumulant to zero and thus closing the hierarchy. This procedure is known as the quasi-normal approximation. It was however shown by Ogura (1963) that the quasi-normal approximation can lead to unphysical negative energy spectra.

Kraichnan (1959a) made a major breakthrough in developing statistical closure theories for turbulence with the formulation of his direct interaction approximation (DIA) for equations having quadratic nonlinearity. He derived his closure theory on the basis of perturbation theory and a heuristic renormalization procedure. The DIA closure was subsequently derived by more formal mathematical methods that have their origins in quantum field theory as discussed in the Introduction. In the DIA closure the propagators, the second-order cumulant and response function, are renormalized but the vertex function is unaltered from its bare value.

In this section we derive the DIA closure theory for two-dimensional turbulence based on the vorticity equation following a simplified approach along the lines of Kraichnan (1959a). We consider homogeneous turbulence in which the mean field

$$\langle\, \zeta_{\mathbf{k}}(t)\, \rangle \;=\; 0\,, \tag{3.1a}$$

and the second-order moment and cumulant are identical and diagonal

$$\begin{aligned}
\langle\, \zeta_{\mathbf{k}}(t)\,\zeta_{-\mathbf{l}}(t')\, \rangle \;&=\; \delta_{\mathbf{kl}}\,\langle\, \zeta_{\mathbf{k}}(t)\,\zeta_{-\mathbf{k}}(t')\, \rangle \\
&=\; \delta_{\mathbf{kl}}\, C_{\mathbf{k}}(t,t')\,.
\end{aligned} \tag{3.1b}$$

We expand the vorticity and response function in perturbation series

$$\zeta_{\mathbf{k}}(t) \;=\; \zeta_{\mathbf{k}}^{(0)}(t) + \lambda\,\zeta_{\mathbf{k}}^{(1)}(t) + \dots\,, \tag{3.2a}$$

$$\hat{R}_{\mathbf{k}}(t,t') \;=\; \hat{R}_{\mathbf{k}}^{(0)}(t,t') + \lambda\,\hat{R}_{\mathbf{k}}^{(1)}(t,t') + \dots\,. \tag{3.2b}$$

Then, from Eqs. (2.3a) and (2.9) we have to zero order in λ

$$\left(\frac{\partial}{\partial t} + D_0(k)\right)\zeta_{\mathbf{k}}^{(0)}(t) \;=\; f_{\mathbf{k}}^0(t)\,, \tag{3.3a}$$

$$\left(\frac{\partial}{\partial t} + D_0(k)\right)\hat{R}_{\mathbf{k}}^{(0)}(t,t') \;=\; \delta(t-t')\,, \tag{3.3b}$$

where we assume that $f_{\mathbf{k}}^0(t)$ and $\zeta_{\mathbf{k}}^{(0)}(t)$ have zero means and are Gaussianly distributed. This leads to some important properties of the third- and fourth-order moments which we shall use in the perturbation analysis. Let $X_i, i = 1,4$ be variables having these same zero mean Gaussian properties; then

$$\langle X_1 X_2 X_3\rangle_G \;=\; 0\,, \tag{3.4}$$

$$\langle X_1 X_2 X_3 X_4\rangle_G =$$
$$\langle X_1 X_2\rangle\langle X_3 X_4\rangle + \langle X_1 X_3\rangle\langle X_2 X_4\rangle + \langle X_1 X_4\rangle\langle X_2 X_3\rangle\,. \tag{3.5}$$

From Eqs. (3.3a) and (3.3b),

$$\hat{R}_{\mathbf{k}}^{(0)}(t,t') \;=\; exp[-D_0(k)(t-t')] \;=\; exp[-\nu_0 k^2(t-t')]\,, \qquad (3.6)$$

$$\zeta_{\mathbf{k}}^{(0)}(t) \;=\; \int_{t_0}^{t} ds\, \hat{R}_{\mathbf{k}}^{(0)}(t,t')\, f_{\mathbf{k}}^0(s)\,. \qquad (3.7)$$

To first order in λ we have from Eq. (2.3a) that

$$\left(\frac{\partial}{\partial t} + D_0(k)\right)\zeta_{\mathbf{k}}^{(1)}(t) =$$
$$\sum_{\mathbf{p}}\sum_{\mathbf{q}} \delta(\mathbf{k}+\mathbf{p}+\mathbf{q})\, K(\mathbf{k},\mathbf{p},\mathbf{q})\, \zeta_{-\mathbf{p}}^{(0)}(t)\, \zeta_{-\mathbf{q}}^{(0)}(t)\,, \qquad (3.8)$$

with solution

$$\zeta_{\mathbf{n}}^{(1)}(t) =$$
$$\int_{t_0}^{t} ds\, \hat{R}_{\mathbf{n}}^{(0)}(t,s) \sum_{\mathbf{m}}\sum_{\mathbf{l}} \delta(\mathbf{n}+\mathbf{m}+\mathbf{l})\, K(\mathbf{n},\mathbf{m},\mathbf{l})\, \zeta_{-\mathbf{m}}^{(0)}(s)\, \zeta_{-\mathbf{l}}^{(0)}(s)\,, \qquad (3.9)$$

with $\mathbf{n} = \mathbf{k}$. Now from Eqs. (2.3a), (3.1b) and (3.2a) we have to order λ^2

$$\left(\frac{\partial}{\partial t} + D_0(k)\right) C_{\mathbf{k}}(t, t') =$$

$$\lambda \sum_{\mathbf{p}} \sum_{\mathbf{q}} \delta(\mathbf{k} + \mathbf{p} + \mathbf{q})\, K(\mathbf{k}, \mathbf{p}, \mathbf{q})\, [\, \langle \zeta_{-\mathbf{p}}^{(0)}(t)\, \zeta_{-\mathbf{q}}^{(0)}(t)\, \zeta_{-\mathbf{k}}^{(0)}(t') \rangle$$

$$+\, 2\lambda\, \langle \zeta_{-\mathbf{p}}^{(1)}(t)\, \zeta_{-\mathbf{q}}^{(0)}(t)\, \zeta_{-\mathbf{k}}^{(0)}(t') \rangle + \lambda\, \langle \zeta_{-\mathbf{p}}^{(0)}(t)\, \zeta_{-\mathbf{q}}^{(0)}(t)\, \zeta_{-\mathbf{k}}^{(1)}(t') \rangle \,]$$

$$+\, \langle f_{\mathbf{k}}^0(t)\, \zeta_{-\mathbf{k}}(t') \rangle$$

$$= \lambda^2 \sum_{\mathbf{p}} \sum_{\mathbf{q}} \delta(\mathbf{k} + \mathbf{p} + \mathbf{q})\, K(\mathbf{k}, \mathbf{p}, \mathbf{q})$$

$$\cdot\, [\, 2 \int_{t_0}^{t} ds \hat{R}_{-\mathbf{p}}^{(0)}(t, s) \sum_{\mathbf{m}} \sum_{\mathbf{l}} \delta(-\mathbf{p} + \mathbf{m} + \mathbf{l})\, K(-\mathbf{p}, \mathbf{m}, \mathbf{l})$$

$$\cdot\, \langle \zeta_{-\mathbf{m}}^{(0)}(s)\, \zeta_{-\mathbf{l}}^{(0)}(s)\, \zeta_{-\mathbf{q}}^{(0)}(t)\, \zeta_{-\mathbf{k}}^{(0)}(t') \rangle$$

$$+ \int_{t_0}^{t'} ds\, \hat{R}_{-\mathbf{k}}^{(0)}(t', s) \sum_{\mathbf{m}} \sum_{\mathbf{l}} \delta(-\mathbf{k} + \mathbf{m} + \mathbf{l})\, K(-\mathbf{k}, \mathbf{m}, \mathbf{l})$$

$$\cdot\, \langle \zeta_{-\mathbf{p}}^{(0)}(t)\, \zeta_{-\mathbf{q}}^{(0)}(t)\, \zeta_{-\mathbf{m}}^{(0)}(s)\, \zeta_{-\mathbf{l}}^{(0)}(s) \rangle \,]$$

$$+\, \langle f_{\mathbf{k}}^0(t)\, \zeta_{-\mathbf{k}}(t') \rangle\,. \quad (3.10a)$$

Here we have used the symmetry properties of the interaction coefficients Eq. (2.5a) and as well Eqs. (3.4) and (3.9). We then use Eq. (3.5) to replace the fourth-order moment by three products of second-order moments. Because of homogeneity (11b) and the delta function (4c) in Eq. (3.10a) one of these three products is zero for each of the two fourth-order moments and the other two can be combined using the symmetry property (5a). Finally we have

$$\left(\frac{\partial}{\partial t} + D_0(k)\right) C_{\mathbf{k}}(t, t') = 4\lambda^2 \int_{t_0}^{t} ds \sum_{\mathbf{p}} \sum_{\mathbf{q}} (\mathbf{k} + \mathbf{p} + \mathbf{q})\, K(\mathbf{k}, \mathbf{p}, \mathbf{q})$$

$$\cdot\, K(-\mathbf{p}, -\mathbf{q}, -\mathbf{k})\, \hat{R}_{-\mathbf{p}}^{(0)}(t, s)\, C_{-\mathbf{q}}^{(0)}(t, s)\, C_{-\mathbf{k}}^{(0)}(t', s)$$

$$+\, 2\lambda^2 \int_{t_0}^{t'} ds \sum_{\mathbf{p}} \sum_{\mathbf{q}} (\mathbf{k} + \mathbf{p} + \mathbf{q})\, K(\mathbf{k}, \mathbf{p}, \mathbf{q})$$

$$\cdot\, K(-\mathbf{p}, -\mathbf{q}, -\mathbf{k})\, C_{-\mathbf{p}}^{(0)}(t, s)\, C_{-\mathbf{q}}^{(0)}(t, s)\, \hat{R}_{-\mathbf{k}}^{(0)}(t', s)$$

$$+\, \langle f_{\mathbf{k}}^0(t)\, \zeta_{-\mathbf{k}}(t') \rangle\,. \quad (3.10b)$$

Kraichnan (1959a) then took the bold step of carrying out the heuristic renormalizations

$$\lambda \;\longrightarrow\; 1\,, \qquad\qquad (3.11a)$$

$$C_{\mathbf{k}}^{(0)}(t,t') \;\longrightarrow\; C_{\mathbf{k}}(t,t')\,, \qquad\qquad (3.11b)$$

$$\hat{R}_{\mathbf{k}}^{(0)}(t,t') \;\longrightarrow\; R_{\mathbf{k}}(t,t')\,, \qquad\qquad (3.11c)$$

where

$$R_{\mathbf{k}}(t,t) \;=\; \langle \hat{R}_{\mathbf{k}}(t,t') \rangle\,. \qquad\qquad (3.12)$$

He realized that to close the equations he also needed the corresponding prognostic for the mean response function. This is obtained as follows. To first order in λ, we have from Eq. (2.9)

$$\left(\frac{\partial}{\partial t} + D_0(k) \right) \hat{R}_{\mathbf{k}}^{(1)}(t,t') \;=\; 2 \sum_{\mathbf{p}} \sum_{\mathbf{q}} \delta(\mathbf{k}+\mathbf{p}+\mathbf{q})\, K(\mathbf{k},\mathbf{p},\mathbf{q})$$
$$\cdot\, \hat{R}_{-\mathbf{p}}^{(0)}(t,t')\, \zeta_{-\mathbf{q}}^{(0)}(t)\,, \qquad\qquad (3.13)$$

with solution

$$\hat{R}_{\mathbf{n}}^{(1)}(t,t') \;=\; 2 \int_{t'}^{t} ds\; \hat{R}_{\mathbf{n}}^{(0)}(t,s) \sum_{\mathbf{m}} \sum_{\mathbf{l}} \delta(\mathbf{n}+\mathbf{m}+\mathbf{l})\, K(\mathbf{n},\mathbf{m},\mathbf{l})$$
$$\cdot\, \hat{R}_{-\mathbf{m}}^{(0)}(s,t')\, \zeta_{-\mathbf{l}}^{(0)}(s)\,, \qquad\qquad (3.14)$$

with $\mathbf{n} = \mathbf{k}$. Here the lower limit is t' since $\hat{R}_{-\mathbf{m}}^{(0)}(s,t')$ vanishes for $s < t'$ because of causality, Eq. (6). Now from Eqs. (2.9), (3.1b) and (3.2b) we have to order λ^2

$$\left(\frac{\partial}{\partial t} + D_0(k)\right) R_{\mathbf{k}}(t,t') =$$

$$2\lambda \sum_{\mathbf{p}} \sum_{\mathbf{q}} \delta(\mathbf{k}+\mathbf{p}+\mathbf{q})\, K(\mathbf{k},\mathbf{p},\mathbf{q})\, [\, \langle \hat{R}^{(0)}_{-\mathbf{p}}(t,t')\, \zeta^{(0)}_{-\mathbf{q}}(t)\rangle$$

$$+\, \lambda\, \langle \hat{R}^{(0)}_{-\mathbf{p}}(t,t')\, \zeta^{(1)}_{-\mathbf{q}}(t)\rangle + \lambda\, \langle \hat{R}^{(1)}_{-\mathbf{p}}(t,t')\, \zeta^{(0)}_{-\mathbf{q}}(t)\rangle \,]$$

$$+\, \delta(t-t')$$

$$= 4\lambda^2 \sum_{\mathbf{p}} \sum_{\mathbf{q}} \delta(\mathbf{k}+\mathbf{p}+\mathbf{q})\, K(\mathbf{k},\mathbf{p},\mathbf{q})$$

$$\cdot \sum_{\mathbf{m}} \sum_{\mathbf{l}} \delta(-\mathbf{p}+\mathbf{m}+\mathbf{l})\, K(-\mathbf{p},\mathbf{m},\mathbf{l})$$

$$\cdot \int_{t'}^{t} ds \langle \hat{R}^{(0)}_{-\mathbf{p}}(t,s)\, \hat{R}^{(0)}_{-\mathbf{m}}(s,t')\, \zeta^{(0)}_{-\mathbf{l}}(s)\, \zeta^{(0)}_{-\mathbf{q}}(t)\rangle$$

$$+\, \delta(t-t')\,. \qquad (3.15a)$$

Here we have used the fact that the zero-order response function is statistically sharp and the mean field vanishes (11a) as well as Eq. (3.14). Again homogeneity (11b) and the symmetry property (5a) leads to

$$\left(\frac{\partial}{\partial t} + D_0(k)\right) R_{\mathbf{k}}(t,t') =$$

$$4\lambda^2 \int_{t'}^{t} ds \sum_{\mathbf{p}} \sum_{\mathbf{q}} \delta(\mathbf{k}+\mathbf{p}+\mathbf{q})\, K(\mathbf{k},\mathbf{p},\mathbf{q})\, K(-\mathbf{p},-\mathbf{q},-\mathbf{k})$$

$$\cdot\, \hat{R}^{(0)}_{-\mathbf{p}}(t,s)\, C^{(0)}_{-\mathbf{q}}(t,s)\, \hat{R}^{(0)}_{\mathbf{k}}(s,t') + \delta(t-t')\,. \qquad (3.15b)$$

The response function equation is also renormalized as in Eqs. (3.11a), (3.11b), (3.11c) and (3.12) leading to the DIA closure equations. The closure has the important properties of conserving kinetic energy and enstrophy, like the spectral vorticity equation (3), and having a generalized Langevin equation representation which exactly reproduces the closure (Kraichnan, 1961; Herring and Kraichnan, 1972; Frederiksen 1999). Martin et al. (1973) note that renormalization of the primitive perturbation theory removes the divergences due to secular behavior and liken it to a *Pade'*-approximant scheme for functions.

4 The DIA and regularized DIA closures

Next, we present the DIA equations in a form that will be useful for further analysis and consider a generalization including an empirical vertex renormalization that we call the regularized DIA (RDIA). We specialize to turbulence which is isotropic (in addition to being homogeneous) where $C_{\mathbf{k}} = C_k$, $R_{\mathbf{k}} = R_k$. That is the second-order moments or cumulants and response functions only depend on k the magnitude of $\mathbf{k}$.

We write the renormalized DIA response function (25b) in the form

$$\frac{\partial R_k(t,t')}{\partial t} + \int_{t'}^{t} ds \, [D_0(k)\delta(t-s)+$$

$$\eta_k(t,s)] \, R_k(s,t') = \delta(t-t') \, , \quad (4.1)$$

where the dissipation has been taken under the integral sign by multiplying by the Dirac delta function. This presentation highlights the fact that the nonlinear damping defined by Eq. (4.5) below modifies the damping specified by the bare dissipation function D_0. Similarly the equation for the renormalized two-time second-order cumulant is

$$\frac{\partial C_k(t,t')}{\partial t} + \int_{t_0}^{t} ds \, [D_0(k) \, \delta(t-s) + \eta_k(t,s)] \, C_k(t',s)$$

$$= \int_{t_0}^{t'} ds \, [F_k^0(t,s) + S_k(t,s)] \, R_k(t',s) \, , \qquad (4.2)$$

where

$$F_k^0(t,t') \;\; = \;\; \langle f_{\mathbf{k}}^0(t) \, f_{-\mathbf{k}}^0(t') \rangle \, . \qquad (4.3)$$

Here, the expectation of the product of the bare forcing and the vorticity spectral component in Eq. (3.10b) has been expressed as an integral over the noise covariance and the response function. This relationship follows simply from zero-order perturbation theory, as in Section 3, and more generally from Novikov's (1965) theorem. The representation (27) shows that the nonlinear noise defined by Eq. (4.6) below modifies the bare noise covariance. We also need the equation for the single-time cumulant viz.,

$$\frac{\partial C_k(t,t)}{\partial t} + 2 \int_{t_0}^{t} ds \, [D_0(k) \, \delta(t-s) + \eta_k(t,s)] \, C_k(t,s)$$

$$= 2 \int_{t_0}^{t} ds \, [F_k^0(t,s) + S_k(t,s)] \, R_k(t,s) , \qquad (4.4a)$$

since

$$\frac{\partial C_k(t,t)}{\partial t} = \lim_{t' \to t} \left\{ \frac{\partial C_k(t,t')}{\partial t} + \frac{\partial C_k(t,t')}{\partial t'} \right\} . \qquad (4.4b)$$

In the above equations, the nonlinear damping

$$\eta_k(t,s) = \sum_{\mathbf{p}} \sum_{\mathbf{q}} B_{\mathbf{kpq}} \, R_p(t,s) \, C_q(t,s) , \qquad (4.5)$$

and nonlinear noise

$$S_k(t,s) = \sum_{\mathbf{p}} \sum_{\mathbf{q}} A_{\mathbf{kpq}} \, C_p(t,s) \, C_q(t,s)$$

$$= \sum_{\mathbf{p}} \sum_{\mathbf{q}} B_{\mathbf{kpq}} \, C_p(t,s) \, C_q(t,s) , \qquad (4.6)$$

where

$$B_{\mathbf{kpq}} = -4\delta \, (\mathbf{k}+\mathbf{p}+\mathbf{q}) \, K(\mathbf{k},\mathbf{p},\mathbf{q}) \, K(-\mathbf{p},-\mathbf{q},-\mathbf{k}) , \qquad (4.7)$$

$$A_{\mathbf{kpq}} = 2\delta \, (\mathbf{k}+\mathbf{p}+\mathbf{q}) \, [K(\mathbf{k},\mathbf{p},\mathbf{q})]^2$$

$$= \frac{1}{2} \, [B_{\mathbf{kpq}} + B_{\mathbf{kqp}}] . \qquad (4.8)$$

The relationship between A and B in Eqs. (4.8) and (4.6) follows from Eq. (2.5b). We also define the nonlinear enstrophy transfer by

$$N_k(t,t) = \int_{t_0}^{t} ds \, [S_k(t,s) \, R_k(t,s) - \eta_k(t,s) \, C_k(t,s)] . \qquad (4.9)$$

The failure of the Eulerian DIA closure to yield the correct inertial ranges, discussed in the Introduction, was shown by Kraichnan (1964b) to be due to spurious convection effects of small-scale eddies by large eddies in the DIA formalism. This may be overcome by the RDIA closure where the interaction coefficient or vertex function is renormalized such that

$$B_{\mathbf{kpq}} \longrightarrow \theta(p - k/\alpha)\,\theta(q - k/\alpha)B_{\mathbf{kpq}} , \qquad (4.10)$$

and this modification is applied to Eqs. (4.1) and (4.2) but not to Eq. (4.4a). Here, θ is the Heaviside step function which vanishes for negative argument and is otherwise unity. It restricts the interactions between large and small wavenumbers, depending on the cut-off ratio α, making the interaction more local and in the process can be shown to improve small-scale aspects of the DIA closure (Frederiksen and Davies, 2002 and references therein).

5   Diagnostics

The prognostic DNS and closure equations of Sections 2 to 4 will be analysed in the following sections by comparing a number of diagnostic quantities. Here we summarize these diagnostics.

We define the total energy, enstrophy, palinstrophy, enstrophy dissipation and palinstrophy production respectively as follows:

$$\mathcal{E}(t) \;=\; \frac{1}{2}\sum_{\mathbf{k}} C_k(t,t)/k^2 \qquad (5.1a)$$

$$\mathcal{F}(t) \;=\; \frac{1}{2}\sum_{\mathbf{k}} C_k(t,t) \qquad (5.1b)$$

$$\mathcal{P}(t) \;=\; \frac{1}{2}\sum_{\mathbf{k}} k^2 C_k(t,t) \qquad (5.1c)$$

$$\eta(t) \;=\; \sum_{\mathbf{k}} \hat{\nu}k^2 C_k(t,t) \qquad (5.1d)$$

$$\mathcal{K}(t) \;=\; \sum_{\mathbf{k}} k^2 N_k(t,t) . \qquad (5.1e)$$

Note that the enstrophy dissipation $\eta(t) = 2\hat{\nu}\mathcal{P}(t)$ where $\mathcal{P}(t)$ is the palinstrophy and $\hat{\nu}$ is the viscosity.

Then the large-scale Reynolds number R_L and skewness S may be defined as

$$R_L(t) = \mathcal{E}/(\hat{\nu}\eta^{1/3}) \tag{5.2}$$

$$S(t) = 2\mathcal{K}/(\mathcal{P}\mathcal{F}^{1/2}) , \tag{5.3}$$

(Herring et al., 1974; Frederiksen and Davies, 2000). The band-averaged energy spectrum, palinstrophy spectrum and enstrophy transfer spectra are defined as

$$E(k_i, t) = \frac{1}{2} \sum_{\mathbf{k}\epsilon\mathcal{S}_i} C_k(t,t)/k^2 \tag{5.4a}$$

$$P(k_i, t) = \frac{1}{2} \sum_{\mathbf{k}\epsilon\mathcal{S}_i} k^2 C_k(t,t) \tag{5.4b}$$

$$N(k_i, t) = \sum_{\mathbf{k}\epsilon\mathcal{S}_i} N_k(t,t) , \tag{5.4c}$$

where the set $\mathcal{S}_i$ is given by

$$\mathcal{S}_i = \left\{ \mathbf{k} \,|\, k_i = \text{Int.}[k + \tfrac{1}{2}] \right\} . \tag{5.5}$$

Here the subscript i is used to indicate that the integer part is taken in Eq. (5.5) so that all $\mathbf{k}$ that lie within a given radius band of unit width are summed over. We may also define the enstrophy flux through wavenumber k_i by

$$Z(k_i, t) = \sum_{j=k_i}^{k_{max}} N(j,t) , \tag{5.6}$$

where k_{max} is the largest retained wavenumber in our circular truncation scheme.

6 Comparisons of DIA and RDIA closures with DNS

Frederiksen and Davies (2000) compared the performance of discrete wavenumber non-Markovian closures with DNS and with the continuous wavenumber DIA closure results of Herring et al. (1974) for very similar initial spectra and large-scale Reynolds numbers (R_L) between ≈ 50 and ≈ 300. Most of these runs used a C63 circular truncation with maximum wavenumber $k_{max} = 63$ and $1 \leq k \leq k_{max}$. They found that the discrete wavenumber closures perform considerably better than the continuous wavenumber closures of Herring et al. (1974) in comparison with DNS as far as the evolved kinetic energy and transfer spectra and skewness are concerned. As well, they compared closure and DNS runs for spectra at the higher resolution of C96, with $k_{max} = 96$, and high evolved large-scale Reynolds number of $R_L \approx 4000$. For the range of Reynolds numbers considered it was found that the discrete wavenumber DIA, SCFT and LET closures are in reasonable agreement with DNS in the energy containing range of the large scales but that these closures, without vertex renormalization, underestimate the enstrophy flux to high wavenumbers, particularly at high Reynolds number, resulting in an underestimation of small-scale kinetic energy.

More recently, Frederiksen and Davies (2002) have performed similar runs with the RDIA closure. They find that on the whole the RDIA closure, with empirical vertex renormalization (Eq. (4.10) and regularization cut-off ratio $\alpha = 6$, performs remarkably well compared with DNS for the spectra considered and for evolved Reynolds numbers R_L ranging between 50 and 4000. The RDIA closure largely overcomes the small-scale systematic biases of the non-Markovian DIA, SCFT and LET closures.

The findings of Frederiksen and Davies (2000, 2002) are exemplified by Figs. 1–4 (redrawn from Frederiksen and Davies, 2002) which present results for runs started from their initial spectrum B. This spectum is specified by

$$C_k(0,0) \;\; = \;\; 1.8 \times 10^{-1} k^2 exp(-\frac{2}{3}k) \; . \tag{6.1}$$

The runs have resolution C63 and start from Gaussian initial conditions for both the closures and DNS. The integrations have been performed between nondimensional times $t = 0$ and $t = 0.8$ with a timestep of $\Delta t = 0.016$ for the closure and $\Delta t = 0.004$ for DNS to ensure computational stability. The viscosity is specified as $\nu = 0.0025$ and the Reynolds number starts at $R_L(0) = 307$ and the final evolved Reynolds number $R_L(0.8) = 280$ for both DNS and RDIA and $R_L(0.8) = 291$ for the DIA closure.

Fig. 1 shows the initial and evolved ($t = 0.8$) kinetic energy spectra $E(k)$ for the RDIA and DIA closures and compares them with DNS results, where the DNS results are the average over 40 realizations. The RDIA closure compares closely with DNS and represents a considerable improvement over

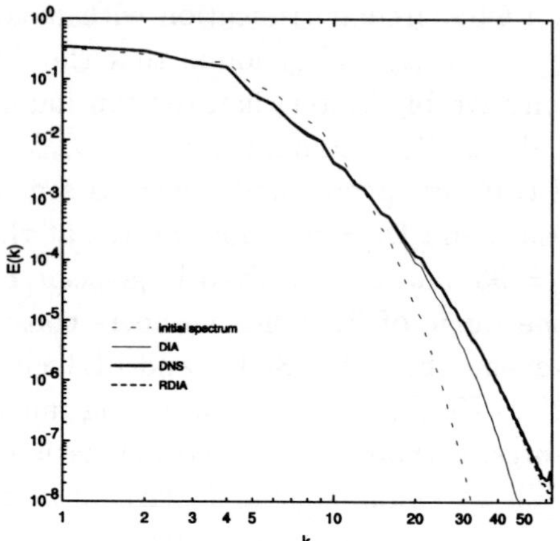

Fig. 1   Initial $(t = 0)$ and evolved $(t = 0.8)$ kinetic energy spectra $E(k)$ for runs starting from initial spectrum B at C63 for DNS and RDIA and DIA closures.

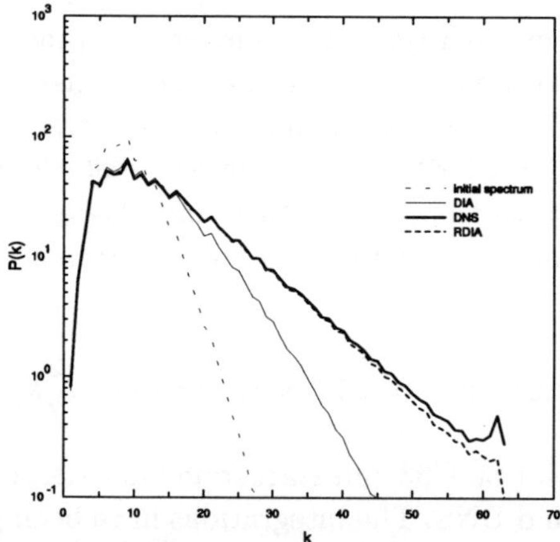

Fig. 2   As in Figure 1 for palinstrophy spectra $P(k)$.

the DIA closure for $k > 20$. The palinstrophy spectra $P(k)$ in Fig. 2 highlight the small scales and again indicate close agreement between the RDIA closure and DNS except at the very smallest scales where $k > 58$. We also note that the RDIA closure results appear to be generally superior to the abridged Lagrangian-history direct interaction (ALHDI) and strain-based abridged Lagrangian-history direct interaction (SBALHDI) closure

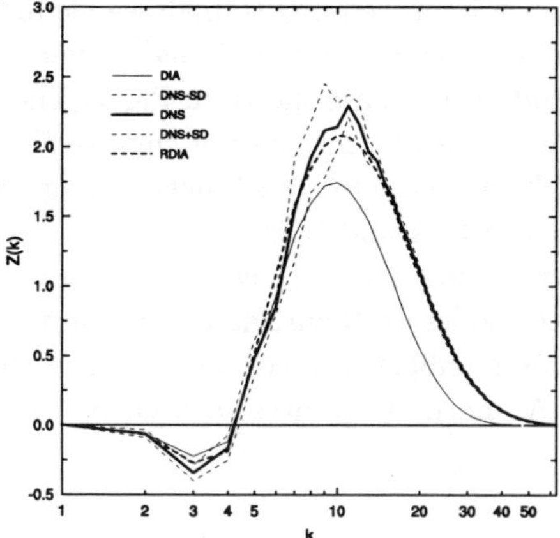

Fig. 3   As in Figure 1 for evolved $(t = 0.8)$ enstrophy flux spectra $Z(k)$.

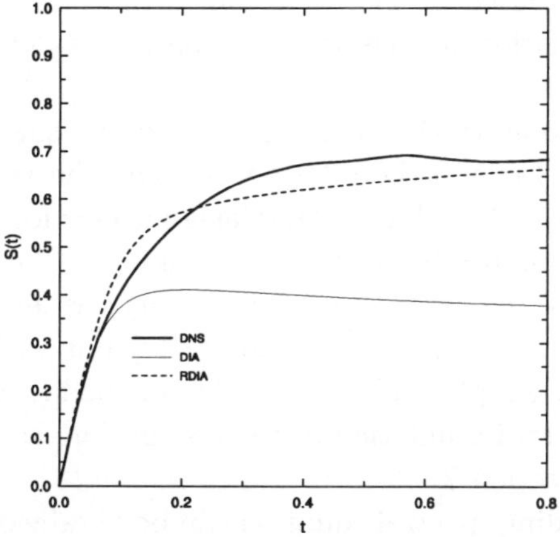

Fig. 4   Time evolution of skewness factor $S(t)$ for closures and DNS for runs starting from initial spectrum B at C63.

results shown in Fig.8b of Herring and Kraichnan (1979) for a closely similar spectrum.

Fig.3 shows the enstrophy flux $Z(k)$ for the RDIA and DIA closures at $t = 0.8$ and compares them with the average DNS results. Also shown are the mean $\pm$ the standard deviation $(DNS \pm SD)$ of four 10-member ensemble averages. We note that the RDIA closure is in good agreement with DNS with a much improved flux to high wavenumbers compared with the DIA closure. The time evolutions of the skewness $S(t)$ for the RDIA and DIA closures and for DNS are shown in Fig.4. The skewness as defined in Eq. (5.3) is a very sensitive measure of small-scale differences between the closures and DNS. We note the much better comparison between RDIA and DNS than between DIA and DNS.

Frederiksen and Davies (2002) show that the results reproduced here for spectrum B with moderate Reynolds number are in fact representative for evolved Reynolds numbers between ≈ 50 and ≈ 4000 in terms of the comparison of RDIA and DIA closures with DNS.

7   Vorticity equation and DIA closure on the sphere

Next, we consider the problem of determining self-consistent subgrid-scale parameterizations based on closures. Since the aim is to apply these to atmospheric circulation models the formulation is developed in spherical geometry.

In spherical geometry the barotropic vorticity equation takes the same form as in Eq. (2.1a) and the relation between the vorticity and stream-function is as in Eq. (2.1b) but the Laplacian is now for spherical geometry. If we non-dimensionalize the equations by taking a (the earth's radius) and Ω^{-1} (the inverse of the earth's angular velocity) as length and time scales then the Jacobian is again given by Eq. (2.1c) but with x replaced by λ_l, the longitude, and y replaced by μ, the sine of the latitude. We shall use a slightly more general formulation of the dissipation, as specified in Section 10, than that in Section 2.

The corresponding spectral equation can be obtained by expanding each of the functions in spherical harmonics; for example

$$\zeta(\lambda_l,\mu,t) = \sum_{m=-T}^{T} \sum_{n=|m|}^{T} \zeta_{mn}(t)P_n^m(\mu)exp(im\lambda_l)\,, \tag{7.1}$$

where m is zonal wavenumber, n is total wavenumber, T is the 'triangular' truncation wavenumber, and P_n^m are orthonormalized Legrendre functions.

Now let

$$P_{\mathbf{k}} \equiv P_n^m \,, \tag{7.2a}$$

$$\mathbf{k} = (m, n) = (m_k, n_k) \,, \tag{7.2b}$$

$$-\mathbf{k} = (-m, n) \,, \tag{7.2c}$$

$$k = n \,. \tag{7.2d}$$

Then the spectral equation is

$$\left(\frac{\partial}{\partial t} + D_0(k) \right) \zeta_{\mathbf{k}}(t) =$$

$$i \sum_{\mathbf{p}} \sum_{\mathbf{q}} \delta(m_k + m_p + m_q) \, K(\mathbf{k}, \mathbf{p}, \mathbf{q}) \, \zeta_{-\mathbf{p}}(t) \, \zeta_{-\mathbf{q}}(t)$$

$$+ \, f_{\mathbf{k}}^0(t) \,, \tag{7.3}$$

$$D_0(k) = \nu_0(k) k(k+1) \,, \tag{7.4}$$

where

$$K(\mathbf{k}, \mathbf{p}, \mathbf{q}) = \frac{1}{2} \left(\frac{p(p+1) - q(q+1)}{p(p+1)q\,(q+1)} \right)$$

$$\cdot \int_{-1}^{1} d\mu \, P_{\mathbf{k}} \left(m_p P_{\mathbf{p}} \frac{dP_{\mathbf{q}}}{d\mu} - m_q P_{\mathbf{q}} \frac{dP_{\mathbf{p}}}{d\mu} \right) \,, \tag{7.5}$$

$$\delta(m_k + m_p + m_q) = \begin{cases} 1 & \text{if } m_k + m_p + m_q = 0 \,, \\ 0 & \text{otherwise} \,. \end{cases} \tag{7.6}$$

In spherical geometry, the DIA closure for isotropic turbulence is again given by the equations in Section 4 but with the interaction coefficient given by Eq. (7.5) and the Kronecker delta function Eq. (7.6) replacing that of Eq. (2.4c) in Eq. (4.7) and Eq. (4.8). The selection rules on the total wavenumbers for non-zero interaction coefficients are given in Eq. (2.3) of Frederiksen and Sawford (1980).

8 EDQNM closure on the sphere

The time-history integrals associated with the non-Markovian DIA closure reflect memory effects associated with turbulent eddies. Other non-Markovian closure theories such as the SCFT and LET closures have very similar performance to the DIA closure. The SCFT and LET closures replace either the second-order two-time cumulant or the response function with the fluctuation dissipation theorem (FDT). Unlike quantum field theories which deal with Hermitian operators, classical systems tend not to be self-adjoint and the FDT in general only applies at canonical equilibrium (Kraichnan, 1959b; Carnevale and Frederiksen, 1983a and references therein). Nevertheless, as noted above it appears to be a reasonable approximation more generally.

The non-Markovian DIA closure can be simplified to a Markovian closure called the eddy-damped quasi-normal Markovian (EDQNM) closure by replacing the equation for the two-time cumulant (Eq. (4.2) by the stationary form of the FDT:

$$C_k(t,t')\theta(t-t') \;=\; R_k(t,t')C_k(t,t) \,, \tag{8.1}$$

and assuming an exponential decay form for the response functions

$$R_k(t,t') \;=\; exp[-\mu_k(t-t')] \,. \tag{8.2}$$

Here, θ is the Heaviside step function which is zero for negative argument and otherwise is unity and the eddy damping is given by the empirical form

$$\mu_k \;=\; \gamma\,[k(k+1)C_k(t,t)]^{\frac{1}{2}} \;+\; \nu_0(k)k(k+1) \tag{8.3}$$

(Orszag,1970; Leith, 1971; Frederiksen and Davies, 1997). This form of the eddy damping is consistent with the -3 enstrophy cascading inertial range power law of two-dimensional turbulence. We find that taking

$$\gamma \;=\; 0.6 \,, \tag{8.4}$$

gives good comparison with DNS.

The single-time cumulant equation (Eq. (4.4a) then reduces to the ordinary differential equation

$$\frac{\partial C_k(t,t)}{\partial t} \;+\; 2\,\nu_0(k)k(k+1)C_k(t,t) - F_0(k;t)$$
$$=\; 2S_k(t) - 2\eta_k(t)C_k(t,t) \,, \tag{8.5}$$

where we have used Eq. (7.4). We have also specialized to white noise forcing for which

$$\langle f_{\mathbf{k}}^0(t)\, f_{-\mathbf{k}}^0(t')\rangle \;=\; F_0(k;t)\delta(t-t') \,, \tag{8.6a}$$

so that, from Eqs. (4.4a) and (4.4b),

$$2 \int_{t_0}^{t} ds \, F_k^0(t,s) \, R_k(t,s) \quad \longrightarrow \quad F_0(k;t) \, . \tag{8.6b}$$

The nonlinear damping and nonlinear noise then reduce to

$$\eta_k(t) \;=\; \sum_{\mathbf{p}} \sum_{\mathbf{q}} B_{\mathbf{kpq}} \, \theta_{kpq} \, C_q(t,t) \, , \tag{8.7}$$

$$S_k(t) \;=\; \sum_{\mathbf{p}} \sum_{\mathbf{q}} A_{\mathbf{kpq}} \, \theta_{kpq} \, C_p(t,t) \, C_q(t,t) \quad (>0)$$

$$\;=\; \sum_{\mathbf{p}} \sum_{\mathbf{q}} B_{\mathbf{kpq}} \, \theta_{kpq} \, C_p(t,t) \, C_q(t,t) \, . \tag{8.8}$$

We note that the nonlinear noise is positive. The time-history integrals can be calculated analytically to determine the triad relaxation time as

$$\theta_{kpq}(t) \;=\; \int_{t_0}^{t} ds \, R_k(t,s) \, R_p(t,s) \, R_q(t,s)$$

$$\;=\; \frac{1 \,-\, exp[-\,(\mu_k + \mu_p + \mu_q)\,(t - t_0)]}{\mu_k + \mu_p + \mu_q} \, , \tag{8.9a}$$

with

$$\theta_{kpq}(\infty) \;=\; (\mu_k + \mu_p + \mu_q)^{-1} \, . \tag{8.9b}$$

Here the summations over $\mathbf{p}$ and $\mathbf{q}$ are determined by

$$\mathcal{T}(T) \;=\; \{\mathbf{p}, \mathbf{q} \,| \quad -T \le m_p \le T \, , \; |m_p| \le p \le T \, ,$$

$$-T \le m_p \le T \, , \; |m_p| \le p \le T \, \} \, . \tag{8.10}$$

9 Subgrid-scale parameterizations

Next, we examine how to establish self-consistent subgrid-scale parameterizations when the resolution is reduced from triangular truncation T to $T_R < T$ where T_R is the triangular truncation wavenumber of the resolved scales. We define the set of resolved scales by

$$\mathcal{R} \;=\; \mathcal{T}(T_R) \, , \tag{9.1a}$$

and the set of subgrid-scales by

$$\mathcal{S} \;=\; \mathcal{T}(T) - \mathcal{R} \,. \tag{9.1b}$$

Then, the nonlinear damping and nonlinear noise due to the resolved scales (respectively subgrid-scales) are given by Eqs. (8.5) and (8.8), with subscript $\mathcal{R}$ (respectively $\mathcal{S}$) for $\mathbf{p}, \mathbf{q}$ in $\mathcal{R}$ (respectively $\mathcal{S}$).

a. The EDQNM Closure

For the EDQNM closure, the equation for (twice) the enstrophy components of the resolved scales can then be written in the form

$$
\begin{aligned}
\frac{\partial C_k(t,t)}{\partial t} \;+\;& 2\left[\nu_0(k)k(k+1) \;+\; \eta_k^{\mathcal{S}}(t)\right] C_k(t,t) \\
-\;& \left[F_0(k;t) \;+\; 2S_k^{\mathcal{S}}(t)\right] \\[2ex]
=\;& 2S_k^{\mathcal{R}}(t) \;-\; 2\eta_k^{\mathcal{R}}(t) C_k(t,t) \,.
\end{aligned}
\tag{9.2}
$$

It is then clear that $\eta_k^{\mathcal{S}}$ modifies the bare viscous damping and $S_k^{\mathcal{S}}$ modifies the random forcing variance. As noted in Eq. (8.8), $S_k^{\mathcal{S}}$ is positive and corresponds to an injection of enstrophy from the subgrid scale eddies to the resolved scales; that is, it represents stochastic backscatter. We therefore define the eddy drain viscosity

$$\nu_d(k) \;=\; [k(k+1)]^{-1}\, \eta_k^{\mathcal{S}} \,, \tag{9.3a}$$

the renormalised viscosity

$$\nu_r(k) \;=\; \nu_0(k) + \nu_d(k) \,, \tag{9.3b}$$

the stochastic backscatter

$$F_b(k) \;=\; 2S_k^{\mathcal{S}} \,, \tag{9.3c}$$

and the renormalised noise variance

$$F_r(k) \;=\; F_0(k) + F_b(k) \,. \tag{9.3d}$$

The injection of enstrophy due to the stochastic backscatter term may contribute to the growth of instabilities, which may be suppressed in lower resolution LES unless the flow is randomly forced or suitably chaotic (Leith, 1990; Piomelli et al., 1991). Most atmospheric circulation models have however not accounted for the stochastic backscatter term, but have tried to account for the differences between the drain and injection terms by an effective or net viscosity. This may be achieved as in Frederiksen and Davies (1997) by defining a (negative) eddy backscatter viscosity ν_b through

$$\nu_b(k) \;=\; -[k(k+1)C_k(t,t)]^{-1} S_k^{\mathcal{S}} \,, \qquad (9.4a)$$

so that,

$$F_b(k) \;=\; -2\nu_b(k)k(k+1)C_k \,. \qquad (9.4b)$$

The net eddy viscosity is defined by

$$\nu_n(k) \;=\; \nu_d(k) + \nu_b(k) \,, \qquad (9.4c)$$

and the renormalised net viscosity by

$$\nu_{rn}(k) \;=\; \nu_n(k) + \nu_0(k) \,. \qquad (9.4d)$$

b. The DIA Closure

For the DIA closure the viscosities again have the relationships defined for the EDQNM closure but with the eddy drain viscosity defined by

$$\nu_d(k) \;=\; [k(k+1)C_k(t,t)]^{-1} \int_{t_0}^{t} ds\, \eta_k^{\mathcal{S}}(t,s)\, C_k(t,s)\, ds \,, \qquad (9.5a)$$

the stochastic backscatter by

$$F_b(k) \;=\; 2 \int_{t_0}^{t} ds\, S_k^{\mathcal{S}}(t,s)\, R_k(t,s)\, ds \,, \qquad (9.5b)$$

and the eddy backscatter viscosity by

$$\nu_b(k) \;=\; -\,[k(k+1)C_k(t,t)]^{-1} \int_{t_0}^{t} ds\, S_k^{\mathcal{S}}(t,s)\, R_k(t,s)\, ds \qquad (9.5c)$$

(Frederiksen and Davies, 1997). Here the integral terms of course come from the integrals in the DIA closure equation (29).

10   Comparisons of DNS and LES with subgrid-scale parameterizations

Frederiksen and Davies, (1997) compared DNS with LES at various resolutions and including dynamic subgrid-scale parameterizations as defined

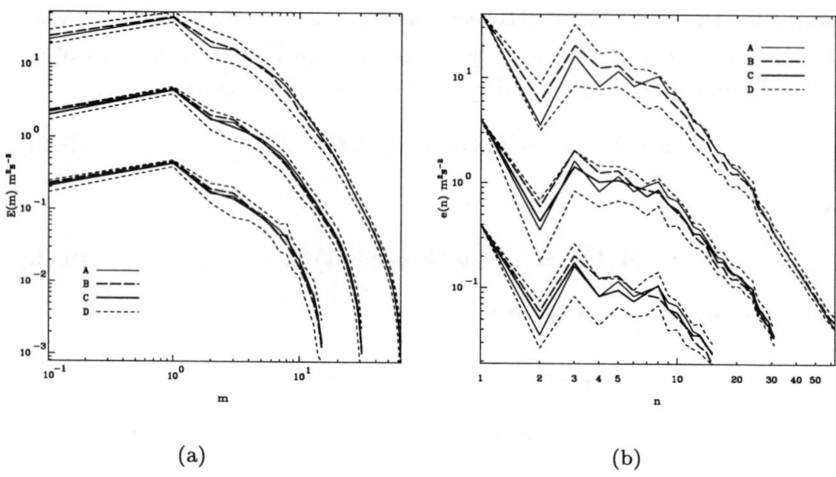

(a)   (b)

Fig. 5   (a) Kinetic energy spectra $\overline{E}(m)$ in $m^2 s^{-2}$ for (A) isotropized January 1979, (B) DNS at T63, (C) LES at T31 or T15, and (D) $\overline{E}(m) \pm \Sigma(m)$ for DNS or LES. The T31 results are scaled by 10^{-1} and T15 by 10^{-2}. Results are for the case using EDQNM-based renormalized stochastic backscatter and renormalized viscosity in LES. Note that results for $m = 0$ are plotted at 10^{-1}. (b) As in (a) for $\overline{e}(n)$ shown as (A), (B), (C) and $\overline{e}(n) \pm \sigma(n)$ shown as (D).

in Section 9. They focussed on DNS simulations which reproduced the observed January 1979 isotropized kinetic energy spectrum as a statistically stationary spectrum in the presence of random forcing and dissipation. The DNS runs have been performed at triangular T63 resolution with the dissipation taken as a linear combination of surface drag and a Laplacian dissipation for which the nondimensional bare viscosity takes the form

$$\nu_0(k) = \begin{cases} \frac{1.014 \times 10^{-2}}{k(k+1)} & \text{for } 2 \le k \le 15 \,, \\ \frac{1.014 \times 10^{-2}}{k(k+1)} + 4.223 \times 10^{-5} & \text{for } 16 \le k \le 63 \,. \end{cases}$$

$$(10.1)$$

The drag corresponds to $7.4 \times 10^{-7} s^{-1}$, or an e-folding decay time of 15.6 days, and the Laplacian contribution corresponds to $1.25 \times 10^5 m^2 s^{-1}$ in dimensional units.

To determine the random forcing variance spectrum needed to balance the dissipation, the steady state EDQNM equation is used as follows. With the enstrophy components specified by the T63 January 1979 isotropic spec-

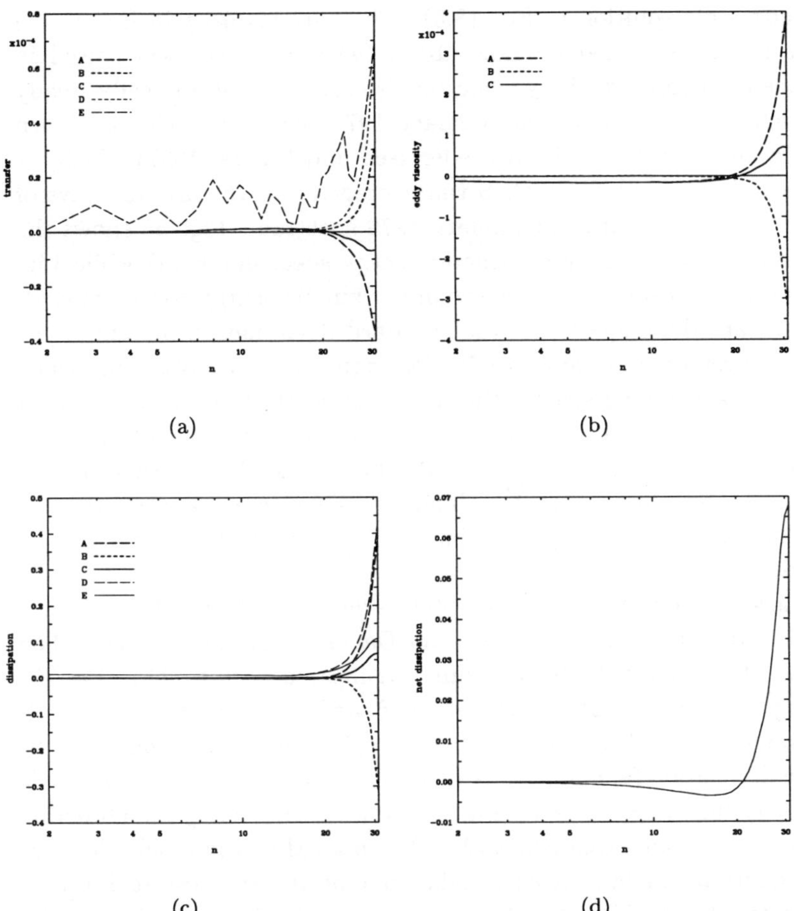

Fig. 6   (a) EDQNM subgrid-scale parametrizations for T31. Transfers corresponding to (A) $-\eta_n^S C_n$, (B) S_n^S, (C) $S_n^S - \eta_n^S C_n$, (D) $F_b(n)$, and (E) $F_r(n)$. (b) As in (a) for viscosities corresponding to (A) $\nu_d(n)$, (B) $\nu_b(n)$, and (C) $\nu_n(n)$. (c) As in (a) for dissipations (A) $\nu_d(n)n(n+1)$, (B) $\nu_b(n)n(n+1)$, (C) $\nu_n(n)n(n+1)$, (D) $\nu_r(n)n(n+1)$, and (E) $\nu_{rn}(n)n(n+1)$. (d) As in (a) for $\nu_n(n)n(n+1)$ with expanded scale.

trum, and the triad relaxation time given by the stationary form in Eq. (8.9b), $F_0(k)$ must be specified by

$$F_0(k) \;=\; \left(2\nu_0(k)k(k+1) + 2\eta_k\right) C_k \;-\; 2S_k \tag{10.2}$$

for a steady state solution to Eq. (8.5). The bare viscosity in Eq. (10.1) and bare forcing specified by Eq. (10.2) (with $\gamma = 0.6$) when used in DNS of the barotropic vorticity equation also produce a statistically steady state which is very close to the January 1979 spectrum. This is shown in Fig. 5a and b (redrawn from Frederiksen and Davies, 1997) which, at T63, compare the results of a DNS run averaged over the last 100 days of a 150-day simulation with the January 1979 spectrum. Fig. 5a shows the dimensional zonal wavenumber kinetic energy spectrum $\overline{E}(m)$ while Fig. 5b shows the dimensional total wavenumber kinetic energy spectrum $\overline{e}(n)$. Also shown are these spectra $\pm$ the standard deviations $\Sigma(m)$ and $\sigma(n)$. The zonal wavenumber spectra are in close agreement at most scales while for the total wavenumber spectra the agreement at the largest scales, where there are few m components to average over and where the eddy turn-over time is long, is not as good as at small scales. The DNS spectra in Fig. 1, or, almost equivalently, the T63 January 1979 spectra, are regarded as the benchmark or "truth" against which LES at lower resolution are to be compared.

The various EDQNM subgrid-scale terms defined in Section 9, are shown in Fig. 6 (redrawn from Frederiksen and Davies, 1997) for the case when the initial T63 January 1979 spectrum is truncated back to T31. Fig. 6a shows $-\eta_k^S$, S_n^S, $S_n^S - \eta_n^S C_n$, $F_b(n) = 2S_n^S$ and $F_r(n) = F_b(n) + F_0(n)$. Fig. 6b depicts the eddy viscosities $\nu_d(n)$, $\nu_b(n)$ and $\nu_n(n)$ while Fig. 6c shows the dissipation functions $\nu_d(n)n(n+1)$, $\nu_b(n)n(n+1)$, $\nu_n(n)n(n+1)$, $\nu_r(n)n(n+1)$ and $\nu_{rn}(n)n(n+1)$. Fig. 6c shows the net dissipation function $\nu_n(n)n(n+1)$ on an expanded scale. We note that these subgrid-scale parameterizations all have a cusp behaviour at the smallest scales as is also characteristic of eddy viscosity parameterizations for three-dimensional turbulence in Cartesian geometry (Kraichnan, 1976; Chasnov 1991). Very similar subgrid-scale parameterizations are obtained when the January 1979 spectrum is truncated back to other lower resolutions as shown at T15 in Fig. 4 of Frederiksen and Davies (1997).

Fig. 5 also compares the spectra of LES using the renormalized viscosity $\nu_r(n) = \nu_d(n) + \nu_0(n)$ and renormalized random forcing $F_r(n) = F_b(n) + F_0(n)$ at resolutions T31 and T15. Again the kinetic energy spectra shown are averaged over the last 100 days of 150-day simulations. The diagrams also show the spectra $\pm$ the standard deviation $\Sigma(m)$ and $\sigma(n)$ and the energy spectra of the T63 DNS, and January 1979 observations, truncated back to T31 or T15. For the zonal wavenumber spectra, there is good agreement between the LES kinetic energies at T31 and T15 and both the DNS and January 1979 results truncated to the resolution of the LES. For the total wavenumber spectra, the agreement between LES and DNS or January 1979 kinetic energies is good at the smaller scales; at the larger scales there are some deviations, presumably because of the few m components to average over and the long eddy turn-over times.

Frederiksen and Davies (1997) show that the generally good agreement between DNS and LES using the EDQNM based renormalized viscosity and renormalized noise forcing is also found if the EDQNM based renormalized net viscosity is used instead or if the corresponding very similar DIA based parameterizations are employed. This is also the situation in the presence of a differential rotation speed characteristic of the earth. They also find that the comparison between DNS and LES is much better than when a number of ad hoc viscosity parameterizations are used in the LES.

11 Discussions and conclusions

We have presented a simple derivation of the DIA closure equations and reviewed recent developments in renormalized closure theory and its application for developing subgrid-scale parameterizations for two-dimensional turbulence. We have focussed on homogeneous and isotropic turbulence with zero mean field. It should however be noted that the theoretical developments presented here may be generalized in various ways. The DIA and EDQNM closures may be derived for the interaction of waves and turbulence (Carnevale and Frederiksen, 1983b and references therein). These closures reduce to the Boltzmann equation of Hasselmann's (1966) resonant wave interaction formalism in the limit of weakly interacting waves. The limit however is singular in the sense that it has an additional constant of motion not inherent in the original field equations or in the DIA and EDQNM closures (Frederiksen and Bell, 1983).

Recently, a tractable quasi-diagonal DIA closure theory has been developed for the interaction of general inhomogeneous two-dimensional flows (including mean fields) with topography (Frederiksen, 1999; O'Kane and Frederiksen, 2002). This closure was also used for formulating a subgrid-scale parameterization for the eddy-topographic force, which describes the interaction of subgrid-scale eddies with resolved scale topography. As well, generalizations of the eddy viscosity and stochastic backscatter parameterizations, considered in this chapter, were established for flow over topography. The closure based eddy viscosity parameterization has also been used to improve atmospheric circulations in more complex global climate models (Frederiksen et al., 2002).

Acknowledgments

It is a pleasure to thank Rowena Ball and Robert Dewar for inviting me to present this material at Dynamic Summer 2002. I wish to thank Steve Kepert for assistance with this work.

References

Bowman,J.C., Krommes,J.A. and Ottaviani,M. "The realizable Markovian closure. I. General theory, with application to three-wave dynamics." *Phys. Fluids B* **5**, 3558 (1993).

Carnevale,G.F. and Frederiksen,J.S. "Viscosity renormalization based on direct-interaction closure." *J. Fluid Mech.* **131**, 289 (1983a).

Carnevale,G.F. and Frederiksen,J.S. "A statistical dynamical theory of strongly nonlinear internal gravity waves." *Geophys. Astrophys. Fluid Dyn.* **23**, 175 (1983b).

Chasnov,J.R. "Simulation of the Kolmogorov inertial subrange using an improved subgrid model." *Phys. Fluids A* **3**, 188 (1991).

Frederiksen,J.S. "Subgrid-scale parameterizations of eddy topographic force, eddy viscosity and stochastic backscatter for flow over topography." *J. Atmos. Sci.* **56**, 1481 (1999).

Frederiksen,J.S. and Bell,R.C. "Statistical dynamics of internal gravity wave-turbulence." *Geophys. Astrophys. Fluid Dyn.* **26**, 257 (1983).

Frederiksen,J.S. and Davies,A.G. "Eddy viscosity and stochastic backscatter parameterizations on the sphere for atmospheric circulation models." *J. Atmos. Sci.* **54**, 2475 (1997).

Frederiksen,J.S. and Davies,A.G. "Dynamics and spectra of cumulant update closures for two-dimensional turbulence." *Geophys. Astrophys. Fluid Dyn.* **92**, 197 (2000).

Frederiksen,J.S. and Davies,A.G. "The regularized DIA closure for two-dimensional turbulence." *Geophys. Astrophys. Fluid Dyn.* (2002) submitted.

Frederiksen,J.S. and Sawford,B.L. "Statistical dynamics of two-dimensional inviscid flow on a sphere." *J. Atmos. Sci.* **37**, 717 (1980).

Frederiksen,J.S., Davies,A.G. and Bell,R.C. "Closure theories with non-Gaussian restarts for truncated two-dimensional turbulence." *Phys. Fluids* **6**, 3153 (1994).

Frederiksen,J.S., Dix,M.R. and Davies,A.G. "The effects of closure-based eddy diffusion on the climate and spectra of a GCM." *Tellus* (2002) In press.

Hasselmann,K. "Feynman diagrams and interaction rules of wave-wave scattering processes." *Rev. Geophys.* **4**, 1 (1966).

Herring,J.R. "Self-consistent-field approach to turbulence theory." *Phys. Fluids* **8**, 2219 (1965).

Herring,J.R. and Kraichnan,R.H. "Statistical models and turbulence." *Lecture Notes in Physics*, J. Ehlers, K. Hepp and H.A. Weidenmuller, Eds., Springer-Velag, 148-194 (1972).

Herring,J.R. and Kraichnan,R.H. "A numerical comparison of velocity-based and strain-based Lagrangian-history turbulence approximations." *J. Fluid Mech.* **91**, 581 (1979).

Herring,J.R., Orszag,S.A., Kraichnan,R.H. and Fox,D.G. "Decay of two-dimensional homogeneous turbulence." *J. Fluid Mech.* **66**, 417 (1974).

Jensen,R.V. "Functional integral approach to classical statistical dynamics." *J. Stat. Phys.* **25**, 183 (1981).

Koshyk,J.N., Hamilton,K. and Mahlman,J.D. "Simulation of the $k^{-\frac{5}{3}}$ mesoscale spectral regime in the GFDL SKYHI general circulation model." *Geophys. Res. Lett.* **26**, 843 (1999).

Kraichnan,R.H. "The structure of isotropic turbulence at very high Reynolds numbers." *J. Fluid Mech.* **5**, 477 (1959a).

Kraichnan,R.H. "Classical fluctuation-relaxation theorem." *Phys. Rev.* **113**, 1181 (1959b).

Kraichnan,R.H. "Dynamics of nonlinear stochastic systems." *J. Math. Phys.* **2**, 124 (1961).

Kraichnan,R.H. "Decay of isotropic turbulence in the direct-interaction approximation." *Phys. Fluids* **7**, 1030 (1964a).

Kraichnan,R.H. "Kolmogorov's Hypothesis and Eulerian Turbulence Theory." *Phys. Fluids* **7**, 1723 (1964b).

Leith,C.E. "Atmospheric predictability and two-dimensional turbulence." *J. Atmos. Sci.* **28**, 145 (1971).

Leith,C.E. "Climate response and fluctuation dissipation." *J. Atmos. Sci.* **32**, 2022 (1975).

Leith,C.E. "Stochastic backscatter in a subgrid-scale model: Plane shear mixing layer." *Phys. Fluids A* **2**, 297 (1990).

Leith,C.E. and Kraichnan,R.H. "Predictability of turbulent flows." *J. Atmos. Sci.* **29**, 1041 (1972).

Martin,P.C., Siggia,E.D. and Rose,H.A. "Statistical dynamics of classical systems." *Phys. Rev. A* **8**, 423 (1973).

McComb,W.D. "A local energy-transfer theory of isotropic turbulence." *J. Phys. A* **7**, 632 (1974).

McComb,W.D. *The Physics of Fluid Turbulence* (Clarendon, Oxford, 1990) 572.pp.

McComb,W.D., Filipiak,M.J. and Shanmugasundaram,V. "Rederivation and further assessment of the LET theory of isotropic turbulence, as applied to passive scalar convection." *J. Fluid Mech.* **245**, 279 (1992).

Millionshtchikov,M. "On the theory of homogeneous isotropic turbulence." *Dokl. Akad. Nauk. SSSR.* **32**, 615 (1941).

Novikov,E.A. "Functionals and the random force-method in turbulence." *Sov. Phys. - JETP.* **20**, 1290 (1965).

Nastrom,G.D. and Gage,K.S. "A climatology of atmospheric wavenumber spectra of wind and temperature observed by commercial aircraft." *J. Atmos. Sci.* **42**, 950 (1985).

Ogura,Y. "A consequence of the zero fourth cumulant approximation in the decay of isotropic turbulence." *J. Fluid Mech.* **16**, 33 (1963).

O'Kane,T.J. and Frederiksen,J.S. "Integro-differential closure equations for inhomogeneous turbulence." *ANZIAM J.* (2002) submitted.

Orszag,S.A. "Analytical theories of turbulence." *J. Fluid Mech.* **41**, 363 (1970).

Phythian,R. "The functional formalism of classical statistical dynamics." *J. Phys. A: Math.* **10**, 777 (1977).

Piomelli,V., Cabot,W.H., Moin,P. and Lee,S. "Subgrid-scale backscatter in turbulent and transitional flows." *Phys. Fluids A* **3**, 1766 (1991).

Rose,H.A. "An efficient non-Markovian theory of non-equilibrium dynamics." *Physica D* **14**, 216 (1985).

Wyld,H.W. "Formulation of the theory of turbulence in an incompressible fluid." *Ann. Phys.* **14**, 143 (1961).

Chapter 7

Low-Dimensional Modelling of Dynamical Systems Applied to Some Dissipative Fluid Mechanics

A. J. Roberts[1]

Department of Mathematics & Computing, University of Southern Queensland, Toowoomba, Queensland 4350, Australia

Abstract. Consider briefly the equations of fluid dynamics — they describe the enormous wealth of detail in all the interacting physical elements of a fluid flow — whereas in applications we want to deal with a description of just that which is interesting. In a wide variety of situations, simple approximate models are needed to perform practical simulations and make forecasts. I overview the *derivation*, from a mathematical description of the detailed dynamics, to accurate and complete low-dimensional models of the interesting dynamics in a system. The development of centre manifold theory and associated techniques puts this modelling process on a firm basis. The geometric viewpoint of dynamical systems theory greatly enriches our approach.

1 An overview of some low-dimensional models

In secondary school we learn of the parabolic flight of a ball. In university physics we learn how it spins. In elasticity courses we learn that a ball deforms as it spins and bounces. In each successive stage of the modelling of the ball we deal with more details of the dynamics: first the position and velocity of the centre of mass, then position, orientation and their velocities, and lastly the position and velocity of every part of the ball. *We will reverse this process*: we will "reduce" equations describing the very detailed dynamics of a system, such as the equations of elasticity, down to simpler "model" equations that deal with the dynamics at a relatively

---

[1]e-mail: `aroberts@usq.edu.au`

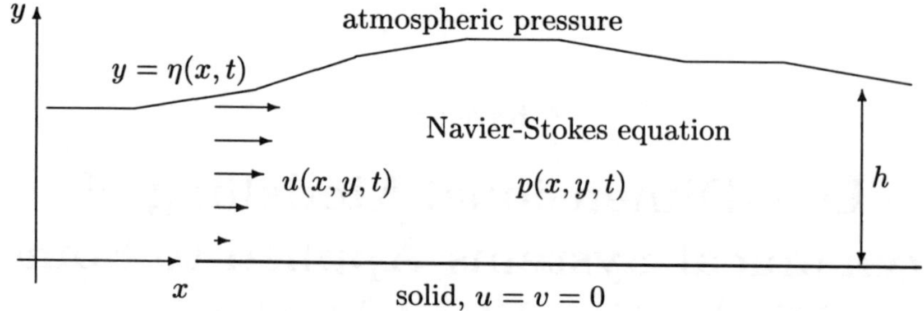

Fig. 1   Schematic diagram of a thin fluid film flowing along a solid bed.

coarse level of description, such as the equations for the centre of mass of a ball.

A model is "simple" if it deals with just a few characteristics of the physical problem. For example, in the flight of the ball we describe just the movement of the centre of mass rather than the movement of each part of the ball. Describing just a few characteristics and so written in terms of just a few parameters, *a simple model is of low-dimension* when compared with the actual physical system. A model is useful if it describes the dynamics of interest with very little extraneous details: we are usually interested in how the ball as a whole moves without needing to know any details about its shape as it spins and deforms. It is the rationale, theory and practice of such reduction in dimensionality that we investigate.

1.1   *Model the flow of a thin fluid film*

For this introduction we leave the flight of a ball and instead introduce many modelling issues in the context of a different physical problem: the flow of a *thin* layer of a viscous fluid drawn in Figure 1. We first derive a model quickly and heuristically and then move on to discuss the modelling features of interest in general. Given the surface of the fluid is located at $y = \eta(x, t)$ above a flat solid substrate, $y = 0$, the classic steps in the modelling are:

(1) The pressure above the thin film due to the atmosphere is essentially constant, say $p = 0$. But due to surface tension the pressure jumps

across the fluid surface by an amount proportional to the curvature, approximately η_{xx}, so the pressure in the fluid at the surface is $p \approx -\sigma\eta_{xx}$ where the constant σ characterises the surface tension.

(2) The Navier-Stokes equation for the vertical velocity of a Newtonian fluid with viscosity μ is

$$\rho(v_t + uv_x + vv_y) = -p_y + \mu(v_{xx} + v_{yy}).$$

Because the fluid is very thin there is effectively no scope for any vertical velocity so we assume $v \approx 0$ and hence the Navier-Stokes equation reduces to $p_y = 0$. The pressure must be constant at each location x, namely $p \approx -\sigma\eta_{xx}$.

(3) The Navier-Stokes equation for the horizontal velocity is

$$\rho(u_t + uu_x + vu_y) = -p_x + \mu(u_{xx} + u_{yy}).$$

In a thin fluid the y-derivatives of u must be much larger than the xt-derivatives and hence we approximate $0 \approx -p_x + \mu u_{yy}$ which is integrated twice to the parabolic velocity profile $u \approx -\frac{1}{\mu}(\eta y - \frac{1}{2}y^2)p_x$.

(4) Conservation of fluid then requires that $\partial_t\eta + \partial_x(\text{flux}) = 0$ where here the integrated horizontal velocity gives the

$$\text{flux} = \int_0^\eta -\frac{1}{\mu}(\eta y - \tfrac{1}{2}y^2)p_x \, dy = -\frac{1}{3\mu}\eta^3 p_x.$$

Since the pressure $p = -\sigma\eta_{xx}$, conservation of mass then requires

$$\eta_t = -\frac{\sigma}{3\mu}\partial_x(\eta^3\eta_{xxx}). \tag{1.1}$$

This differential equation for the film's thickness models the spreading of the fluid film. It is a sort of nonlinear hyper-diffusion equation for the thickness.

One immediate characteristic of the above modelling process is that the arguments are highly specific to the problem: the skills involved are not straightforwardly transferable; further, a slight change in the fluid equations, a pressure dependent constitutive relation for example, will ruin the chain of logic. We will discuss a systematic and widely applicable approach to such modelling, an approach based upon dynamical systems theory.

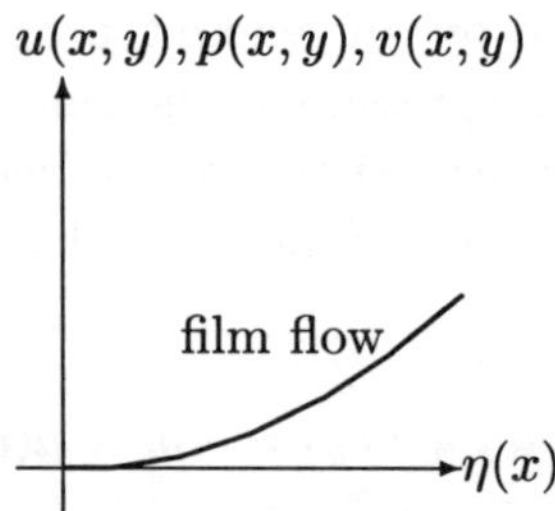

Fig. 2   Schematic diagram of the fluid velocity and pressure fields being uniquely given in the model by the free surface shape η.

1.2   *Issues illustrated by thin film modelling*

Let us look at modelling issues in the context of the thin film model.

(a)   A model is of low dimension

The model is expressed in terms of an evolving film thickness η: for any given η the velocity and pressure fields are uniquely determined, $p = -\sigma\eta_{xx}$, $v = 0$ and $u = -\frac{\sigma}{\mu}(\eta y - \frac{1}{2}y^2)\eta_{xxx}$, drawn schematically in Figure 2. Of all the possible velocity and pressure fields that may instantaneously exist for any given fluid thickness $\eta(x)$, the model only resolves one. This is why a model is a simplification: a model evolves on a low-dimensional subset of states.

(b)   Relevant models are exponentially attractive

In the thin fluid film model, viscosity quickly damps out any lateral shearing motion, and the substrate hinders any vertical motion. Very quickly all that remains is any variations in the film thickness η, which then force a slow long-term evolution according to the model (1.1). Typically we investigate dissipative systems where most modes decay exponentially. Thus the rationale for the modelling is an exponentially quick collapse of the state space *from almost all initial conditions* under the original dynamics. One may then perfectly adequately describe the long term evolution in terms of just a few parameters: in the example of Figure 3, after the quick transients have decayed, we need just one parameter to measure location along the apparent one-dimensional curve. Note that we must parametrise the *curve*; it is not adequate to simply neglect the decaying modes. In general, the decaying modes are not zero, they are forced by the dynamics and, as we shall see, such forcing can feedback signif-

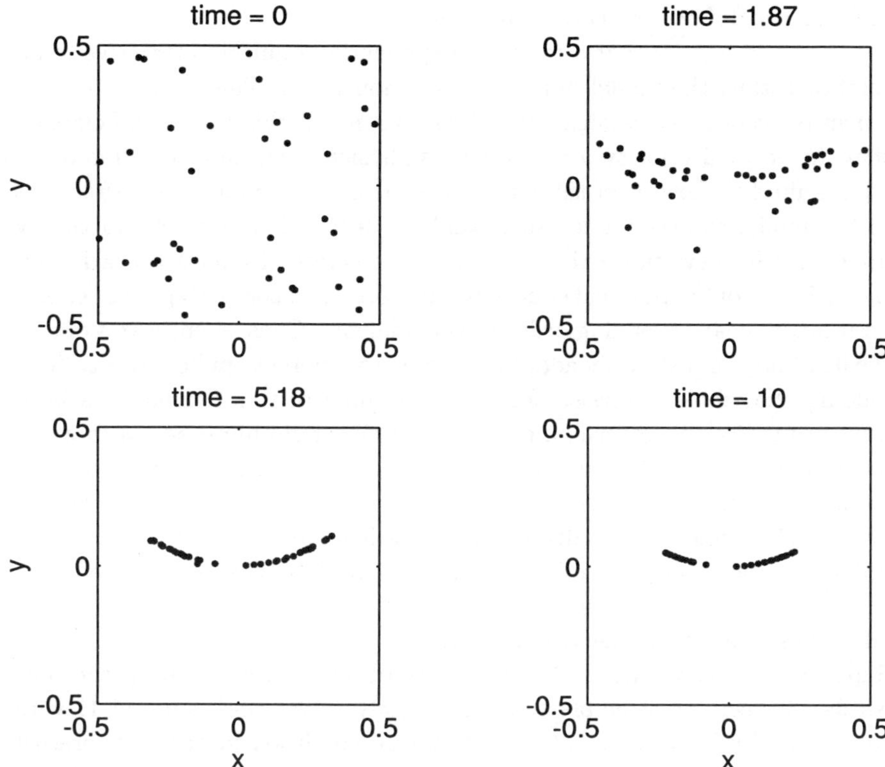

Fig. 3   Forty random initial conditions evolve in time according to the example (2.1). Observe how all these initial conditions collapse onto a one-dimensional set of states in which *all* the long-term dynamics take place.

icantly upon the dynamics. For example, in a thin film flow it is the weak shearing flow forced by film thickness variations that flatten the fluid film. We use centre manifold theory, introduced in §2, to justify and construct dynamical models based upon the picture of exponential decay to a curved set of states such as that seen in Figure 3. The use of centre manifold theory for modelling was initiated by Coullet & Spiegel [21] (who in §1 give an historical perspective) and Carr & Muncaster [11; 12]. The rationale of exponential collapse that underpins centre manifold theory has been also invoked by other methods (such as Haken's synergetics [42, e.g.]), but the geometric picture and associated analytical techniques make centre manifold theory a powerful approach to modelling.

(c)   A model solves the equations

We have just argued that a model is exponentially quickly attractive to all solutions; thus the model must be a solution itself. This is an important tenent of accurate dynamical modelling. But in the thin fluid film I claimed that the vertical velocity $v = 0$, whereas physically we know $v \neq 0$ in order for the fluid to flow from thick regions to thin—there must be corrections to the model to account for such weaker effects. Centre manifold theory provides a framework for the systematic correction of a model based upon the residuals of the physical equations. Importantly, competing small effects need not appear at leading order in the analysis. Consequently you obtain the flexibility to justifiably adapt the model to different applications *without* redoing the whole analysis. Just one example of this flexibility is later used to modify the governing equations of thin fluid films, §4, so that the model incorporates the extra dynamical degree of freedom to resolve wave dynamics on the film. Such flexibility is also used in holistic discretisations such as (1.8) where the width of the numerical stencil is controlled by the order of error in an artificial parameter. Flexibility is extremely useful.

(d)   Forecast from an initial state

Suppose you place a drop of water on the table and watch it spread outwards. It seems reasonable that one simply solves the model (1.1) with the initial fluid thickness $\eta_0(x)$ as its initial condition in order to forecast the details of the spread. However, now consider the colliding paintbrush problem: two sheets of fluid of the same constant small thickness η_0 collide together—what initial condition should be provide to the model (1.1)? It cannot be that $\eta(x,0) = \eta_0$ because physically we know that before the two fluid layers slow down by viscosity, they will pile together to form a hump at the collision point. Somehow we need to resolve this, preferably without having to solve for the details of the initial flow. The geometric picture of evolution near the centre manifold suggests analysis, §5.2, that provides correct initial conditions to use with a model to make correct long-term forecasts. The algebra is based upon how neighbouring trajectories evolve, identifying which ones approach each other exponentially quickly and thus have the same long-term evolution. For example, the initial condition $\eta(x,0) = h_0(x)$ for the fluid film model (1.1) is to first order accuracy [99]

$$h_0 \approx \eta_0 + \partial_x \overline{u_0 \left(\tfrac{1}{2}y^2 - \eta_0 y \right)}, \tag{1.2}$$

where $u_0(x, y)$ is the initial lateral velocity and the over-bar denotes integration across the fluid depth. Thus to forecast the flow following the collision of two thin films, for which the lateral velocity u_0 changes sign across the collision point, we must reasonably start the model from an initial condition with a hump. For another example, in the long-term dispersion down a channel (§3), despite writing the model in terms of cross stream averages, we do discern the difference between dumping contaminant into the slow moving flow near the bank and into the fast core flow. Normal form transformations, §5.1, are also important tools to illustrate the projection of initial conditions onto the centre manifold model.

In addition, normal form transformations show limitations in the so-called slow manifold models, §5.3, that are formed by removing the dynamics of fast oscillations. For example, the motion of the centre of mass of the elastic ball. In contrast to dissipative systems, through nonlinear interaction, neglected fast waves may resonate and cause inevitable errors to accumulate over time.

1.3 *Examples of low-dimensional dynamical models*

The general modelling challenge is to start with an accurate and reliable description of the dynamics of a system of interest, a "Theory Of Everything" (TOE) of interest, then analyse it systematically and extract routinely the simple, low-dimensional dynamical models which are relevant in given situations:

$$\text{TOE} \quad \mapsto \quad \text{model}.$$

Specific examples of this process of modelling are:

- equations of elasticity $\mapsto$ rigid body motion (mentioned above), schematically

$$\frac{\partial u}{\partial t} = q \,, \; \rho \frac{\partial q}{\partial t} = \nabla \cdot \sigma \quad \mapsto \quad \dot{x} = v \,, \; m\dot{v} = \text{forces}, \qquad (1.3)$$

  where x and v denote the position and velocity of the centre of mass of the body;
- heat or mass transfer in a pipe of channel $\mapsto$ longitudinal advection and

dispersion (§3), schematically

$$\frac{\partial c}{\partial t} + \boldsymbol{q} \cdot \boldsymbol{\nabla} c = \kappa \nabla^2 c \quad \mapsto \quad \frac{\partial C}{\partial t} + U \frac{\partial C}{\partial x} = D \frac{\partial^2 C}{\partial x^2} \tag{1.4}$$

for the cross-sectional average concentration $C(x,t)$ where remarkably the effective coefficient of longitudinal diffusion $D \propto 1/\kappa$;

- Navier-Stokes equations $\mapsto$ thin films, sheets and jets of fluids (§4), schematically

$$\boldsymbol{\nabla} \cdot \boldsymbol{q} = 0, \quad \rho \left[\frac{\partial \boldsymbol{q}}{\partial t} + \boldsymbol{q} \cdot \boldsymbol{\nabla} \boldsymbol{q} \right] = -\boldsymbol{\nabla} p + \mu \nabla^2 \boldsymbol{q}$$

$$\mapsto \quad \frac{\partial \eta}{\partial t} = -\tfrac{1}{3} \boldsymbol{\nabla} \cdot (\eta^3 \boldsymbol{\nabla} \nabla^2 \eta) \tag{1.5}$$

for the thickness η of a fluid layer spreading over a flat substrate;

- fluid and heat flow equations $\mapsto$ Ginzburg-Landau equation for the complex amplitude if the spatial pattern of convection, schematically

$$\boldsymbol{\nabla} \cdot \boldsymbol{q} = 0, \quad \rho \left[\frac{\partial \boldsymbol{q}}{\partial t} + \boldsymbol{q} \cdot \boldsymbol{\nabla} \boldsymbol{q} \right] = -\boldsymbol{\nabla} p + \mu \nabla^2 \boldsymbol{q} + \rho \boldsymbol{g},$$

$$\frac{\partial T}{\partial t} + \boldsymbol{q} \cdot \boldsymbol{\nabla} T = \kappa \nabla^2 T$$

$$\mapsto \quad \frac{\partial A}{\partial t} = rA + \frac{\partial^2 A}{\partial x^2} - A^3 \tag{1.6}$$

for the complex amplitude $A(x,t)$ of the convective rolls that occur when heating is sufficiently strong $(r > 0)$.

- elasticity equations $\mapsto$ beam theory of torsion, bending and longitudinal waves, schematically [80]

$$\frac{\partial \boldsymbol{u}}{\partial t} = \boldsymbol{q}, \quad \rho \frac{\partial \boldsymbol{q}}{\partial t} = \boldsymbol{\nabla} \cdot \boldsymbol{\sigma}$$

$$\mapsto \quad \rho \frac{\partial^2 \phi}{\partial t^2} = \mu \frac{\partial^2 \phi}{\partial x^2}, \quad \rho \frac{\partial^2 \eta}{\partial t^2} = \frac{Ea^2}{4} \frac{\partial^4 \eta}{\partial x^4} \quad \text{and}$$

$$\rho \frac{\partial^2 \zeta}{\partial t^2} = E \frac{\partial^2 \zeta}{\partial x^2} + \frac{E \nu^2 a^2}{2} \frac{\partial^4 \zeta}{\partial x^4} \tag{1.7}$$

for the mean twist $\phi(x,t)$, mean sideways deflection $\eta(x,t)$ and mean longitudinal displacement $\zeta(x,t)$;

- partial differential equations $\mapsto$ holistic finite element approximations [85], for example Burgers' equation

$$\frac{\partial u}{\partial t} + u\frac{\partial u}{\partial x} = \frac{\partial^2 u}{\partial x^2}\,,$$

$$\mapsto \quad \dot{u}_j + u_j\frac{u_{j+1} - u_{j-1}}{2h} = \left(1 + \frac{u_j^2}{12}\right)\frac{u_{j+1} - 2u_j + u_{j-1}}{h^2} \quad (1.8)$$

for the values u_j of the field on a grid with spacing h;
- some Markov chains $\mapsto$ quasi-stationary distribution [70];
- atmospheric models $\mapsto$ quasi-geostrophic approximation (§5.3).

The modelling process we investigate is represented by the arrows in the above examples. We aim to describe a complete approach based upon centre manifold theory, techniques and concepts that enables ready identification of when a model exists, how quickly it is valid, its systematic construction, and the provision of appropriate initial conditions.

(e) Four modelling issues we do not explore herein

(1) One attribute of basic centre manifold theory is that it deals with autonomous dynamical systems. However, in the presence of a time dependent forcing the system is pushed away from the centre manifold and so the geometric projection of initial conditions may be used to determine what forcing is appropriate in the model [22]. One can show, for example, that a model may be very sensitive to a forcing which more primitive approaches would neglect. There is also many interesting issues in the modelling of noisy dynamical systems [15].

(2) Many useful models are expressed in terms of partial differential equations in space and time. Such partial differential equations must have boundary conditions. For example, models for dispersion in a channel (1.4) or for beam theory (1.7) require boundary conditions at the inlet and outlet of the channel or the ends of the beam respectively. Similarly, boundary conditions are needed for the model (1.1) at the extremes of a thin fluid film. Arguments based upon the *spatial evolution* away from the boundary, applied to both the full system and the model, give a rationale which provides correct boundary conditions. To leading order these boundary conditions are typically those obtained from physical heuristics. But, there are corrections accounting for more subtle features of the dynamics [78].

(3) Computing details of the centre manifold and its dynamic model often
involves considerable algebra. This is especially true for centre manifolds
of more than just a few dimensions as the algebraic complexity of the
model may increase combinatorially. Computer algebra can be used to
minimise the human labour involved. After all:

> "It is unworthy of excellent persons to lose hours like slaves in the
> labour of calculation"...Gottfried Wilhelm von Leibniz.

Algorithms for computer based algebra packages which are relatively simple and reliable to implement are reported in [82; 84].

(4) Throughout we focus on continuous time dynamics, flows, as expressed
through differential equations. Similar concepts and analysis can be also
developed for discrete maps but I will not elaborate on these.

2   Rational theory underlies modelling

Simple models must have low dimension corresponding to the few variables
in the model. Since a model must have something to say about the original
physics, the state space of the model must be able to be embedded in the
high-dimensional state space of the TOE. As drawn in Figure 2, typically
we imagine that the state space of a model forms a low-dimensional smooth
manifold within the state space of the TOE.

If the model is to accurately capture the dynamics of the TOE, then the
manifold must be made of some of the trajectories of the TOE. The detail
lost in apparently ignoring all the dynamics outside the states described by
the model is an inevitable consequence of forming a simple, low-dimensional
model. The process of analysing the TOE and creating a low-dimensional
model is sometimes termed *coarse graining* [62, e.g.] because of the loss of
fine detail in forming a model. In this section I introduce centre manifold
theory as a basis for the rational modelling of dynamical systems, as was
first recognised by Coullet & Spiegel [21] and Carr & Muncaster [11; 12].
An interesting application to nonlinear diffusion shows how the theory may
also illuminate some similarity solutions. But to unleash the full potential
of the theory we need a trick originally used to unfold bifurcations. The
trick is used in later sections in a variety of ways.

2.1   *Exponential collapse gives a rationale*

If many modes of the TOE decay exponentially, then all that is left after the transients decay are the *relatively* slowly evolving modes of long-term importance. The evolution of these few significant modes effectively forms a low-dimensional dynamical system on a low-dimensional set of states in state space. Through the rapid exponential decay all neighbouring trajectories are quickly attracted to these low-dimensional dynamics and so they form an accurate low-dimensional model of the TOE.

This idea also lies behind the construction of so-called *inertial manifolds* by Temam [105] and others. Functional analysis is used to construct inertial manifolds that contain the attractor of the dynamics (see Foias et al. [35; 34] for example). But what if we are not just interested in the ultimate attractor, but are curious about long-lasting transients? However, before the eventual death of the Universe become an overriding issue to humans there are many problems of interest to study, albeit transient. A more prosaic example is the dispersion of material in an infinitely long channel or pipe: the ultimate attractor is a completely dispersed contaminant of effectively zero concentration everywhere; however, as to be seen in §3, we are very interested in modelling the long time spreading of a contaminant as it is carried downstream.

One case where the distinction between exponential decay and long-lasting importance is made with absolutely clarity is in the neighbourhood of an equilibrium or fixed point. Without loss of generality we take the reference equilibrium at the origin. Let the *linearised* dynamics be $\dot{u} = \mathcal{L}u$, then the eigenvalues of the linear operator $\mathcal{L}$ determine the dynamics in the neighbourhood:[2]

- modes with $\mathrm{Re}(\lambda) < 0$ decay exponentially;
- modes with $\mathrm{Re}(\lambda) = 0$ do not decay, are long-lasting and form the basis of a low-dimensional model.

An elementary example from [73] is

$$\dot{x} = -xy, \quad \text{and} \quad \dot{y} = -y + x^2 - 2y^2 \,. \tag{2.1}$$

Very quickly all trajectories approach a curved "subspace" in state space, called the centre manifold, $y = x^2$ in this example as seen in Figure 3. Thus

---

[2] Any modes with $\mathrm{Re}(\lambda) > 0$ contradicts the dissipative nature of the dynamics.

no matter what the initial condition, the only states of long-term interest are those of the centre manifold. The evolution on the centre manifold, here $\dot{x} = -x^3$, then forms a low-dimensional model of *all* the dynamics in the system.[3] We base our modelling on the linear picture of the dynamics near a relevant equilibrium (fixed point).

In the above, I introduced the criterion that $\mathrm{Re}(\lambda) = 0$ for determining the modes that form the low-dimensional model. Arguments relax this restrictive equality. An argument could be made that modes with $\mathrm{Re}(\lambda) < -\mu < 0$ decay quickly, whereas modes with $\mathrm{Re}(\lambda) \geq -\mu$ are at least *longer-lasting* and form the basis of a low-dimensional model—here it would be the modes with $\mathrm{Re}(\lambda) < -\mu$ that "collapse" the state space. For an elementary example, also from [73], consider

$$\dot{x} = \epsilon x - xy, \quad \text{and} \quad \dot{y} = -y + x^2 - 2y^2, \tag{2.2}$$

which has the exponentially attractive manifold

$$y = \frac{x^2}{1 + 2\epsilon}, \tag{2.3}$$

as seen in Figure 4, on which the model evolution is $\dot{x} = \epsilon x - x^3/(1 + 2\epsilon)$. For $\epsilon \geq 0$ this is termed a *centre-unstable manifold*, a concept used by Armbruster et al. [1] to investigate the Kuramoto-Sivashinksy dynamics, by Cheng & Chang [16] for subharmonic instabilities of waves, and by Chow & Lu [18] to compare with the method of averaging. For general parameter ϵ, not too large in magnitude, (2.3) describes an *invariant manifold* [115, §1.1C], which when exponentially attractive may be used to create accurate models [76; 77], such as that of dispersion in simple channels [111]. Invariant manifolds, based upon modes with $\mathrm{Re}(\lambda) > -\mu$ for some threshold decay rate μ, may improve the numerical solution of spatio-temporal PDEs through what others have called the *nonlinear Galerkin method* [57; 104; 32; 55; 33]. The disadvantage of these more general invariant manifolds for modelling is that the consequent algebraic analysis in constructing the manifold is much more difficult. The rationale of exponential collapse also underlies the so-called "intrinsic low-dimensional manifolds" of Maas & Pope [56; 72] in their application to chemical reactions. In §2.3 we discuss a method

---

[3] As an exercise you might like to try your hand at deducing the model for $\dot{x} = xy + x^3$ and $\dot{y} = -y - 2x^2$. We will return to the interesting dynamics of this problem later in §4.2.

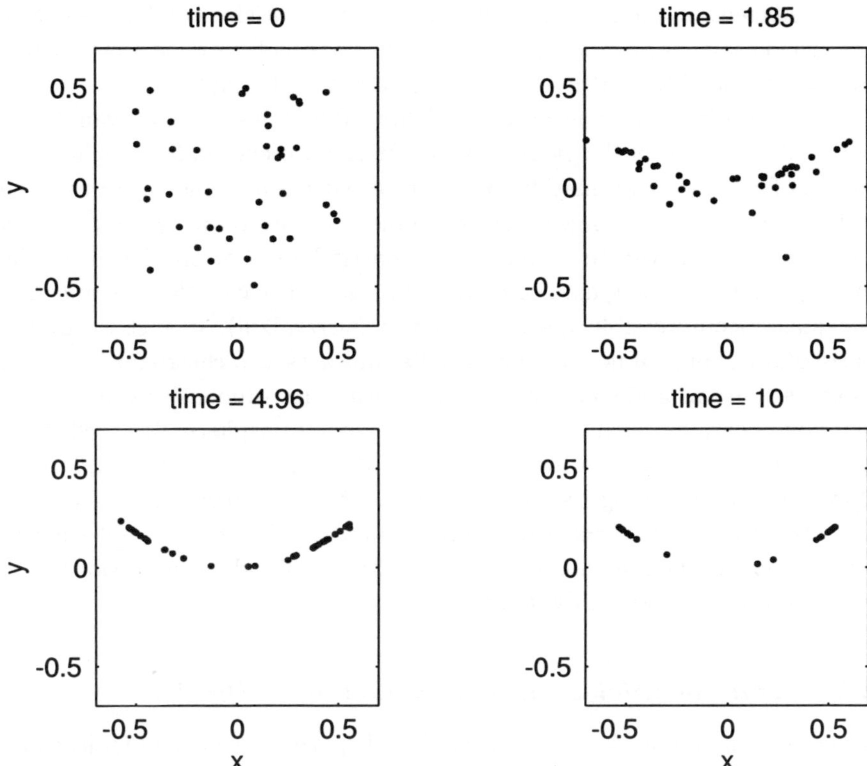

Fig. 4 Forty random initial conditions evolve in time according to the example (2.2) with parameter $\epsilon = 0.2$. Again observe how all these initial conditions collapse onto the so-called centre-unstable manifold, $\mathcal{M}_{cu}$, a one-dimensional set of states in which *all* the long-term dynamics take place.

to approximate more general invariant manifold models while maintaining the relatively simple algebra associated with centre manifolds.

Some problems with algebraic collapse to a low-dimensional set of states may also be encompassed in this framework—we introduce the example of a nonlinear diffusion soon in §2.2.

A completely different approach, albeit justified by the same exponential collapse, is to use symmetry considerations to sketch out possible structurally stable, low-dimensional models. See [30] for example, or the review by Crawford & Knobloch [26] on symmetry in fluid dynamics. The limitation of this approach is that one can practically never determine quantitative coefficients in the model. Such an approach is satisfactory for predicting the broad range of phenomena possible, but it cannot produce a model able to make detailed forecasts about a specific physical situation.

A quite different concept in modelling dynamics is what van Kampen [108] calls that of the *guiding centre*. In the presence of fast oscillations, such as fast waves, we may be only interested in the long-term evolution of the dynamics of the mean. For example, we are primarily interested in the motion of the centre of mass of an elastic ball. For another example, incompressible fluid dynamics ignores fast sound waves (for instance see the quasi-incompressible approximation of Durran [29]). In such a case the particular dynamical flow without oscillations acts as a centre for flow with rapid oscillations and so is considered to form a low-dimensional model. In many applications, such as beam theory [80] or atmospheric flows [53, e.g.], the model dynamics are known as the slow dynamics on the *slow manifold*. However, the modelling issues associated with this concept are much more delicate and I discuss them only briefly in §5.3. Mostly we will concentrate upon exponential collapse to the centre manifold as the rationale for forming low-dimensional dynamical models.

2.2   *Centre manifolds—including nonlinear diffusion*

The centre manifold, $\mathcal{M}_c$, is the curved "subspace" to which all trajectories in the neighbourhood of a fixed point are attracted exponentially quickly. Being composed of trajectories of the TOE and being low-dimensional, the evolution on $\mathcal{M}_c$ qualifies as forming a model of the TOE.

Three theorems claim:

(1) $\mathcal{M}_c$ *exists* for the previously mentioned structure of eigenvalues, provided the nonlinear terms are not "badly" behaved;

(2) $\mathcal{M}_c$ is *relevant* as all solutions in the neighbourhood are attracted exponentially to a solution on $\mathcal{M}_c$;

(3) $\mathcal{M}_c$ may be *constructed* to any desired degree of accuracy (asymptotically speaking).

Detailed proofs of these theorems have, for example, been given in the excellent little book by Carr [10], and more recently by Vanderbauwhede & Iooss [110; 109] and Shilnikov et al. [96, Chapt. 5] with a useful extension in [51]. In this subsection we see how such theorems support modelling.

2.2.1   *Existence*

We address dynamical systems, the TOE, in the general form

$$\dot{u} = \mathcal{L}u + f(u)\,, \tag{2.4}$$

where $u(t)$in$I\!\!R^n$, the linear operator $\mathcal{L}$ has m eigenvalues with zero real-part (counted according to their multiplicity), the remaining eigenvalues have strictly negative real-part, bounded above by $-\mu < 0$, and f is quadratically nonlinear at the origin, of $\mathcal{O}(u^2)$ where $u = \|u\|$. This general form (2.4) is appropriate for most applications where we prefer to perform analysis with meaningful physical variables.

Linearly, according to $\dot{u} = \mathcal{L}u$, all solutions collapse exponentially quickly onto the ,

$$\mathcal{E}_c = \operatorname{span}\{e_1,\ldots,e_m\}\,,$$

where e_j are the m eigenvectors (or generalised eigenvectors) of the m eigenvalues with zero real-part (counted according to multiplicity). Theory asserts that at least near the origin the nonlinear terms just "bend" this invariant subspace and modify the dynamics on the subspace, as shown in Figure 5.

Theorem 7.1   *(existence) sufficiently near the origin, in some neighbourhood U, there exists an m-dimensional invariant manifold for (2.4), $\mathcal{M}_c$, with tangent space $\mathcal{E}_c$ at the origin, in the form $u = v(s)$ (that is, locations on $\mathcal{M}_c$ are parametrised by a set of variables s, often called the). The flow on $\mathcal{M}_c$ is governed by the m-dimensional dynamical system*

$$\dot{s} = \mathcal{G}s + g(s)\,, \tag{2.5}$$

where $\mathcal{G}$ is the restriction of $\mathcal{L}$ to $\mathcal{E}_c$, and the nonlinear function g is determined from $\mathcal{M}_c$ and (2.4). Because of the nature of the eigenvalues of $\mathcal{L}$ this invariant manifold is called a centre manifold *of the system.*

A centre manifold is, in some neighbourhood of the origin, at least as differentiable as the nonlinear terms f. However, it may not be analytic even though f is analytic. Also, a centre manifold need not be unique, but the differences between the possible centre manifolds are of the same order as the differences we set out to ignore in establishing the low-dimensional model [73; 76]. Non-uniqueness, when it arises, is irrelevant for modelling.

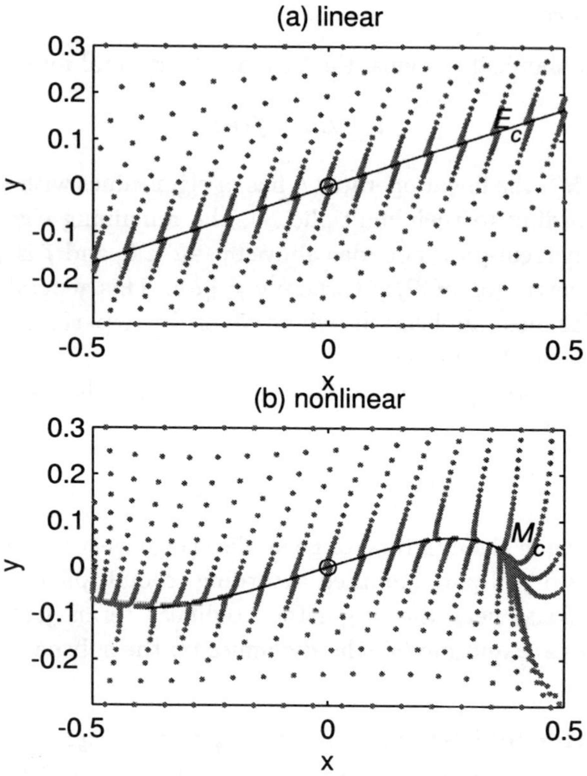

Fig. 5   An example of the exponential collapse of (a) linear dynamics onto the centre subspace $\mathcal{E}_c$, each dot on a trajectory a fixed time Δt apart, to compare with (b) showing that the nonlinear dynamics "bend" the subspace to the centre manifold $\mathcal{M}_c$ and produce a slow evolution thereon.

In many applications, such as fluid dynamics, we need theory dealing with not only infinite dimensional TOE's, but also infinite dimensional centre manifolds. Extensions of the above theorem to infinite dimensional dynamics are extant but limited. The most general theory I am aware of is currently due to Gallay [36], but it suffers from restrictions upon the nonlinear terms, they have to be bounded, which limits its rigorous application. Scarpellini [92] apparently places significantly less restrictions upon the nonlinearities in the dynamical equations, but while he addresses infinite dimensional centre manifolds, the results are severely constrained by requiring finite dimensional stable dynamics. Hărăgus [44;

45] has developed theory supporting infinite dimensional models, such as the Korteweg-de Vries equation, but only by placing extremely limiting restrictions upon the time derivatives in the TOE.

(f)   There is no spectral gap on an infinite domain!

for example, consider the nonlinear diffusion problem for a field $u(x,t)$ on the infinite domain $-\infty < x < \infty$:

$$u_t = u_{xx} + f(u).\qquad(2.6)$$

Here f is some nonlinear reaction or advection term or terms; we consider some examples later. Linearly, that is neglecting f, a solution $u = \exp(ikx + \lambda t)$ leads to the classic continuous spectrum $\lambda = -k^2$ as wavenumber k varies continously in an infinite domain. See that there is no clear division between decaying and long-lasting modes! There are decaying modes with decay rate arbitrarily close to zero. It would apear that centre manifold theory is of no use in this problem.

However, we know that linearly, a *compact release* leads to the similarity solution of an algebraic decaying Gaussian. We seek to determine the emergence of this simple structure not only for the linear problem, but also for some nonlinear problems.

(g)   The Wayne transform

[113; 112] is consistent for some nonlinearities f:

- first slow down time by writing the dynamics in terms of $\tau = \log t$ which turns algebraic decay in t to exponential decay in τ;
- second, beacuse the initial condition is one of a compact release, assumed localised near $x = 0$ without loss of generality, we scale space $\xi = x/\sqrt{t}$ to capture the ultimate spreading nature;
- last, extract the leading algebraic decay of the field u by transforming to seeking v such that $u = t^{-1/2}v(\xi,\tau)$

The nonlinear diffusion equation (2.6) transforms to

$$v_\tau = \tfrac{1}{2}v + \tfrac{1}{2}\xi v_\xi + v_{\xi\xi} + t^{3/2}f(u).\qquad(2.7)$$

For example, for Burgers' equation the nonlinearity is $f = -uu_x$, which in the aboved transformed equations becomes $t^{3/2}f = -vv_\xi$.[4] The utility of

---

[4]As an exercise: how might one handle nonlinearities $f = (u_x)^2$? or $f = u^2$?

the Wayne transform is that linearly the eigenmodes are $v = H_n(\xi)\exp(-\xi^2/2 + \lambda_n t)$ for the nth Hermite polynomial. The corresponding linear spectrum is now discrete: $\lambda_n = -n/2$! Thus a centre manifold model exists for the nonlinear diffusion (2.6), $v = v(a,\xi)$, parametrised by the amplitude $a(\tau)$ of the Gaussian $\exp(-\xi^2/2)$.

The utility of the Wayne transform, applied to a nonlinear diffusion problem in [98] for example, is very suggestive for other infinite domain problems. Perhaps many other similarity solutions can be placed on an interesting dynamical basis using this transform.

2.2.2   *Relevance*

Based on the rationale of neglecting rapidly decaying transients, our intention is to consider (2.5) as a simple model system for the TOE (2.4); simpler in the sense that it has lower dimensionality, m instead of n. However, there is a subtle point which is often overlooked. We must be assured that the actual solutions of the model (2.5) do indeed correspond to solutions of the full system (2.4). Exponential attraction to an invariant manifold is *not* enough to assure us that the solutions on the manifold actually model accurately all the nearby dynamics.[5]

Theorem 7.2   *(relevance) The neighbourhood U may be chosen so that all solutions of (2.4) staying in U tend exponentially to some solution of (2.5). That is, for all solutions $\boldsymbol{u}(t)$inU for all $t \geq 0$ there exists a solution $s(t)$ of the model (2.5) such that*

$$\boldsymbol{u}(t) = \boldsymbol{v}\left(s(t)\right) + \mathcal{O}\left(e^{-\mu' t}\right) \quad as \quad t \to \infty,$$

where $-\mu'$ may be estimated *as $-\mu$, the upper bound on the negative eigenvalues of the linear operator $\mathcal{L}$.*

This theorem is crucial to modelling; it asserts that for a wide variety of initial conditions the dynamics of the TOE decays exponentially quickly to a solution which can be simulated by the low-dimensional model. Somewhat pessimistically, the theorem requires initial conditions to be sufficiently small, namely in a neighbourhood U. But consider the system shown in

---

[5]For example, consider the system $\dot{x} = x + xy$ and $\dot{y} = -y$ which has the exponentially attractive invariant manifold $y = 0$. There is generally an unavoidable $\mathcal{O}(1)$ discrepancy between the full system, with solutions $x(t) \sim Ae^t + B$, and the low-dimensional model, $\dot{x} = x$ with $x(t) \sim Ae^t$, despite the exponential collapse onto $y = 0$.

Figure 3. Observe that all trajectories within the picture asymptote exponentially to the centre manifold. Thus the conclusions of this theorem are correct for at least *all* initial conditions within the figure. In practice the neighbourhood U may be quite large.

This relevance appears in the term *asymptotic completeness* for *inertial manifolds* [88, e.g.]. Very loosely, there it is argued that provided trajectories on the low-dimensional model do not approach each other as fast as the approach to the inertial manifold, then all nearby trajectories approach a solution of the model system. The dynamics of approximations relevant to modelling have also been usefully addressed in theory by Pliss & Sell [69].

Note that this strong relevance theorem is very useful. In particular, it is much better than the frequent practise of performing a self consistent algebraic analysis and claiming the results are useful simply because of internal self consistency.

(h)   The similarity in nonlinear diffusion is attractive

consider further the nonlinear diffusion (2.6). By the relevance theorem we know the centre manifold model for the evolution of the amplitude a of the Gaussian shape of v is approached something like

$$\Delta v = \mathcal{O}\big(exp(-\tau/2')\big) = \mathcal{O}\big(t^{-1/2'}\big) \,,$$

where the rate of attraction to the model $1/2' \approx 1/2$. Thus the *nonlinear* similarity solution for u is approached algebraically, roughly like $\mathcal{O}\big(t^{-1'}\big)$. For problems where the Wayne transform is applicable we will know that the similarity solution is relevant to the long-term dynamics of the system.

2.2.3   *Approximation*

We need to find an equation to solve which gives the centre manifold $\mathcal{M}_c$ and the model evolution thereon. This is obtained straightforwardly by substituting the assumed functional relations, that $\boldsymbol{u}(t) = \boldsymbol{v}(\boldsymbol{s}(t))$ where $\dot{\boldsymbol{s}} = \mathcal{G}\boldsymbol{s} + \boldsymbol{g}(\boldsymbol{s})$, into the TOE (2.4) and using the chain rule for time derivatives to obtain

$$\mathcal{L}\boldsymbol{v}(\boldsymbol{s}) + \boldsymbol{f}\left(\boldsymbol{v}(\boldsymbol{s})\right) = \frac{\partial \boldsymbol{s}}{\partial \boldsymbol{v}}\left[\mathcal{G}\boldsymbol{s} + \boldsymbol{g}\left(\boldsymbol{s}\right)\right] . \tag{2.8}$$

This is the equation to be solved for the centre manifold $\mathcal{M}_c$ (the right-hand side is just $d\boldsymbol{u}/dt$).

An extra condition is that $\mathcal{E}_c$ is the tangent space of $\mathcal{M}_c$ at the origin. Put more crudely, we require that v is quadratically close to $\mathcal{E}_c$. This condition ensures that the constructed manifold truly contains the whole of the critical centre modes, and nothing but the centre modes. Without it, the solution of (2.8) could be based on an almost arbitrary mixture of linear modes. Indeed, other invariant manifolds of note satisfy (2.8) but are tangent to different subspaces as $s \to 0$.

It is typically impossible to find exact solutions to (2.8). However, in applications we may approximate $\mathcal{M}_c$ to any desired accuracy. For functions $\phi : I\!\!R^m \to I\!\!R^n$ (imagine that ϕ approximates the shape v) and $\psi : I\!\!R^m \to I\!\!R^m$ (imagine that ψ approximates the evolution $\dot{s} = \mathcal{G}s + g(s)$) define the *residual* of (2.8)

$$\mathsf{R}(\phi, \psi) = \frac{\partial \phi}{\partial s} \psi(s) - \mathcal{L}\phi(s) - f\left(\phi(s)\right) , \qquad (2.9)$$

and observe that $\mathcal{M}_c$ satisfies $\mathsf{R}(v(s), \mathcal{G}s + g(s)) = 0$.

Theorem 7.3   *(approximation) If the tangent space of $\phi(s)$ at the origin is $\mathcal{E}_c$, and the residual $\mathsf{R}(\phi, \psi) = \mathcal{O}(s^p)$ as $s \to 0$ for some $p > 1$ (where $s = |s|$) then $v(s) = \phi(s) + \mathcal{O}(s^p)$ and $\dot{s} = \mathcal{G}s + g(s) = \psi(s) + \mathcal{O}(s^p)$ as $s \to 0$.*

That is, if we can satisfy the physical equations (2.8) to some order of accuracy then the centre manifold model is obtained to the same order of accuracy.

When problems are specified in the special separated form (5.1) the centre manifold may be determined simply by iteration [10, e.g.], in essence we performed just one iteration to obtain the thin fluid model (1.1). However, typically a solution is sought in the form of an asymptotic power series in s as developed by Coullet & Spiegel [21] for the general form (2.4) (and reinvented by Leen [50]). Such a power series solution looks very like the Lyapunov-Schmidt method [43, e.g.] for determining the nontrivial fixed points near a simple bifurcation; in contrast though, centre manifold theory additionally determines the dynamics, see [2; 101; 64; 17] for comparisons. More recently an iterative algorithm using the physical form (2.4) of the equations have been developed with the virtue that it is readily implemented and interpretted in computer algebra, see [82; 84]. This algorithm is very flexible and is recommended.

Centre manifold models need not be anchored to just the one equilibrium. Some very interesting models are constructed for the dynamics in the neighbourhood of a manifold of equilibria [10, e.g.]. Such models may be uniformly valid across the whole set of equilibria. One example is the thin fluid film model (1.1) where a fluid at rest of constant but *arbitrary* thickness provides the reference equilibria for a model of the evolution of the fluid film's thickness.

2.3 *A trick rescues the theory from obscurity*

Requiring that $\mathrm{Re}(\lambda) = 0$ precisely as the criterion for the modes of the dynamical model is far too restrictive in practice. Fortunately, an extension enables enormous applications.

It is easiest to see the trick in an example: to (2.2) adjoin the trivial dynamical equation $\dot{\epsilon} = 0$. Based on the linearisation about the equilibrium at the origin in the ϵxy-space (note that in this space ϵx is a nonlinear term), there then exists a two-dimensional centre manifold, parametrised by ϵ and x. The relevance and approximation theorems may similarly be applied to construct the centre manifold for small x and ϵ, and to validate the model, that

$$\dot{\epsilon} = 0, \quad \text{and} \quad \dot{x} = \epsilon x - x^3 + \mathcal{O}\!\left(s^4\right),$$

where $s = |(\epsilon, x)|$. Fixing upon any small value for ϵ then gives a model, actually involving only the evolution of x, from which one may predict, for example, the presence of a pitchfork bifurcation.

This example shows that we may apply centre manifold theory to problems which do not precisely fit the linear structure required by the theory. By adjoining trivial dynamical equations we may, in essence, perturb eigenvalues with small real-part, either negative or positive, so that the "interesting" modes then have eigenvalues that precisely satisfy $\mathrm{Re}(\lambda) = 0$ and hence are included in the centre manifold. This trick is not only used to unfold bifurcations, as above, it is used also to partially justify the long-wave approximation (§3), and to approximate "hard" problems by writing them as perturbations of easier problems (§4).

Because parameters are so important in application, it is useful to quote a more powerful and flexible approximation theorem. Consider the extended dynamical system with ℓ parameters ϵ:

$$\dot{\epsilon} = \mathbf{0}, \quad \text{and} \quad \dot{u} = \mathcal{L}u + f(\epsilon, u), \tag{2.10}$$

with $\mathcal{L}$ as before. For functions $\phi : \mathbb{R}^\ell \times \mathbb{R}^m \to \mathbb{R}^n$ approximating the centre manifold, and $\psi : \mathbb{R}^\ell \times \mathbb{R}^m \to \mathbb{R}^m$ approximating the evolution thereon, we have the following more general theorem [51].

Theorem 7.4   *(parameter approximation)* *If the tangent space of* $\phi(\epsilon, s)$ *at the origin is* $\mathbb{R}^\ell \times \mathcal{E}_c$, *and the residual of (2.10) is* $\mathrm{R}(\phi, \psi) = \mathcal{O}(s^p + \epsilon^q)$ *as* $(\epsilon, s) \to 0$ *for some* $p > 1$ *and* $q \geq 1$, *then* $v(\epsilon, s) = \phi(\epsilon, s) + \mathcal{O}(s^p + \epsilon^q)$ *and* $\dot{s} = \mathcal{G}s + g(\epsilon, s) = \psi(\epsilon, s) + \mathcal{O}(s^p + \epsilon^q)$ *as* $(\epsilon, s) \to 0$. *Arguments also justify the same statement but with errors* $\mathcal{O}(s^p, \epsilon^q)$.

Consequently centre manifold models are justifiably constructed to a rich variety of asymptotic errors. We may choose after construction the truncation appropriate to the use of the model. Centre manifold theory generates flexible models.

For example, we later expand some power series to relatively high order, 10th to 20th order perhaps, in a parameter and show convergence in this parameter. Other parameter and variable dependencies are only determined to low order. The approximation theorem justifies such flexibility.

2.4   *Summary*

In this section we have discussed a great theoretical basis of low-dimensional dynamical modelling. Using this basis the sequence of steps in the analysis of a dynamical system is the following.

(1) Find a useful equilibrium (or many of them).
(2) Investigate the linear spectrum.
(3) Creatively adapt the linear spectrum to fit centre manifold theory.
(4) Construct the model (perhaps using computer algebra) to some truncation as justified by the Approximation theorem.
(5) Then existence, accuracy and relevance are supported by the theory.

3   Slow space variations—dispersion in a channel

Centre manifold analysis has often been applied to reduce dynamics down to a "handful" of ODEs, typically the centre manifold is of dimension 1–3. Also typically, a normal form transformation (see §5) is then used to classify the dynamics obtained on the centre manifold (although Ipsen et al. [47] directly seek a centre manifold parametrised to be in normal form). This

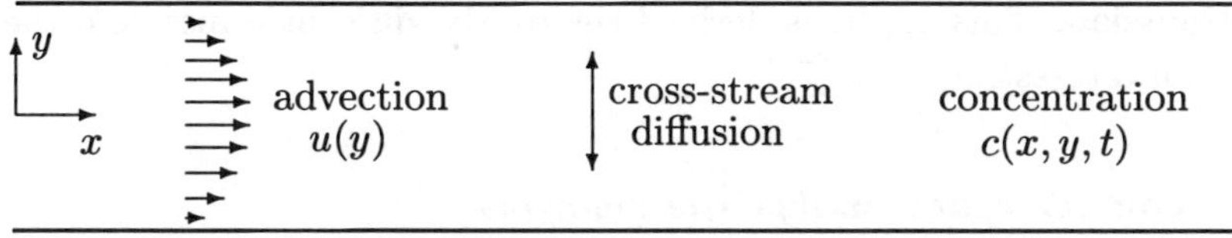

Fig. 6 Schematic diagram of the dispersion of a contaminant with concentration field $c(x,y,t)$ in a channel with advection $u(y)$ and cross-stream diffusion.

established approach is described in various books dealing with dynamical systems, see Carr [10], Guckenheimer & Holmes [40, Ch.3], Wiggins [115, Ch.2], Iooss & Adelmeyer [46, Ch.1], or Kuznetsov [49, Ch.3–5]. However, there is much greater scope for the use of centre manifold analysis in low-dimensional modelling.

Of particular interest are "infinite dimensional" models such as (1.1) for thin fluid flow. We investigate a simpler example first—namely, the dispersion of material along a long channel. The analysis is relatively simple because this dispersion is linear in the concentration and leads to a rigorous analysis in Fourier space. However, physical variables are much easier to understand. Lastly a novel trick in the application of theory leads to a "holistic discretisation" of the dispersion along the channel.

3.1 *Taylor modelled advection-dispersion in a pipe*

In a long, thin channel or pipe, as shown schematically in Figure 6, the dominant process is diffusion of contaminant, concentration $c(x,y,t)$, across the thin channel. This leads to decay of all modes except the constant mode (across the channel). But at different stations the constant will typically be different, and so the cross-sectional mean concentration C will depend upon downstream distance x. The shearing caused by the varying advection velocity, $u(y)$ then causes these variations in x to move and interact leading to the Taylor model [102; 103]

$$\frac{\partial C}{\partial t} = -U\frac{\partial C}{\partial x} + D\frac{\partial^2 C}{\partial x^2} + \cdots,\tag{3.1}$$

where U is simply the mean downstream velocity, but the effective diffusivity D is a nontrivial function of the velocity field. The Taylor model predicts the long-term concentration is in the shape of a moving and spreading Gaussian. This model is derived rigorously via centre manifold theory using an interesting trick [74].

3.2 *Fourier space analysis is rigorous*

Consider dispersion in the example channel shown in Figure 6 which has the non-dimensional governing equation

$$\frac{\partial c}{\partial t} = -\mathcal{P}u(y)\frac{\partial c}{\partial x} + \nabla^2 c, \tag{3.2}$$

where, for definiteness, the advecting velocity is here $u(y) = 3(1 - y^2)/2$ in the channel, non-dimensionally $-1 < y < 1$, and where $\mathcal{P}$, the Peclet number, is the downstream velocity relative to the rate of diffusion across the channel and is typically quite large. The Peclet number $\mathcal{P} \approx 10^2\text{--}10^4$ in many applications. Now take the Fourier transform of (3.2), so that $c = \int_{-\infty}^{\infty} \hat{c}e^{ikx}dx$, to obtain the Fourier space equation

$$\frac{\partial \hat{c}}{\partial t} = -ik\mathcal{P}u(y)\hat{c} - k^2\hat{c} + \frac{\partial^2 \hat{c}}{\partial y^2}. \tag{3.3}$$

Then utilise the trick introduced in §2.3 to unfold bifurcations, but here to introduce the approximation of a *large-scale, slowly-varying* spatial dependence. The trick is to adjoin to (3.3) the trivial equation that the wavenumber k is constant:

$$\frac{\partial k}{\partial t} = 0. \tag{3.4}$$

Adjoining this trivial equation bases the centre manifold analysis upon the small wavenumber, large-scale dynamics along the channel. The advection term $-ik\mathcal{P}u(y)\hat{c}$ and downstream diffusion, $k^2\hat{c}$, are then "nonlinear" in k and $\hat{c}$. Linearly (in this new sense), cross-channel diffusion causes concentrations $\hat{c}$ to decay exponentially quickly to a constant with respect to y, that is $\hat{c} \to \hat{C}$: the modes are $\cos(n\pi(y + 1)/2)\exp(\lambda_n t)$ for eigenvalues $\lambda_n = -n^2\pi^2/4$ for $n = 0, 1, 2, \ldots$. Centre manifold theory applied to (3.3–3.4) thus guarantees the existence of a relevant model parametrised by wavenumber k and the cross-stream average concentration $\hat{C}$.

The detail in the steps to construct the model are similar to those for many other perturbation methods. By the approximation theorem we seek to satisfy the transformed advection-diffusion equation (3.3) to some order of accuracy by substituting

$$\hat{c} \;=\; \hat{C} + ikv_1(y)\hat{C} + (ik)^2 v_2(y)\hat{C} + \cdots,$$
$$\text{such that} \quad \hat{C}_t \;=\; ikg_1\hat{C} + (ik)^2 g_2\hat{C} + \cdots, \tag{3.5}$$

where the model is linear in $\hat{C}$ as the original problem is linear in c. Substituting and equating powers of wavenumber k and transformed concentration $\hat{C}$ leads to a hierarchy of equations, the first two being:

$$g_1 \;=\; -\mathcal{P}u(y) + \frac{\partial^2 v_1}{\partial y^2}\,;$$

$$g_1 v_1 + g_2 \;=\; -\mathcal{P}u(y)v_1 + 1 + \frac{\partial^2 v_2}{\partial y^2}\,.$$

Integrating the first across the channel and using the boundary conditions shows $g_1 = -\mathcal{P}$, whence solving the differential equation gives the perturbation to the concentration profile to be $v_1 = -\mathcal{P}(7 - 30y^2 + 15y^4)/120$. Then integrating the second across the channel and using the boundary conditions shows $g_2 = 1 + 2\mathcal{P}^2/105$. Thus the Fourier transform of the cross-channel average $\hat{C}$ will in the long-term evolve according to

$$\frac{\partial \hat{C}}{\partial t} = -ik\mathcal{P}\hat{C} - \left(1 + \frac{2\mathcal{P}^2}{105}\right)k^2\hat{C} + \mathcal{O}(k^3)\,. \tag{3.6}$$

Since the dispersion is linear in concentration c, the residual has only a component in wavenumber k and not in concentration $\hat{C}$, and hence by the approximation theorem the only error is in the wavenumber k. Taking the inverse Fourier transform leads to the Taylor model (3.1) with the particular mean advection velocity $U = \mathcal{P}$ and effective diffusivity $D = 1 + 2\mathcal{P}^2/105$. See that the effective diffusivity *along the channel* is remarkably enhanced, through the Peclet number $\mathcal{P}$, by the shear in the velocity profile.

The Taylor model (3.1) approximates the dominant branch of the spectrum of the full problem as shown in Figure 7. The small wavenumber analysis here is based upon the real eigenvalues obtained when wavenumber $k = 0$. In essence we have solved a perturbed eigenvalue problem for the leading branch of eigenvalues.

(i)   Extensions:

the centre manifold analysis of dispersion in a pipe has also been extended by Balakotaiah & Chang [3] to the dynamics in a chromatograph chemical reactor where chemical reactions occur either in the flow or with the walls of

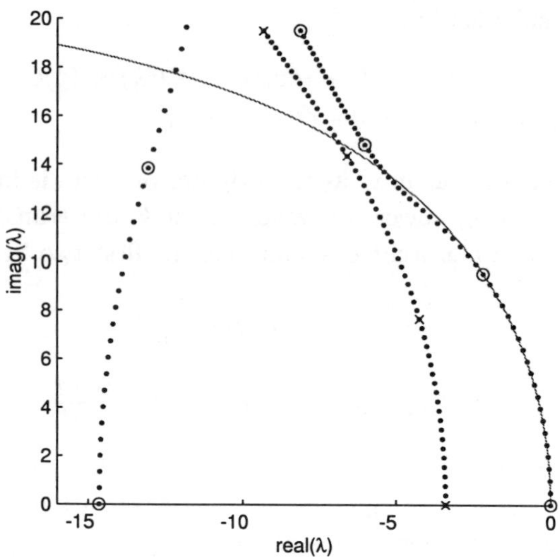

Fig. 7   Spectrum of advection-diffusion (3.2) as a function of wavenumber: the dots are the complex eigenvalues plotted for wavenumbers $\Delta k = 0.5$ apart starting from the real axis when $k = 0$; the three apparent branches correspond to different structure across the channel; the solid curve is that of the Taylor model (3.1) showing a good agreement between the original and model for wavenumbers up to about 10.

the pipe. It has also been applied to the modelling of separation of macro-molecules in a field flow fractionation channel by Suslov & Roberts [99; 100]. Rosencrans [89] has generalised the analysis to channels which have periodic and significant variation in their width along the channel. The resulting advection-dispersion model describes the evolution of the concentration averaged over each periodic cell of the channel.

Observe that the Taylor model (3.1), while certainly simpler than the TOE (3.2), is still of "infinite" dimension—there is an infinite freedom in $C(x)$ that is governed by the model. Here this inifinity comes from the inverse Fourier transform of (3.6) which combines together the "infinite" number of one dimensional models (3.6) that are valid for small k. In such "infinite" dimensional centre manifolds there is a much richer field of possible and applicable models to investigate than in centre manifolds of just a few dimensions.

It is straightforward for the centre manifold analysis to compute higher order models of the dispersion, involving terms such as C_{xxx} and C_{xxxx} for example, that show the evolution of the skewness and kurtosis of the concentration distribution. Indeed, Mercer & Roberts extended the analysis of the dispersion in a channel [60] and pipe [61] to very high order and shown, using a generalisation of the Domb-Sykes plot [60, Appendix], that these models actually converge for non-dimensional downstream wavenumber $|k|$ less than roughly 10, see Figure 7, depending upon the particular problem in hand (similar bounds also have been computed for beam theory, see [80]). Thus here we are assured that the model for dispersion resolves spatial details down to roughly $2\pi/10$ times the downstream advection distance in a cross-stream diffusion time. This quantitative limit on the resolution is some 10 times better than that estimated by Taylor [103]. The Relevance Theorem also assures us that the model will resolve temporal dynamics longer than a cross-stream diffusion time, the time scale of approach to the centre manifold. Both of these limits to the resolution of the model can only be improved by retaining more dynamic modes in the model, as was done by Watt & Roberts [111] in investigating an invariant manifold model based upon the two or three gravest modes in the channel. That we can sometimes find these *quantitative* bounds to the domain of the validity of a low-dimensional model is part of the power of centre manifold theory.

3.3 *Physical variables are understood*

The above analysis in Fourier space is rigorous. However, upon inspecting the details of the algebra it is apparent that the wavenumber, in the combination ik, just acts as a place holder for the spatial derivative ∂_x. The adjoining of the trivial dynamical equation (3.4), that $\partial k/\partial t = 0$, just focuses attention upon small wavenumber. Precisely the same effect is achieved in physical variables simply by treating spatial derivatives ∂_x as "small." A rationale for doing this is outlined in [74] based upon the local dynamics in any reach of the domain, but the operator ∂_x is unbounded and extant theory [36] cannot be applied directly. However, an analysis based upon this idea has the advantage that one deals in physical variables, rather than Fourier transformed quantities, and it easily generalises to more difficult problems such as thin fluid films (1.1) and as seen in the next section.

Treating ∂_x as small is based upon a local expansion in space [74]. Consider the channel near any station $x = X$. We may reasonably expand the concentration field in a Taylor series in its longitudinal dependence:

$$
\begin{aligned}
c(x,y,t) &= u_0(y,t) + u_1(y,t)(x-X) + u_2(y,t)\tfrac{1}{2}(x-X)^2 + \cdots \\
&= \sum_{n=0}^{\infty} u_n(y,t)T_n(x),
\end{aligned}
$$

where $T_n(x) = (x-X)^n/n!$ are basis functions for smoothly varying fields in the channel near $x = X$. Substitute this change of basis into (3.2), equate coefficients of the basis T_n to determine

$$
\frac{\partial u_n}{\partial t} = \frac{\partial^2 u_n}{\partial y^2} - \mathcal{P}u(y)u_{n+1} + u_{n+2} \quad n = 0,1,2,\dots. \tag{3.7}
$$

In this basis the right-hand side is upper triangular with ∂_y^2 down the diagonal. Hence all eigenvalues are negative except for a zero eigenvalue associated with each u_n. In a suitable inner product space we then claim the concentration field exponentially quickly approaches states $u_n \approx C_n$—representating a concentration field constant across the channel but with an arbitrary Taylor series description along the channel:

$$
c(x,y,t) = C_0 + C_1(x-X) + C_2\tfrac{1}{2}(x-X)^2 + \cdots.
$$

See that $C_n = \partial^n C/\partial x^n$ at the station $x = X$. Now seek corrections to this description. The corrections come from the coupling between the derivatives C_n through the off-diagonal terms on the right-hand side of (3.7). Seek

$$
u_n = C_n + u'_n(y,\boldsymbol{C}) \quad \text{such that} \quad \frac{\partial C_n}{\partial t} = g'_n(\boldsymbol{C}).
$$

Upon substituting into (3.7), straightforwardly deduce $g'_n = -\mathcal{P}C_{n+1} + C_{n+2}$ and $u'_n = \mathcal{P}v_1(y)C_{n+1}$: these represent the leading order change in the concentration due to the different advection velocities interacting with cross-stream diffusion. A further iteration suffices to find the shear dispersion as then

$$
\frac{\partial C_n}{\partial t} = g_n(\boldsymbol{C}) = -\mathcal{P}C_{n+1} + DC_{n+2} + \cdots,
$$

for the earlier given dispersion coefficient D. Recalling $C_n = \partial^n C/\partial x^n$, the $n = 0$ case of this equation simply reduces to the Taylor model (3.1). This derivation proceeds with more physical quantities. But it may be simplified. We only needed the $n = 0$ case at the end. All the other

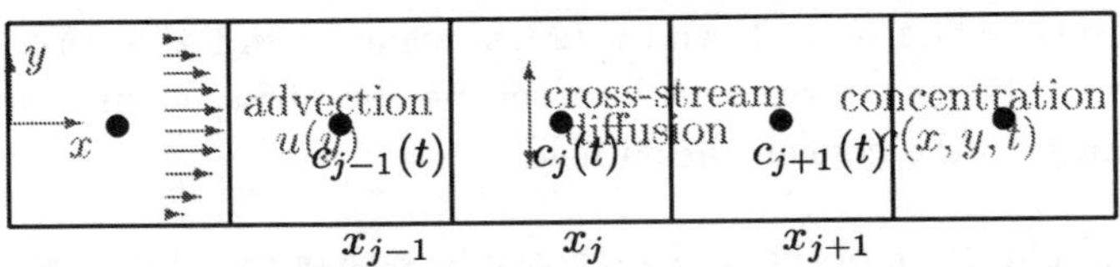

Fig. 8   The processes of advection and diffusion of contaminant $c(x, y, t)$ in a channel, overlaid by an artificial division into elements of finite length h. The mid point of each element are at $(x_j, 0)$ where the concentration is the evolving grid value $c_j(t)$.

equations could be obtained from spatial differentiation and by recognising that differentiation may be interchanged:

$$\frac{\partial}{\partial t}\left(\frac{\partial^n C}{\partial x^n}\right) = \frac{\partial^n}{\partial x^n}\left(\frac{\partial C}{\partial t}\right) .$$

We again conclude that simply treating longitudinal derivatives ∂_x as small is sound.

3.4   *Holistic discretisation resolves subgrid scale processes*

A novel and interesting approach to neumerical models uses centre manifold theory [85; 86]. One application is to create a numerical model of shear dispersion along the channel *direct* from the 2D advection-diffusion equation. We use the term "holistic" because rather than discretise eqations term by term, we instead construct a discretisation that incorporates interactions between all terms by solving the equations and hence resolving physical processes on the subgrid scale.

We first adapt the physical problem. As shown in Figure 8, divide the channel into elements of length h and we seek the concentration in the channel as a function of the grid values c_j and space: $c = v_j(x, y, c)$. These finite elements are created by introducing rather curious artificial internal boundary conditions (IBCs) linking the concentration in neighbouring elements [87]:

$$v_j(x_{j+1}, y) - v_j(x_j, y) = \gamma\left[v_{j+1}(x_{j+1}, y) - v_j(x_j, y)\right] . \tag{3.8}$$

See that when

- $\gamma = 0$, all elements are isolated from each other and the above IBC approximates an insulating condition at the boundary $x = x_j + h/2$;

- conversely, when $\gamma = 1$, the $v_j(x_j, y)$ cancel, and this IBC essentially requires sufficient agreement between neighbouring elements to restore continuity to the concentration field.

This particular form of the IBCs is chosen to satisfy the above two principles and to additionally ensure the discretisation is consistent as $h \to 0$ [87].

For definiteness we apply this technique to the varying diffusivity problem (so that algebraic results can be shown)

$$\frac{\partial c}{\partial t} = -\mathcal{P}\tfrac{3}{2}(1-y^2)\frac{\partial c}{\partial x} + \boldsymbol{\nabla} \cdot \left[(1-y^2)\boldsymbol{\nabla}c\right] . \tag{3.9}$$

For interest, the (continuum) corresponding Taylor model is

$$\frac{\partial C}{\partial t} = -\mathcal{P}\frac{\partial C}{\partial x} + \left(\frac{2}{3} + \frac{\mathcal{P}^2}{30}\right)\frac{\partial^2 C}{\partial x^2} + \left(-\frac{2\mathcal{P}}{45} + \frac{\mathcal{P}^3}{630}\right)\frac{\partial^3 C}{\partial x^3} + \cdots . \tag{3.10}$$

We will see some of these shear dispersion terms arise naturally in the discretisation. But we completely bypass this model in constructing the discretisation of (3.9).

(j) Apply centre manifold theory:

adjoin $\frac{\partial \gamma}{\partial t} = \frac{\partial \mathcal{P}}{\partial t} = 0$ to base the analysis on the dynamics when the advection is relatively weak and the elements in the channel are completely uncoupled. Thus linearly, diffusion is the dominant mechanism in each element and all subgrid structure decays exponentially quickly to a different constant in each element. The decay corresponds to negative eigenvalues, whereas the constant field corresponds to the existence of a zero eigenvalue associated with each element. Hence there exists a centre manifold parametrised by, say, the grid value $c_j = c(x_j, 0)$, the coupling γ and the advection parameter, the Peclet number $\mathcal{P}$. The centre manifold evolution of the grid values c_j then forms a numerical discretisation.

By the relevance theorem, the rate of decay to the model is will be of the order of the cross-element diffusion time. Thus the model is relevant at finite h. *This is novel:* the model is not justified in the small h limit but is fully justified by centre manifold theory at finite h.

(k) Construction:

in terms of centre differences $\mu\delta c_j = (c_{j+1} - c_{j-1})/2$ and $\delta^2 c_j = c_{j+1} - 2c_j + c_{j-1}$, the concentration field within the jth element on the centre manifold model is

$$
\begin{aligned}
c \;=\; & c_j + \gamma \left[\xi + \frac{\mathcal{P}}{4h} y^2 \right] \mu \delta c_j \\
& + \gamma \left[\frac{1}{2}\xi^2 - \frac{1}{6h^2} y^2 + \frac{\mathcal{P}h}{4}(\xi^3 - \xi) \right] \delta^2 c_j \\
& + \mathcal{O}\!\left(\gamma^2, \mathcal{P}^2\right),
\end{aligned}
$$

where $\xi = (x - x_j)/h$ is scaled to the length of the elements. See that this field resolves both downstream and across stream structures within each element. The field may be constructed by iteration; I outline the first iterate. We know a first approximation to the field in each element is just the grid value $c = c_j$. Let's seek a correction: $c = c_j + v'_j$ where the grid values will now evolve according to $\frac{\partial c_j}{\partial t} = g'_j$. Substitute into the governing advection-diffusion equation (3.9) to obtain the forced elliptic equation

$$
g'_j + \underbrace{\sum_k \frac{\partial v'_j}{\partial c_k} g'_k}_{\text{neglected}} = \underbrace{-\mathcal{P}u(y)\frac{\partial v'_j}{\partial x}}_{\text{neglected}} + \nabla \cdot \left[(1 - y^2)\nabla v'_j \right] .
$$

The negligible terms are small as products of corections and small asymptotically small factors such as $\mathcal{P}$ or γ. The boundary conditions (3.8) are of the form

$$
v'_j(x_{j+1}, y) - v'_j(x_j, y) = \gamma\left[c_{j+1} - c_j\right],
$$

generating a coupling between adjacent elements. The solution v'_j of this problem gives the $\mathcal{O}(\gamma)$ terms in the above concentration field within each element. *We resolve subgrid fields by solving the* PDE *(3.9)*

(1)   Evolution:

the discretisation is formed by further iteration to $\mathcal{O}(\gamma^2)$ in the element coupling. Shear dispersion appears at this order. We find

$$\frac{\partial c_j}{\partial t} = \frac{2}{3h^2}\left(\gamma\delta^2 - \tfrac{\gamma^2}{12}\delta^4\right)c_j + (\gamma - \gamma^2)\frac{\mathcal{P}^2}{8}\delta^2 c_j + \gamma^2\frac{\mathcal{P}^2}{30h^2}\delta^2 c_j$$
$$- \frac{\mathcal{P}}{h}\left(\gamma\mu\delta - \tfrac{\gamma^2}{6}\mu\delta^3\right)c_j - \gamma^2\frac{2\mathcal{P}}{45h^3}\mu\delta^3 c_j$$
$$+ \gamma^2\left(\frac{2}{135} + \frac{\mathcal{P}^2 h^2}{72} - \frac{\mathcal{P}^2 h^4}{20}\right)\frac{1}{h^4}\delta^4 c_j + \mathcal{O}(\mathcal{P}^3,\gamma^3) :$$
$$= \; x\text{-diffusion} + \text{stabilisation} + \text{shear dispersion}$$
$$+ \text{advection} + \text{skewness term}$$
$$+ \text{kurtosis} + \text{h.o.t.}$$

See that when evaluated at $\gamma = 1$:

- the first term on the right-hand side is an $\mathcal{O}(h^4)$ estimate of the longitudinal diffusion;
- the second term, if truncated to errors $\mathcal{O}(\gamma^2)$, would stablise the discretisation for large advection $(\mathcal{P})$, but here truncated to errors $\mathcal{O}(\gamma^3)$ disappears to leave
- the third term which approximates shear dispersion as appeared in the one-dimensional model (3.10);
- whereas the fourth term (the first on the second line above) is an $\mathcal{O}(h^4)$ estimate of the longitudinal advection at mean velocity $\mathcal{P}$;
- the fifth term contributes to the skewness of the solution as does a corresponding term in (3.10);
- and lastly the sixth term affects the kurtosis.

A remarkable fact: the shear dispersion appears here in a model created at finite h whereas it previously has only been derived by the large length scale, slowly varying assumption. Shear dispersion appears here because we resolve sugrid physical processes through the use of centre manifold theory.

3.5   *Summary*

- Infinite dimensional models for large-scale dynamics are justified to some extent. The trick is to just treat ∂_x as small parameter.

- The analysis is rigorous in linear problems such as shear dispersion when phrased in terms of small wavenumber k. Remarkably the asymptotic series converges!
- Numerical discretisations may be constructed using the centre manifold approach. By resolving subgrid scale processes the resultant models should perform well on finite size grids. See the web page `http://www.sci.usq.edu.au/staff/aroberts/holistic.html`
- Both of the examples seen in this section show different ways to manipulate the spectrum for application of centre manifold theory. Be creative in applications.

4   Cross-sectional averaging is unsound—thin film flows

In this section we briefly revisit the thin film model in light of the the ideas developed in the previous section by the application of centre manifold theory to shear dispersion. Then we recognise that a model with better temporal resolution is often needed. But the usual method of cross-sectional averaging the fluid equations can fail as is clearly demonstrate in a simple example. Instead include inertia into a fluid film model by basing the analysis on an imaginatively forced linear problem.

4.1   *Revisit the thin film model*

We return to the dynamics of thin films of fluid which are important in many industrial, environmental and biological processes. If surface tension σ is the only driving force then the simplest long-wave model is (1.1). To analyse the Navier-Stokes equations, the TOE, to derive such a model [90], recognise that across the thin fluid film, viscosity acts quickly to damp almost all cross-film shear flow. The distinction between the longitudinal and cross fluid dynamics is exactly analogous to that of dispersion in a pipe; we may proceed with a similar analysis. However, the equations are very different (see Figure 1). In particular, this problem has many nonlinearities: not only is the advection in the Navier-Stokes equation described by a nonlinear term, but also the thickness of the fluid film is to be found as part of the solution and its unknown location is another source of nonlinearity. Here, the practical course of analysis is to deal with physical variables and treat ∂_x as a "small" parameter which is negligible to the leading "linear" approximation. Treating ∂_x as small is equivalent to the long-wave approximation.

Assuming no longitudinal variations at all, $\partial_x = 0$, a linear analysis of the equations shows that there is one critical mode in the cross-fluid dynamics, corresponding to conservation of fluid, all others are lateral shear modes which decay exponentially due to viscosity. The critical mode is associated with conservation of the fluid and consequently it is natural to express the low-dimensional model in terms of the film thickness $\eta(x,t)$. An interesting aspect is the fact that although this is a nonlinear problem, conservation of fluid applies no matter how thick the fluid layer, and the state of no flow is an equilibrium for all constant η. Thus the analysis is valid for arbitrarily large variations in the thickness of the film, just so long as the variations are sufficiently slow in space and time. A relatively simple toy example of this is also discussed in [74]. These are examples, as mentioned in §2.2.3, of a centre manifold analysis based upon a manifold of equilibria rather than being anchored to just one fixed point.

Treating ∂_x as a small perturbing operator, the systematic centre manifold formalism straightforwardly derives models such as (1.1). In this problem each fluid equation neatly gives an update for one physical variable. For example, suppose we have some approximation, $\tilde{u}$ and $\tilde{v}$, for the velocity field and consider finding corrections, to $\tilde{u} + u'$ and $\tilde{v} + v'$, using the continuity equation $u_x + v_y = 0$: first substitute to deduce

$$\frac{\partial u'}{\partial x} + \frac{\partial v'}{\partial y} = -\frac{\partial \tilde{u}}{\partial x} - \frac{\partial \tilde{v}}{\partial y} = \text{(residual)};$$

second realise that ∂_x is a small operator so u'_x can be neglected here; hence lastly we just integrate the residual with respect to y to obtain the next correction to the vertical velocity v'. Similarly, though more complicated, for other equations and variables. Thus higher order models may be straightforwardly constructed if necessary [82], viz

$$\eta_t = -\frac{\sigma}{\mu}\partial_x \left[\tfrac{1}{3}\eta^3\eta_{xxx} + \tfrac{3}{5}\eta^5\eta_{xxxxx} + 3\eta^4\eta_x\eta_{xxxx} \right.$$
$$\left. + \eta^4\eta_{xx}\eta_{xxx} + \tfrac{11}{6}\eta^3\eta_x^2\eta_{xxx} - \eta^3\eta_x\eta_{xx}^2\right] + \mathcal{O}\!\left(\partial_x^7\right).$$

These new sixth order terms generally improve accuracy. However, since the leading order $\partial_x(\eta^3\eta_{xxx})$ is structural stable nonlinear diffusion, the higher order corrections are usually unlikely to be significant. Probably the best use of such terms is to provide a *good* estimate the error of the leading order model. A well know "rule of thumb" when summing asymptotic series is that the error is approximately the first neglected term: the above sixth-order terms quantitatively estimate the error in the model $\eta_t \propto \partial_x(\eta^3 \eta_{xxx})$. More interesting is to model the 2D spread of a 3D fluid over a curved substrate [90] including complicating effects of gravity and fluid inertia to produce an extensive model.

One significant limitation in the use of the whole family of models discussed above is that they only resolve dynamics significantly slower than the time scale of viscous decay, $T \approx 0.4\eta^2/\nu$, of the gravest shear mode, $u_1 \propto$ $\sin(y\pi/2\eta)$. For thicker fluid films, interesting dynamics may occur on this time scale. Another class of models is needed in such a situation. Traditionally these are obtained by integrating or averaging the horizontal momentum equation over the depth of the fluid to obtain an evolution equation for both the mean horizontal velocity $\bar{u}(x,t)$ and the fluid thickness $\eta(x,t)$ [71; 14, e.g.]. We explore such averaging next.

4.2 *Cross-sectional averaging can fail*

Unfortunately, simple modelling methods of cross-sectional averaging do not just get the coefficients wrong, they can fail dramatically. Abstractly, cross-sectional averaging is a projection of the dynamical equations onto some linear subspace of the state space, namely the space of functions which are constant over cross-sections. Typically, linear considerations suggest that this would be a useful procedure. But consider the simple example [49, p153]

$$\dot{x} = xy + x^3, \quad \text{and} \quad \dot{y} = -y - 2x^2, \tag{4.1}$$

whose trajectories are plotted in Figure 9. Linearly, $\dot{x} = 0$ and $\dot{y} = -y$ and so solutions exponentially quickly approach the subspace $y = 0$. Thus, one might plausibly argue projecting the dynamics onto $y = 0$ to give $\dot{x} = x^3$ will form a reasonable model of the long-term dynamics of the system from which one would deduce that $x = 0$ is an unstable fixed point. But this is not so. The centre manifold is straightforwardly constructed by iteration: Starting with approximations $\dot{x} = 0$ on $y = 0$:

- y-equation with the above substituted except for the linear y on the right-hand side then gives $y \approx -2x^2$ whence the x equation becomes $\dot{x} \approx -x^3$;

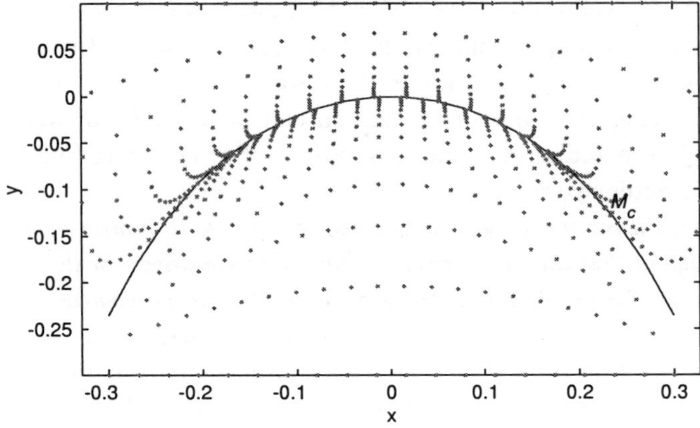

Fig. 9   Trajectories of the dynamical system (4.1), plotted as dots $\Delta t = 0.4$ apart, showing that although trajectories move away from the origin on $y = 0$, nonetheless the nonlinear dynamics on the centre manifold, $\mathcal{M}_c$, shows that the origin is stable.

- in a second iteration the y-equation then gives $y \approx -2x^2 - 4x^4$ whence the x-equation becomes $\dot{x} \approx -x^3 - 4x^5$.

Evidently the centre manifold is $y = -2x^2 - 4x^4 + \mathcal{O}(x^6)$ whence we deduce the correct model of the long-term dynamics to be $\dot{x} = -x^3 + \mathcal{O}(x^4)$ and hence the origin is actually stable, as seen in Figure 9. *Cross-sectionally averaging dynamical equations is unsound as a modelling paradigm.*

I briefly give another example of averaging deficiencies. Cross-sectionally averaging the advection-diffusion equation (3.2) leads to an advection-diffusion model (3.1) with the correct mean velocity $\mathcal{P}$ but with a woefully incorrect effective diffusion of 1 instead of $D = 1 + 2\mathcal{P}^2/105$ (recall the Peclet number $\mathcal{P}$ is typically large). We need a more systematic approach to modelling dynamics than that provided by simple averaging and filtering.

4.3   *Include inertia into a fluid film model*

Returning to the dynamics of a fluid film, observe that in the horizontal shear modes, $u_n \propto \sin(n y \pi / 2\eta)$ there is a relatively large *spectral gap* between the gravest mode u_1 and the next mode u_2; the eigenvalues are $\lambda_1 = -2.5\nu/\eta^2$ and $\lambda_2 = -22.2\nu/\eta^2$ respectively. Surely we should be able to gather the u_1 mode into the centre manifold. Because of the ratio of these eigenvalues, such a model would resolve dynamics on a time scale

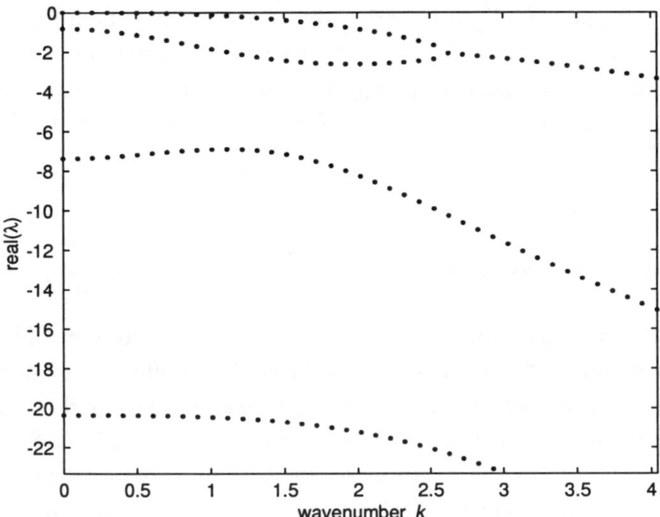

Fig. 10   Real part of the eigenvalues for the viscous decay of small waves, wavenumber k on a thin fluid film. The Reynolds number is Re = 3. Each branch corresponds to more cross-film structure in the horizontal shear. See the first two branches interact, merging at $k \approx 2.5$ to form decaying waves.

which is an order of magnitude shorter than those resolved by (1.1). See the eigenvalues for the linear problem as a function of lateral wavenumber k in Figure 10: see the leading branch and the second branch, corresponding to the slowest decaying lateral shear mode, interact to become decaying waves at wavenumber $k \approx 2.5$. The other branches decay much faster. A model resolving the two leading branches will be valid over a large parameter regime.

(m)   Creatively adapt:
the trick of §2.3 allows us to apply centre manifold techniques to obtain a coupled model for the evolution of $\bar{u}$ and η which resolves transients on a shorter time scale [81]. Manipulate the non-dimensional horizontal component of the Navier-Stokes equation (1.5) to

$$\frac{\partial u}{\partial t} + u\frac{\partial u}{\partial x} + v\frac{\partial u}{\partial y} = -\frac{\partial p}{\partial x} + \nu\nabla^2 u \quad + (1-\gamma)\nu\left(\frac{\pi}{2\eta}\right)^2 u\,, \qquad (4.2)$$

and adjoining $\dot{\gamma} = 0$ changes the spectrum. When $\gamma = 0$, the introduced artificial forcing makes u_1 a critical mode, along with η, and hence there exists a centre manifold parametrised by $\bar{u}$, measuring the amplitude of u_1, the film thickness η, and the artificial parameter γ. Setting $\gamma = 1$ recovers the original PDEs and so pursuing the analysis and subsequently setting $\gamma = 1$ leads to an approximate model for the original dynamics. The model is found [81] to be

$$\eta_t \;=\; -\partial_x\left(\bar{u}\eta\right), \tag{4.3}$$

$$\bar{u}_t \;\approx\; 0.8238\,\sigma\eta_{xxx} - 1.504\,\bar{u}\bar{u}_x - 2.467\frac{\nu}{\eta^2}\bar{u} - 0.1516\frac{\bar{u}^2}{\eta}\eta_x - \nu\bar{u}_{xx}, \tag{4.4}$$

as in depth averaged equations but quantitatively different. Three coefficients are of note: firstly, $2.467 = \pi^2/4$ is the coefficient of decay of the gravest horizontal shear mode; secondly, 1.504 resolves conflicting opinions about the coefficient of the advection term; and lastly, $0.8238 = \pi^2/12$ is significantly less than the value 1 obtained by all others based upon cross-film averaging—nonetheless 0.8238 is correct as being the direction cosine of the angle between the pressure forcing and the profile u_1 of the viscous shear flow.

Let's overview some of the initial key steps in deriving the model equation (4.4) while glossing over much tedious detail. First linearising (4.2) by discarding products of velocities, products with γ and any x-derivatives (as these are assumed slow) we obtain

$$\frac{\partial u}{\partial t} = \nu u_{yy} + \nu\left(\frac{\pi}{2\eta}\right)^2 u,$$

which has the neutral mode $u = \bar{u}(x,t)\frac{\pi}{2}\sin\zeta$ where $\zeta = \pi y/(2\eta)$ is a scaled vertical coordinate ranging from 0 (the substrate) to $\pi/2$ (the surface). Now seek small modifications by substituting

$$u = \bar{u}(x,t)\tfrac{\pi}{2}\sin\zeta + u', \quad v \approx 0, \quad \text{and} \quad p \approx -\sigma\eta_{xx},$$

the last from earlier arguments to obtain (1.1). Substitute into (4.2) and now keep the leading order terms in the modifications to obtain

$$\bar{u}_t\tfrac{\pi}{2}\sin\zeta + \bar{u}\bar{u}_x\tfrac{\pi^2}{4}\sin^2\zeta = \sigma\eta_{xxx} + \nu\bar{u}_{xx}\tfrac{\pi}{2}\sin\zeta - \gamma\nu\bar{u}\tfrac{\pi^3}{8\eta^2}\sin\zeta + \nu\left[u'_{yy} + \tfrac{\pi^2}{4}u'\right].$$

This is to be solved for the modification u' across the film. But the operator involving u' on the right is singular and so the other terms must lie in the range of this operator by choosing $\bar{u}_t$. This solvability is achieved by integrating across the film—but we must weight the integral by $\sin\zeta$, namely

$\int_0^{\pi/2} \cdots \sin\zeta\, d\zeta$, not a constant, because $\sin\zeta$ is the relevant homogeneous solution of the operator. Performing this integral we obtain

$$\frac{\partial \bar{u}}{\partial t} + \tfrac{4}{3}\bar{u}\bar{u}_x = \tfrac{8}{\pi^2}\sigma\eta_{xxx} + \nu\bar{u}_{xx} - \gamma\nu\frac{\pi^2}{4\eta^2}\bar{u}. \tag{4.5}$$

When evaluated at the physically relevant $\gamma = 1$ this (4.5) is roughly the reported model (4.4). Higher order refinements will transform (4.5) into (4.4). These refinements are obtained by actually determining the modification u' and corresponding modifications to the pressure and vertical velocity, then updating the evolution from the consequent residual of the governing equations, and repeating to the desired order [81], preferably using computer algebra.

The artificial adaptation (4.2) of the horizontal momentum equation is not unique. Exactly the same eventual model (4.4) is obtained [83] by appropriate artificial manipulation of the tangential stress boundary condition at the free-surface.

Similar ideas are being employed to analyse the dynamics of a turbulent river or flood. In initial work by Mei et al. [58; 59] we took the k-ϵ model of turbulent flow as the TOE, and perturbed some critical coefficients so that a centre manifold inspired model could be constructed based upon the water depth η and the depth averages of the horizontal velocity $\bar{u}$, the turbulent energy $\bar{k}$, and the turbulent dissipation $\bar{\epsilon}$. Further analysis on this interesting model is in progress; it promises a sophisticated and reliable model of flood, river and estuarine dynamics.

However, remember we need more theory on infinite dimensional centre manifolds to provide rigorous support for these sorts of interesting nonlinear, slowly-varying, long-wave models.

4.4 *Summary*

- A high an order expansion usually does not make up for omitting dynamical modes of interest. Sure it may be accurate but there will always be fundamental limits on the temporal resolution. Instead one often seeks to involve some of the weaker decaying dynamical modes.
- However, cross-sectional averaging—in essence a simple linear projection— is unsound, it can fail both quantitatively and qualitatively.
- Instead, again, one may creatively modify the linear spectrum to apply centre manifold theory with more neutral modes.

5   A normal form illuminates modelling principles

Here the normal form of the complete dynamics of the TOE clearly shows both the low-dimensional model (Elphick et al. [31] and Shilnikov et al. [96]), the nature of transients, the correct provision of initial conditions (Cox & Roberts [24]), and that resonances weaken "slow" models of non-dissipative dynamics.

Commonly, as intimated at the start of §3, a normal form transformation may be applied to the dynamics *after* reduction from the TOE to the centre manifold. The normal form then assists in the classification of the dynamics. Recent work by Wittenberg & Holmes [117] indicates though that the normal form of the low-dimensional dynamics *on a centre manifold* may be a poor model: the predicted dynamics of the normal form of the Brusselator dynamics [117] were significantly poorer than that of the centre manifold model itself. I consider that the normal form of the low-dimensional dynamics, though well established for dynamical taxonomy [49, e.g.,Ch.3–5], is inappropriate for modelling: the normal form transformation is not used in a coordinate free fashion—The method assumes that it is good to use polynomials as basis functions to describe the dynamics. Perhaps other basis functions for the normal form could serve better when modelling dynamics. But the purpose of this section is different. Here we discuss how normal forms illuminate issues of modelling dynamics.

5.1   *Normal forms clearly show the model*

In this section we discuss normal form transformations that separates the long-lasting centre dynamics from the decaying stable dynamics and hence display the dynamical model. For simplicity of discussion suppose that the dynamics of the TOE are given in the *separated form*:

$$
\begin{aligned}
\dot{x} &= Ax + f(x,y)\,, \\
\dot{y} &= By + g(x,y)\,,
\end{aligned}
\tag{5.1}
$$

where the eigenvalues of the matrix A have real-part zero, the eigenvalues of the matrix B have strictly negative real-part, and f and g are quadratically nonlinear functions at the origin, of $\mathcal{O}(x^2 + y^2)$ where $x = |x|$ and $y = |y|$. Simple examples of this separated form are (2.1) and (4.1). This separated form has been most often used for theoretical statements as we do here for some modelling issues.

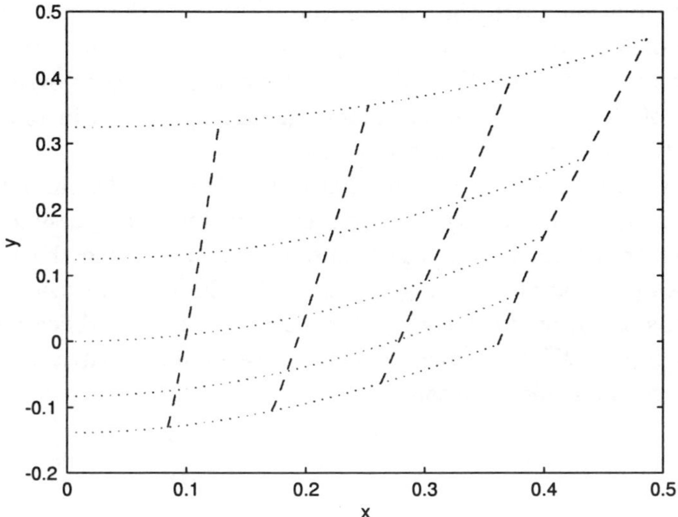

Fig. 11   The normal form transformation of (2.1): the dashed lines are the constant X isochrons; the dotted lines are constant Y; with $\Delta X = \Delta Y = 0.1$. The centre manifold is the dotted curve ($Y = 0$) emanating from the origin.

Consider as an example the system (2.1).[6]   The nonlinear but near identity coordinate transformation

$$\begin{aligned} x &= X + XY + \tfrac{3}{2}XY^2 + \cdots \\ y &= Y + X^2 + 2Y^2 + 4Y^3 + \cdots, \end{aligned} \tag{5.2}$$

shown in Figure 11, separates the dynamics to

$$\dot{X} = -X^3, \quad \text{and} \quad \dot{Y} = -(1 + 2X^2 + \cdots)Y. \tag{5.3}$$

In these new coordinates see that $Y = 0$ is exponentially quickly attractive to all solutions and clearly forms the centre manifold: however $X(t)$ evolves, the decaying solutions are just

$$Y(t) = Y(0) \exp\left[-\int (1 + 2X^2 + \cdots)dt\right].$$

Further, (5.3) shows the evolution of X is independent of Y. That is, $X(t)$ evolves independently of whether the system is on or off the centre manifold,

---

[6]The other example system (4.1) has a normal form that you may like to find approximately by substituting $x = X + aXY + bXY^2$ and $y = Y + cX^2$ such that $\dot{X} = \alpha X^3$ and $\dot{Y} = -Y + \beta X^2 Y$, and neglecting quartic and higher order terms.

and so all solutions with the same initial $X(0)$ have the same subsequent evolution of $X(t)$ and so exponentially quickly approach each other as $Y(t)$ decays to the centre manifold $Y = 0$. These properties of existence and relevance of the centre manifold model are clearly shown in (5.3) by the near identity coordinate transform (5.2).

Such a normal form transformation works its magic by simplifying the differential equations through putting as many terms as possible into the coordinate transform. That the resultant dynamics have the form seen in (5.3) is appreciated by trying to modify it. Suppose we seek to modify the x transformation to $x + x'$ with a corresponding modification to the evolution $\dot{X} = -X^3 + g'$ where x' and g' are small. Substitute into (2.1) and use the chain rule to obtain

$$\frac{\partial x}{\partial X}(-X^3 + g') + \frac{\partial x}{\partial Y}\dot{Y} + \frac{\partial x'}{\partial X}(-X^3 + g') + \frac{\partial x'}{\partial Y}\dot{Y} = -(x + x')y \,.$$

Now recognise $\frac{\partial x}{\partial X}(-X^3) + \frac{\partial x}{\partial Y}\dot{Y} = \dot{x}$, neglect small quantities and approximate $\frac{\partial x}{\partial X} \approx 1$ and $\dot{Y} \approx -Y$ wherever they multiply small modifications to obtain

$$g' - Y\frac{\partial x'}{\partial Y} = -\dot{x} - xy = (\text{residual}) \,.$$

The right-hand side is the residual of the current transform. See that any term involving a power of Y can be absorbed into the coordinate transform x'—it need not appear in the evolution through g'. However, no (smooth) term purely involving powers of X can be generated by $Y\frac{\partial x'}{\partial Y}$ and so any such term in the residual must be incorporated into the evolution of $\dot{X}$, as we have with $-X^3$, to contribute to the evolution on the centre manifold. This explains why the X-equation has no Y dependence.

Now suppose we seek to modify the y-transformation to $y + y'$ and the Y-evolution to $\dot{Y} + h'$ for small y' and h'. Similar substitution into (2.1) and approximation leads to the following equation for the modifications:

$$h' + y' - Y\frac{\partial y'}{\partial Y} = (\text{residual}) \,.$$

See that $y' = Y$ is a homogeneous solution of the "homological operator" $y' - Y\frac{\partial y'}{\partial Y} = 0$ and so all terms except those linear in Y are incorporated into the coordinate transform through y'. This leaves just those terms linear in Y to appear in the transformed dynamics (5.3); thus $Y = 0$

naturally appears as the invariant and attractive manifold in the normal form. This pattern generalises.

For the general separated system (5.1) we also seek a similarly useful, nonlinear, but near identity, coordinate transformation

$$x = X + \Xi(X, Y), \quad \text{and} \quad y = Y + \Psi(X, Y), \tag{5.4}$$

(such as the example shown in Figure 11) so that the TOE is transformed to

$$
\begin{aligned}
\dot{x} &= Ax + f(x, y) \\
\dot{y} &= By + g(x, y)
\end{aligned}
\mapsto
\begin{aligned}
\dot{X} &= AX + F(X) \\
\dot{Y} &= [B + G(X, Y)]Y
\end{aligned}
\tag{5.5}
$$

This is always possible given the pattern of eigenvalues for the existence of a centre manifold [96, Theroem 5.5]. In the form (5.5), observe that $Y = 0$ is clearly the invariant and exponentially attractive centre manifold $\mathcal{M}_c$. The normal form transformation works its magic by decoupling the long-lasting centre modes, X, from the quickly decaying modes Y (though the decaying modes may depend nonlinearly upon the centre modes).

Although the normal form of the TOE is a very useful conceptual tool, it does not provide a practical method for actually constructing a low-dimensional model. The reason is that it involves considerable wasted algebra in the transformation of the ultimately neglected, quickly decaying modes. If there are many decaying mode, as is usual in interesting physical systems (there are an infinite number of decaying modes in shear dispersion and in thin fluid films), then the normal form transformation may be practically impossible to construct, whereas methods based upon the Approximation Theorem 7.3 will be manageable [21; 50; 82].

5.2   *Initial conditions are long-lasting*

Many models are used to make definite predictions of later times given that the system starts from a given initial state—the forecast problem. So, suppose you know the initial state of the system for the TOE: what is the corresponding initial condition for the model?

For example, if a pollutant is released into a river from a given site, we wish to know what will be the dispersion of this specific cloud of pollutant. If the pollutant is released in the middle of a channel it will be initially carried downstream quicker than if it is released at the side; in the long-term the pollutant clouds will have different mean locations. The initial condition

$$C(x,0) = \overline{c_0} + \partial_x(\overline{v_1(y)c_0}) + \mathcal{O}(\partial_x^2 c_0) \tag{5.6}$$

derived in [60; 61; 111] for the Taylor model (3.1) in channels and pipes, where the overline is the cross-sectional average, reflects such differences in the release: for example, a release in mid-channel corresponds to where the structure function $v_1(y)$ is negative and hence the term in the longitudinal derivative c_{0x} effectively shifts the release downstream for the model. Similarly, the long-term variance deficit in the spread of the contaminant is also predicted given the next order in the correct initial condition projection (5.6). As commented in the introduction, the initial condition (1.2) for the lubrication model (1.1) of a thin film of fluid flow has also been derived [99].

With the geometric picture of centre manifolds we have created a mechanism to provide correct initial conditions for low-dimensional dynamical models. Consider further the general normal form (5.5). But also, both on and off $\mathcal{M}_c$, the $X(t)$ evolution is completely independent of the decaying modes Y. Hence all solutions of the TOE from initial conditions with the same $X(0)$ have precisely the same $X(t)$ evolution for all time, and so they all tend to the same solution on the centre manifold. In the example (2.1), all solutions starting on any given dashed curve $X = \text{const}$ in Figure 11 exponentially quickly the corresponding solution on $\mathcal{M}_c$. Thus the normal form (5.5) shows that the projection of initial conditions onto the centre manifold is along constant X. Constant X are termed *isochronic manifolds* [75], *isochrons* [116; 39] or *leaves of the foliation* [96, p279]. *The correct initial condition for the model is* not *the definition of the model's parameters evaluated at the physical initial condition.*

Remarkably, this issue of providing the correct initial conditions for a low-dimensional dynamical model has received very little attention in the past. In many cases this is because attention has focussed on the typical dynamics inherent in a model. However, even when interested in making definitive forecasts, generally people have simply assumed that the provision either is according to the linear dynamics or is simply by the evaluation of the "amplitudes" in the model. That these assumptions are not always sound nonetheless has been occasionally recognised in the phenomenon of "initial slip" [38; 41; 37; 107]. As developed in [75; 24] a useful procedure for determining correct initial conditions for fore-

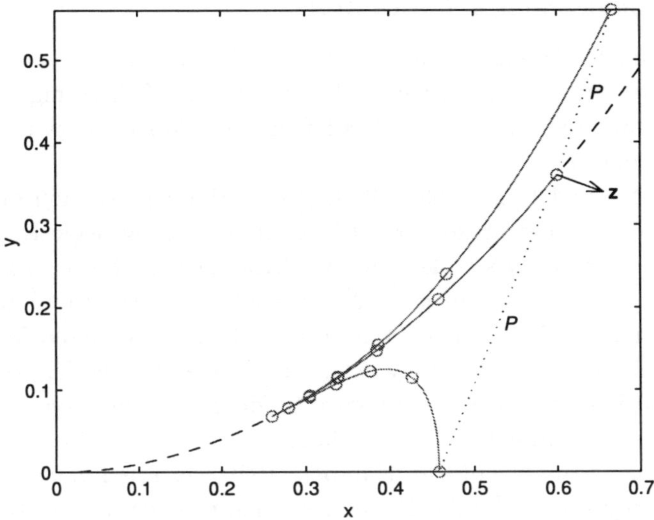

Fig. 12   Trajectories (solid) of (2.1) from three different initial conditions, all with the same long-term dynamics on the centre manifold $\mathcal{M}_c$ (dashed) to an exponentially small error. The circles are plotted at $\Delta t = 1$ apart. The modelling issue is to find the projection P (dotted) from any given initial condition off $\mathcal{M}_c$ onto one for the model on $\mathcal{M}_c$. P is described by its normal vector z at $\mathcal{M}_c$; here $z = (1, -x/(1 + 2x^2))$.

casting with a low-dimensional model is based on the geometric picture of a centre manifold in the state space.

Both the Relevance Theorem 7.2 and the normal form (5.5) assure us that there is indeed a particular solution of the low-dimensional model on $\mathcal{M}_c$ which is approached exponentially by every trajectory of the TOE (provided the trajectory starts close enough to $\mathcal{M}_c$). This is illustrated in Figure 12 where trajectories of the dynamical system (2.1) from three different initial conditions all have the same long-term evolution *to a difference which decays exponentially quickly.* The modelling task is to find the projection P from any given initial state off $\mathcal{M}_c$, say u_0, onto a state on $\mathcal{M}_c$, say $v(s_0)$, so that the long-term evolution, from s_0 in the model, will be the same to an exponentially small error. Some algebra [84] determines this projection by finding orthogonal vectors z to the isochrons as shown in Figure 12.

5.3   *The slow manifold is central*

Consider briefly the flight of an elastic ball. In such a purely elastic body, elastic waves "ring" perpetually within the body. If these high frequency vibrations are ignored, then what is left is the *relatively* slow dynamics of rigid body motion.

Muncaster and Cohen [63; 19] suggested the construction of the low-dimensional manifold of slow, rigid-body dynamics by neglecting the fast modes. The extremely simple example of the motion of a one-dimensional elastic body is discussed in [80, §2]. In contrast to the rapid collapse to the centre manifold, the slow dynamics on the slow manifold form a low-dimensional model because they act as a "centre" for the fast oscillations of neighbouring trajectories, see Figure 13 for an example from Lorenz [53]. This principle of neglecting fast oscillations is completely equivalent to the *guiding centre* principle of Van Kampen [108], which is frequently invoked in plasma physics, see [6; 20; 13; 48] for some recent work. In the context of the quantum dynamics of a nucleus, Troupe and Rosensteel [106] discuss many of the basic issues of low-dimensional slow manifold models under the name of *collective models* or *collective dynamics*.

The concept of a slow manifold is generally applicable to Hamiltonian dynamical systems because of their typical spectrum. Examples are beam models (1.7) and are also found in atmospheric dynamics [91], water waves [25; 66], and plasma physics [6; 67; 68; 114].

You may have noted that all the specific examples of centre manifolds discussed in earlier sections have also been slow manifolds as they are based upon zero-eigenvalue modes. However, in this section I specifically focus on situations where the slow dynamics occurs among fast oscillations. That is, we take the spectrum of the linear dynamics to consist entirely of some zero eigenvalues and some, perhaps infinitely many, purely imaginary eigenvalues.

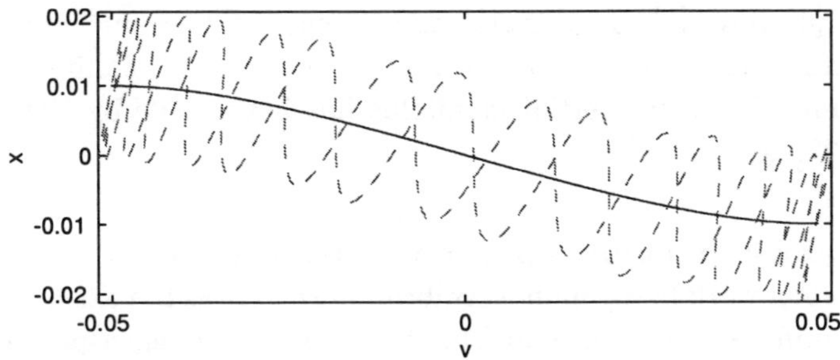

Fig. 13   A comparison of the trajectories on (solid) and off (dashed) the slow manifold $\mathcal{M}_0$ for the Lorenz system (5.7). Observe that the fast oscillations off the slow manifold reasonably track the evolution on $\mathcal{M}_0$.

(n)   Quasi-geostrophy is a slow manifold:

The concept of geostrophy is important in atmospheric dynamics, see recent work in [8; 7]. Lorenz [53] introduced the five mode dynamical system

$$
\begin{aligned}
\dot{u} &= -vw + bvz \\
\dot{v} &= uw - buz \\
\dot{w} &= -uv \\
\dot{x} &= -z \\
\dot{z} &= x + buv\,,
\end{aligned}
\tag{5.7}
$$

to illuminate the nonlinear slow manifold of quasi-geostrophy. Observe that x and z oscillate quickly, while u, v and w evolve slowly, at least near the origin. A couple of trajectories of this system are plotted in Figure 13. The slow manifold of this 5-dimensional TOE is 3-dimensional based upon the three 0 eigenvalues associated with u, v and w in the linear dynamics. Theory by Sijbrand [97, §6–7] applied to (5.7) demonstrates existence of the slow manifold, $\mathcal{M}_0$,

$$
x = -buv + \mathcal{O}\!\left(s^4\right), \quad z = b(u^2 - v^2)w + \mathcal{O}\!\left(s^5\right),
\tag{5.8}
$$

where $s = |(u,v,w)|$, to some level of smoothness. Lorenz later argued [52] that truly slow dynamics do not exist in (5.7) because if it did then there must be an infinite number of singularities in the slow manifold. However, Cox & Roberts [23] have shown that the singularities are exponentially weak and so are negligible for small enough amplitude flow. Recently, Bokhove &

Shepherd [4], Boyd [5], Camassa [9], and Lorenz [54] have further elucidated the structure and dynamics associated with the slow manifold of (5.7), and a connection with inertial manifolds has been made by Debussche & Temam [27; 28].

(o)   Modulation occurs on a sub-centre manifold:

building upon work by Lyapunov, Sijbrand [97] proves the existence of *sub-centre manifold*s. That is, manifolds based upon the eigenspace of a pair of pure imaginary modes; slow manifolds corresponding to eigenvalues that are precisely 0 are a special case. His theorems are directly relevant to the existence and construction of nonlinear normal modes of oscillation as investigated by Shaw et al. [95; 93; 94; 65]. It may be that such theorems also provide the basis for a justification of modulation equations describing the slow space-time evolution of the amplitude and phase of nonlinear dispersive waves. A simple example being the nonlinear Schrödinger equation usually derived using the method of multiple scales. A sub-centre manifold inspired approach to the modulation and interaction of nonlinear waves has been elucidated in [79].

(p)   Resonance destroys relevance:

but there is a further twist in slow manifold models: there is no strong relevance theorem for the low-dimensional dynamics on the slow manifold. There is no rigorous assurance that they do indeed model the TOE. Instead, Cox & Roberts [23; 24] used normal forms, (§5.1) to show that the dynamics on and off the slow manifold generally differ by an amount of $\mathcal{O}(R^2)$, where R measures the amplitude of the fast oscillations, it measures the distance off $\mathcal{M}_0$. That is, there is some unavoidable slip between the model and the TOE. Thus, in general, *one can only expect a slow manifold model to be accurately predictive for a time* $o(1/R^2)$.

The Lorenz model (5.7) exhibits such quadratic resonance. Upon making the near identity, nonlinear coordinate transform

$$u \;=\; U - bVX - \frac{b^2}{4}U(X^2 - Z^2) - bV(U^2 - V^2)X - \frac{b^3}{12}VX(X^2 + 3Z^2),$$

$$v \;=\; V + bUX - \frac{b^2}{4}V(X^2 - Z^2) + bU(U^2 - V^2)X + \frac{b^3}{12}UX(X^2 + 3Z^2),$$

$$w \;=\; W - b(U^2 - V^2)Z + 4bUVWX + b^2UVXZ, \tag{5.9}$$

$$x \;=\; X - bUV - bUV(U^2 - V^2 + 4W^2) - \frac{b^3}{2}UV(X^2 + Z^2),$$

$$z \;=\; Z + b(U^2 - V^2)W,$$

the dynamics is described to quintic errors by

$$\begin{aligned}
\dot{U} &= -VW + b^2VW(U^2 - V^2), \\
\dot{V} &= UW - b^2UW(U^2 - V^2), \\
\dot{W} &= -UV + \frac{b^2}{2}UV(X^2 + Z^2), \\
\dot{X} &= -Z, \\
\dot{Z} &= X + b^2(U^2 - V^2)X - 2b^3UVX^2.
\end{aligned} \tag{5.10}$$

There are three important details:

- first, see from the last two equations in (5.10) that $X = Z = 0$ is precisely the slow manifold $\mathcal{M}_0$ of the dynamics—$\mathcal{M}_0$ reduces to (5.8) upon setting $X = Z = 0$ in the coordinate transform (5.9);
- second, the evolution upon $\mathcal{M}_0$ is given by the first three equations in (5.10) with $X = Z = 0$;
- but thirdly, the evolution on and off $\mathcal{M}_0$ is irretrievably different as the $\dot{W}$ equation has a component in fast wave amplitude $R^2 = X^2 + Z^2$.

Thus, in contrast to centre manifold models, there is inevitably some slip between a slow model and its original system.

The difference is that here the coordinate transform cannot completely eliminate the fast variables from the slow modes. Suppose we try by modifying the w transform to $w + w'$ with the aim of eliminating the awkward terms in the evolution $\dot{W}$ by modifying it to $\dot{W} + g'$ for some small corrections w' and g'. The $\dot{w} = -uv$ equation then becomes, using the chain rule,

$$\frac{\partial w}{\partial U}\dot{U} + \frac{\partial w}{\partial V}\dot{V} + \frac{\partial w}{\partial W}(\dot{W} + g') + \frac{\partial w}{\partial X}\dot{X} + \frac{\partial w}{\partial Z}\dot{Z}$$
$$+ \frac{\partial w'}{\partial U}\dot{U} + \frac{\partial w'}{\partial V}\dot{V} + \frac{\partial w'}{\partial W}(\dot{W} + g') + \frac{\partial w'}{\partial X}\dot{X} + \frac{\partial w'}{\partial Z}\dot{Z} \;\; = \;\; -uv :$$

the first line is $\dot{w} + \frac{\partial w}{\partial W}g' \approx \dot{w} + g'$ as to leading order $w \approx W$; the first three terms in the second line are all negligible because $\dot{U}$, $\dot{V}$ and $\dot{W}$ are quadratically small; however, recognise $\dot{X} \approx -Z$ and $\dot{Z} \approx X$. Thus the equation reduces to

$$g' - Z\frac{\partial w'}{\partial X} + X\frac{\partial w'}{\partial Z} = -\dot{w} - uv = \text{(residual)}. \qquad (5.11)$$

See that we cannot use g' to eliminate $X^2 + Z^2$ terms from $\dot{W}$ because $X^2 + Z^2$ is a homogeneous solution of the homological operator $-Z\frac{\partial w'}{\partial X} + X\frac{\partial w'}{\partial Z}$.[7] Conversely, when such terms involving the square of the fast wave amplitude $X^2 + Z^2$ are generated in the residual, as they are, we cannot put them into the coordinate transform via w', instead they have to be placed in g' and thus remain in the dynamics.

This resonance arises solely due to the linear structure of fast waves, here $\dot{X} = -Z$ and $\dot{Z} = X$, and thus the problem arises generally. Thus for slow manifold models which neglect fast oscillations, that is guiding centre models, the general normal form coordinate transform (5.4) will similarly exhibit the slow manifold as $\mathbf{Y} = \mathbf{0}$ [23; 24]. The coordinate transform will also remove terms linear in $\mathbf{Y}$ from $\dot{\mathbf{X}}$. However, in general, terms quadratic in $\mathbf{Y}$ cannot be eliminated from $\dot{\mathbf{X}}$—there is generally a resonant forcing of the slow modes $\mathbf{X}$ by the fast oscillations or waves $\mathbf{Y}$ which is quadratic in the oscillation amplitude. Thus the evolution of the TOE is generally slightly but unavoidably different to that of the slow model. An further example of such quadratic resonance is the phenomenon of Stokes drift generated by the "fast" water waves superimposed upon the dynamics of large-scale currents. These resonances limit the accuracy of low-dimensional models based upon the concept of a slow manifold.

---

[7]You may like to confirm that terms in XZ or $X^2 - Z^2$ can be absorbed into the coordinate transform by involving such terms in w'.

6 Conclusion

The most important aspect of this work is that we have sought to base modelling upon dynamical arguments in the geometry of state space. The aim is to take a so-called "Theory Of Everything" (TOE), a complete but far too detailed description of the system, and produce a "coarse grained" model that captures all the dynamics, and little else, that are of interest in a particular application. Although the geometric arguments are supported by centre manifold theory, we have seen that many applications go beyond the current rigorous support.

This approach to modelling has many advantages:

- the signature of the model is straightforwardly derived from linearisation (§2.2.1, §5.3), but one can be inventive (§2.3 and §4);
- the model may be systematically refined (§2.2.3, §2.3, §3), perhaps using computer algebra to handle most of the details [82; 84];
- one may be very flexible about the relative magnitude of "small" parameters, resulting in a model which is later justifiably tuned to specific applications;
- it provides correct initial conditions for forecasts (§5.2), initial conditions that guarantee fidelity between the predictions of the model and the long-term evolution of the full dynamics;
- may account for smooth or stochastic forcing [22; 15];
- establishes an approach to determine correct boundary conditions for the models [78].

How to decide upon an accurate TOE on which to base the analysis is another problem—a model is only guaranteed if it is based upon an accurate description of the physical dynamics.

References

D. Armbruster, J. Guckenheimer, and P. Holmes. Kuramoto sivashinsky dynamics on the centre-unstable manifold. *Siam J. Appl Math*, 49:676–691, 1989.

P. Ashwin, I. Moroz, and M. Roberts. Bifurcations of stationary, standing and travelling waves in triply diffusive convection. *Phys. D*, 81(4):374–397, 1995.

V. Balakotaiah and H. C. Chang. Dispersion of chemical solutes in chromatographs and reactors. *Phil Trans R Soc Lond A*, 351:39–75, 1995.

O. Bokhove and T. G. Shepherd. On the hamiltonian balanced dynamics and the slowest invariant manifold. *J. Atmos Sci*, 53(2):276–297, 1996.

J.P. Boyd. The slow manifold of a five-mode model. *J. Atmospheric Sci.*, 51(8):1057–1064, 1994.

A.J. Brizard. Nonlinear gyrokinetic vlasov equation for toroidally rotating axisymmetric tokamaks. *Phys. Plasmas*, 2(2):459–471, 1995.

G. Brunet. On global nonlinear geostrophic balance of an isolated distribution of vorticity using a new time invariant. *J. Atmospheric Sci.*, 50(14):2304–2309, 1993.

A.B.G. Bush, J.C. McWilliams, and W.R. Peltier. Origins and evolution of imbalance in synoptic-scale baroclinic wave life cycles. *J. Atmospheric Sci.*, 52(8):1051–1069, 1995.

R. Camassa. On the geometry of an atmospheric slow manifold. *Physica D*, 84:357–397, 1995.

J. Carr. *Applications of centre manifold theory*, volume 35 of *Applied Math. Sci.* Springer-Verlag, 1981.

J. Carr and R. G. Muncaster. The application of centre manifold theory to amplitude expansions. I. ordinary differential equations. *J. Diff. Eqns.*, 50:260–279, 1983.

J. Carr and R. G. Muncaster. The application of centre manifold theory to amplitude expansions. II. Infinite dimensional problems. *J. Diff. Eqns*, 50:280–288, 1983.

J.R. Cary and S.G. Shasharina. Probability of orbit transition in asymmetric toroidal plasma. *Phys. Fluids B*, 5(7):2098–2121, 1993.

H. C. Chang. Wave evolution on a falling film. *Annu. Rev. Fluid Mech.*, 26:103–136, 1994.

Xu Chao and A. J. Roberts. On the low-dimensional modelling of Stratonovich stochastic differential equations. *Physica A*, 225:62–80, 1996.

M. Cheng and H. C. Chang. Subharmonic instabilities of finite amplitude monochromatic waves. *Phys Fluids A*, 4:505–523, 1992.

P. Chossat and M. Golubitsky. Hopf bifurcation in the presence of symmetry, center manifold and liapunov-schmidt reduction. In *Oscillations, bifurcation and chaos (Toronto, Ont., 1986)*, volume 8 of *CMS Conf. Proc.*, pages 343–352. Amer. Math. Soc., Providence, RI, 1987.

S. N. Chow and K. N. Lu. C^k centre unstable manifolds. *Proc Roy Soc Edin A*, 108:303–320, 1988.

H. Cohen and R. G. Muncaster. The theory of psuedo-rigid bodies. *Springer Tracts In Natural Philosophy*, 33, 1988.

D.R. Cook, A.N. Kaufman, A.J. Brizard, H. Ye, and E.R. Tracy. Two-dimensional reflection of magnetosonic radiation by gyroballistic waves: an analytic theory. *Phys. Lett. A*, 178(5-6):413–418, 1993.

P. H. Coullet and E. A. Spiegel. Amplitude equations for systems with competing instabilities. *SIAM J. Appl. Math.*, 43:776–821, 1983.

S. M. Cox and A. J. Roberts. Centre manifolds of forced dynamical systems.

J. Austral. Math. Soc. B, 32:401–436, 1991.

S. M. Cox and A. J. Roberts. Initialisation and the quasi-geostrophic slow manifold. Technical report, University of Adelaide, 1994.

S. M. Cox and A. J. Roberts. Initial conditions for models of dynamical systems. *Physica D*, 85:126–141, 1995.

W. Craig and M. D. Groves. Hamiltonian long-wave approximations to the water-wave problem. *Wave motion*, 19(4):367, 1994.

J. D. Crawford and E. Knobloch. Symmetry and symmetry-breaking bifurcations in fluid dynamics. *Annu Rev Fluid Mech*, 23:341–387, 1991.

A. Debussche and R. Temam. Inertial manifolds and slow manifolds. *Appl. Math. Lett.*, 4(4):73–76, 1991.

A. Debussche and R. Temam. Inertial manifolds and the slow manifolds in meteorology. *Differential Integral Equations*, 4(5):897–931, 1991.

D. R. Durran. Improving the anelastic approximation. *J. Atmos. Sci.*, 46:1453–1461, 1989.

M. P. Edwards. Classical symmetry reductions of nonlinear diffusion-convection equations. *Phys Letts A*, 190:149, 1994.

C. Elphick, E. Tirapeugi, M. E. Brachet, P. Coullet, and G. Iooss. A simple global characterisation for mormal forms of singular vector fields. *Physica D*, 29:95–127, 1987.

C. Foias, M. S. Jolly, I. G. Kevrekidis, G. R. Sell, and E. S. Titi. On the computation of inertial manifolds. *Phys. Lett. A*, 131:433–436, 1988.

C. Foias, M. S. Jolly, I. G. Kevrekidis, and E. S. Titi. On some dissipative fully discrete nonlinear galerkin schemes for the kuramoto–sivashinsky equation. *Phys. Lett. A*, 186(1):87, 1994.

C. Foias, G. R. Sell, and R. Temam. Inertial manifolds for nonlinear evolution equations. *J. Diff. Eqn*, 73:309–353, 1988.

C. Foias, G. R. Sell, and R. Teman. Inertial manifolds for dissipative differential equations. *Comptes Rendus Serie I*, 301:139–141, 1985.

Th. Gallay. A center-stable manifold theorem for differential equations in banach spaces. *Commun. Math. Phys*, 152:249–268, 1993.

U. Geigenmüller, U. M. Titulaer, and B. U. Felderhof. Systematic elimination of fast variables in linear systems. *Physica A*, 119:41–52, 1983.

H. Grad. Asymptotic theory of the Boltzmann equation. *Phys. Fluids*, 6:147–181, 1963.

J. Guckenheimer. Isochrons and phaseless sets. *J. Math. Biol.*, 1:259–273, 1975.

J. Guckenheimer and P. Holmes. *Nonlinear oscillations, dynamical systems, and bifurcations of vector fields*. Springer-Verlag, 1983.

F. Haake and M. Lewenstein. Adiabatic drag and intial slip in random processes. *Phys. Rev. A*, 28:3060–3612, 1983.

H. Haken. *Synergetics, An introduction*. Springer, Berlin, 1983.

J.K. Hale and H. Kocak. *Dynamics and bifurcations*, volume 3 of *Texts in Applied Mathematics*. Springer-Verlag, New York, 1991.

M. Hărăguş. Model equations for water waves in the presence of surface tension.

Eur. J. Mech, 15:471–492, 1996.

M. Hărăguş. Reduction of pdes on unbounded domains application: unsteady water waves problem. *J. Nonlinear Sci.*, 8:353–374, 1998.

G. Iooss and M. Adelmeyer. *Topics in Bifurcation Theory*. World Sci, 1992.

M. Ipsen, F. Hynne, and P. G. Sorensen. Systematic derivation of amplitude equations and normal forms for dynamical systems. Technical report, [http://arXiv.org/abs/chao-dyn/9803006], 1998.

S. S. Khirwadkar, P. S. Pathak, S. Chaturvedi, and P. I. John. 2-d guiding centre simulation of toroidal elctron clouds. *J. Comput. Phys.*, 132:291–298, 1997.

Y. A. Kuznetsov. *Elements of applied bifurcation theory*, volume 112 of *Applied Mathematical Sciences*. Springer-Verlag, 1995.

T.K. Leen. A coordinate-independent center manifold reduction. *Phys. Lett. A*, 174(1-2):89–93, 1993.

Zhenquan Li and A. J. Roberts. A flexible error estimate for the application of centre manifold theory. Technical report, [http://arXiv.org/abs/math.DS/0002138], February 2000.

E. Lorenz and Krishnamurty. On the non-existence of a slow manifold. *J. Atmos. Sci.*, 44:2940–2950, 1987.

E. N. Lorenz. On the existence of a slow manifold. *J. Atmos. Sci.*, 43:1547–1557, 1986.

E.N. Lorenz. The slow manifold—what is it? *J. Atmospheric Sci.*, 49(24):2449–2451, 1992.

M. Luskin and G. R. Sell. Approximation theories for inertial manifolds. *Math Model & Num Anal*, 23:445–461, 1989.

U. Maas and S. B. Pope. Simplifying chemical kinetics: intrinsic low-dimensional manifolds in composition space. *Combustion and Flame*, 88:239–264, 1992.

M. Marion and R. Temam. Nonlinear Galerkin methods. *SIAM J. Numer. Anal.*, 26(5):1139–1157, 1989.

Z. Mei and A. J. Roberts. Equations for turbulent flood waves. In A. Mielke and K. Kirchgässner, editors, *Structure and dynamics of nonlinear waves in fluids*, pages 342–352. World Sci, 1995.

Z. Mei, A. J. Roberts, and Zhenquan Li. Modelling the dynamics of turbulent floods. Technical report, [http://arXiv.org/abs/patt-sol/9907001], June 1999. submitted to SIAM J Appl Math.

G. N. Mercer and A. J. Roberts. A centre manifold description of contaminant dispersion in channels with varying flow properties. *SIAM J. Appl. Math.*, 50:1547–1565, 1990.

G. N. Mercer and A. J. Roberts. A complete model of shear dispersion in pipes. *Jap. J. Indust. Appl. Math.*, 11:499–521, 1994.

R. G. Muncaster. Invariant manifolds in mechanics i: The general construction of coarse theories from fine theories. *Arch. Rat. Mech. Anal.*, 84:353–373, 1983.

R. G. Muncaster. Invariant manifolds in mechanics ii: Zero-dimensional elastic bodies with directors. *Arch. Rat. Mech. Anal.*, 84:375–392, 1983.

W. Nagata. A perturbed hopf bifurcation with reflection symmetry. *Proc. Roy. Soc. Edinburgh Sect. A*, 117(1-2):1–20, 1991.

A. H. Nayfeh and S. A. Nayfeh. On nonlinear modes of continuous systems. *J. Vibration Acoustics*, 116:129–136, 1994.

C. Nore and T. G. Shepherd. A hamiltonian weak-wave model for shallow water flow. *preprint*, 1996.

S.F. Nosov. The hamiltonian formalism in the drift theory of the motion of charged particles in a magnetic field. *Kinemat. Fiz. Nebesn. Tel*, 8(6):14–31, 1992.

D. Pfirsch and P.J. Morrison. The energy-momentum tensor for the linearized maxwell-vlasov and kinetic guiding center theories. *Phys. Fluids B*, 3(2):271–283, 1991.

V. A. Pliss and G. R. Sell. Approximation dynamics and the stability of invariant sets. *J. Diff. Eqns*, 149:1–51, 1998.

P. K. Pollett and A. J. Roberts. A description of the long-term behaviour of absorbing continuous time markov chains using a centre manifold. *Advances Applied Probability*, 22:111–128, 1990.

Th. Prokopiou, M. Cheng, and H. C. Chang. Long waves on inclined films at high Reynolds number. *J. Fluid Mech.*, 222:665–691, 1991.

C. Rhodes, M. Morari, and S. Wiggins. Identification of low order manifolds: validating the algorithm of Maas and Pope. *Chaos*, 9:108–123, 1999.

A. J. Roberts. Simple examples of the derivation of amplitude equations for systems of equations possessing bifurcations. *J. Austral. Math. Soc. B*, 27:48–65, 1985.

A. J. Roberts. The application of centre manifold theory to the evolution of systems which vary slowly in space. *J. Austral. Math. Soc. B*, 29:480–500, 1988.

A. J. Roberts. Appropriate initial conditions for asymptotic descriptions of the long term evolution of dynamical systems. *J. Austral. Math. Soc. B*, 31:48–75, 1989.

A. J. Roberts. The utility of an invariant manifold description of the evolution of a dynamical system. *SIAM J. Math. Anal.*, 20:1447–1458, 1989.

A. J. Roberts. Low-dimensionality—approximations in mechanics. In J. Noye and W. Hogarth, editors, *Computational Techniques and Applications: CTAC-89*, pages 715–722. Hemisphere, 1990.

A. J. Roberts. Boundary conditions for approximate differential equations. *J. Austral. Math. Soc. B*, 34:54–80, 1992.

A. J. Roberts. A sub-centre manifold description of the evolution and interaction of nonlinear dispersive waves. In L. Debnath, editor, *Nonlinear waves*, chapter 9, pages 127–156. World Sci, 1992.

A. J. Roberts. The invariant manifold of beam deformations. part 1:the simple circular rod. *J. Elas.*, 30:1–54, 1993.

A. J. Roberts. Low-dimensional models of thin film fluid dynamics. *Phys. Letts. A*, 212:63–72, 1996.

A. J. Roberts. Low-dimensional modelling of dynamics via computer algebra. *Comput. Phys. Comm.*, 100:215–230, 1997.

A. J. Roberts. An accurate model of thin 2d fluid flows with inertia on curved surfaces. In P. A. Tyvand, editor, *Free-surface flows with viscosity*, volume 16 of *Advances in Fluid Mechanics Series*, chapter 3, pages 69–88. Comput Mech Pub, 1998.

A. J. Roberts. Computer algebra derives correct initial conditions for low-dimensional dynamical models. *Comput. Phys. Comm.*, 126(3):187–206, 2000.

A. J. Roberts. Holistic discretisation ensures fidelity to Burgers' equation. *Applied Numerical Modelling*, 37:371–396, 2001.

A. J. Roberts. Holistic discretisation illuminates and enhances the numerical modelling of differential equations. In V. V. Kluev and N. E. Mastorakis, editors, *Topics in Applied and Theoretical Mathematics and Computer Science*, pages 81–89. WSES Press, 2001.

A. J. Roberts. A holistic finite difference approach models linear dynamics consistently. *Mathematics of Computation*, to appear, 2001.

J. C. Robinson. The asymptotic completeness of inertial manifolds. *Nonlinearity*, 9:1325–1340, 1996.

S. Rosencrans. Taylor dispersion in curved channels. *SIAM J. Appl. Math.*, 57:1216–1241, 1997.

R. Valery Roy, A. J. Roberts, and M. E. Simpson. A lubrication model of coating flows over a curved substrate in space. *J. Fluid Mech.*, to appear, 2001.

R. Salmon. New equations for nearly geostrophic flow. *J. Fluid Mech.*, 153:461–477, 1985.

B. Scarpellini. Center manifolds of infinite dimensions. i. main results and applications. *Z. Angew. Math. Phys.*, 42(1):1–32, 1991.

S. W. Shaw. An invariant manifold approach to nonlinear normal modes of oscillation. *J. Nonlinear Sci.*, 4:419–448, 1994.

S. W. Shaw. An invariant manifold approach to nonlinear normal modes of oscillation. *J. Nonlinear Sci*, 4:419–448, 1994.

S. W. Shaw and C. Pierre. Normal modes for non-linear vibratory systems. *J. Sound Vibration*, 164(1):85–124, 1993.

L. P. Shilnikov, A. L. Shilnikov, D. V. Turaev, and L. O. Chua. *Methods of qualitative theory in nonlinear dynamics. Part I*, volume 4 of *World Scientific Series on Nonlinear Science, Series A*. World Sci, 1998.

J. Sijbrand. Properties of centre manifolds. *Trans. Amer. Math. Soc.*, 289:431–469, 1985.

S. A. Suslov and A. J. Roberts. Similarity, attraction and initial conditions in an example of nonlinear diffusion. *J. Austral. Math. Soc. B*, 40(E):E1–E26, October 1998. [Online] http://jamsb.austms.org.au/V40/E007 [14 Oct 1998].

S. A. Suslov and A. J. Roberts. Advection-dispersion in symmetric field-flow fractionation channels. *J. Math. Chem.*, 26:27–46, 1999.

S. A. Suslov and A. J. Roberts. Modelling of sample dynamics in rectangular asymmetrical flow field-flow fractionation channels. *Analytical Chemistry*, 72(18):4331–4345, 2000.

P. Takac. Invariant 2-tori in the time-dependent ginzburg-landau equation. *Nonlinearity*, 5(2):289–321, 1992.

G. I. Taylor. Dispersion of soluble matter in solvent flowing slowly through a tube. *Proc. Roy. Soc. Lond. A*, 219:186–203, 1953.

G. I. Taylor. Conditions under which dispersion of a solute in a stream of solvent can be used to measure molecular diffusion. *Proc. Roy. Soc. Lond. A*, 225:473–477, 1954.

R. Temam. Do inertial manifold apply to chaos? *Physica D*, 37:146–152, 1989.

R. Temam. Inertial manifolds. *Mathematical Intelligencer*, 12:68–74, 1990.

J. Troupe and G. Rosensteel. Algebraic nonlinear collective motion. *Annals of Physics*, 270:126–154, 1998.

U.M. Titulaer. On the extraction of macroscopic descriptions from mesoscopic ones. In *Nonequilibrium statistical mechanics (Puerto Iguazu, 1989)*, pages 1–33. World Sci. Publishing, Teaneck, NJ, 1990.

N. G. van Kampen. Elimination of fast variables. *Physics Reports*, 124:69–160, 1985.

A. Vanderbauwhede. Centre manifolds. *Dynamics Reported*, pages 89–169, 1989.

A. Vanderbauwhede and G. Iooss. Center manifold theory in infinite dimensions. *Dynamics Reported*, 1:125–163, 1988.

S. D. Watt and A. J. Roberts. The accurate dynamic modelling of contaminant dispersion in channels. *SIAM J. Appl Math*, 55(4):1016–1038, 1995.

C. E. Wayne. Invariant manifolds and the asymptotics of parabolic equations in cylindrical domains. In Lu & Pan Bates, Chow, editor, *Differential equations and applications*, pages 314–325. International Press, 1997.

C. E. Wayne. Invariant manifolds for parabolic partial differential equations on unbounded domains. *Arch. Rat. Mech. Anal.*, to appear, 1997.

R.B. White. Canonical hamiltonian guiding center variables. *Phys. Fluids B*, 2(4):845–847, 1990.

S. Wiggins. *Introduction to applied nonlinear dynamical sytems and chaos*. Springer-Verlag, 1990.

A. Winfree. Patterns of phase compromise in biological cycles. *J. Math. Biol.*, 1:73–95, 1974.

R. W. Wittenberg and P. Holmes. The limited effectives of normal forms: A critical review and extension of local bifurcation studies of the Brusselator PDE. *Physica D*, 100:1–40, 1997.

Chapter 8

Vortices and Spatial Solitons in Optical Resonators, and the Relations to Other Fields of Physics

C. O. Weiss, K. Staliunas, M. Vaupel, V. B. Taranenko, G. Slekys, and Ye. Larionova

PhysikalischTechnische Bundesanstalt, Braunschweig, Germany

Abstract.   These lectures describe experiments on pattern formation and spatial solitons in nonlinear optical resonators.   The fundamental mathematical model for such resonators is the nonlinear Schrödinger equation, which connects fluids, superfluids, superconductors, particle physics, and chemistry. Pattern formation, hydrodynamics analogies, phase- and intensity- solitons are shown to occur in passive and active nonlinear resonators of various types. Properties of the different types of (dissipative) spatial solitons are discussed. Recent experiments on spatial solitons in semiconductor microresonators pave the way to applications of spatial solitons in telecommunications and in parallel optical information processing. We also show the connections of nonlinear optics to a wide variety of fields from biology to particle physics.

1 Introduction

We discuss the localized structures found experimentally to exist in nonlinear optical resonators. These are primarily vortices. Vortices allow to draw close analogies of resonator optics, with (super-) fluids. Vortices exist in laser type resonators since in these the phase is a free variable. Vortices in lasers with a slowly relaxing material inversion show many similarities with "chemical spiral waves" i.e. vortices in "excitable media". Bright and dark dissipative solitons are characteristic of other nonlinear resonators. They have internal energy flows and can "self replicate" and "die" irreversibly. Thus they are reminiscent of simplest biological structures. Resonators

315

which obtain internal light amplification by a degenerate wave mixing process possess phase-solitons.

All these localized structures can interact, and depending on parameters can form bound states or "molecules" and can condense into "crystalline" space-filling structures, of square geometry for vortices (because of the existence of two kinds of vortices: right-handed, left-handed); and of hexagonal structures for other solitons. The conceptual unification of the concept of vortices on the one hand and bright (dark) solitons on the other hand arises when considering e.g. higher order bright solitons. These often contain a central vortex. As an example the first higher order bright soliton resembles a Laguerre mode with a central vortex.

Many links appear to exist between optics and particle physics, some concerning phenomena usually thought of as being of "quantum" origin.

A number of the results support the conjecture that formation of "structure" or "coherence" in general is a consequence of a bottleneck to the microscopic dissipation. This suggests that very diverse phenomena could be understood from a common mechanism: optical short pulse formation [1], formation of patterns — from the simplest periodic ones to biological structures; pair creation of particles out of vacuum; 1/f noise (subsuming "self organized criticality" [2]); laser emission and Bose condensation. (As well as earthquakes and stock market crashes...). In addition to these general physics aspects, optical localized structures may prove useful in parallel optical processing of information from simple telecommunication problems to problems such as pattern recognition, where typically digital traditional computers prove deficient. The notion of such a usefulness of resonator solitons is supported by the recent demonstration, manipulation, and investigation of solitons in semiconductor microresonators [3].

2   Vortices in lasers

Approaching structure formation in laser resonators from a laser physicist's point of view, the most plausible occurrence of vortices is in Gauss-Laguerre Modes. In [4] it was found that by phase-locking two spatially orthogonal Gauss-Hermite first order modes (TEM01) a first order Gauss-Laguerre Mode with a vortex at the center forms (Fig. 1). The phase difference of the "constituent" Hermite modes arranges itself in the locking at plus or minus p/2, corresponding to vortices of opposite handedness. Phase-locking in the

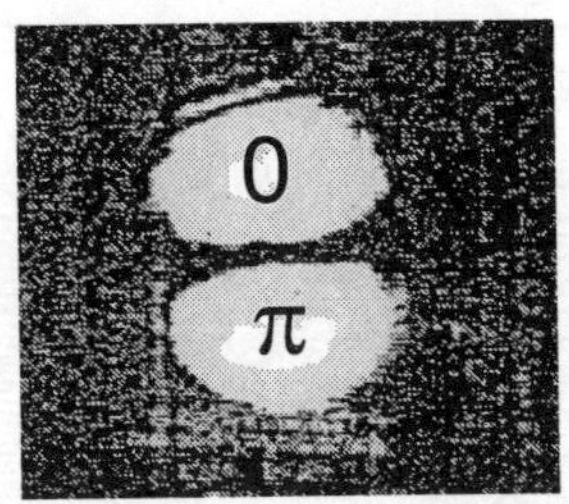

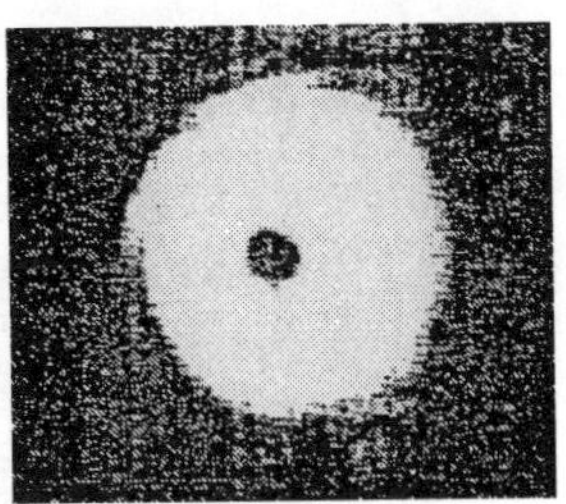

Fig. 1   Phase-locking of the TEM01 and TEM10- modes at 90° phase difference produces a field with a vortex (phase-singularity) at the center (TEM01*).

same fashion higher order Gauss-Hermite modes produces "vortex-crystals" (Fig. 2) [5]. Phase-locking in the analogous fashion higher order Laguerre "flower-modes" leads to multiply-charged vortices (Fig. 5) [6]. Vortices in laser resonators are always bistable in their handedness [7]. A laser in which several Gauss-Laguerre Modes are excited which are not phase-locked generally shows vortices circling around the optical axis. Vortices are then usually moving on concentric circles [8], where the direction of motion on each circle is independent of the direction of motion of the vortices on other circles, and the direction of motions is bistable due to the bistability of the handedness of vortices [7]. Video examples of circling vortices in laser experiments can be viewed at [9]. The dynamics of the field structure can in these cases be understood as the "beating" (or interference) of all modes excited. Non-phase locked modes in general lead to a dynamics which can involve periodic pair-creation and pair-annihilation of vortices of opposite handedness [10].

This is the laser physicist's picture of pattern formation through (i) excitation of particular single transverse modes (ii) simultaneous excitation of more than one transverse mode, possessing different optical frequencies or (iii) being phase-locked. This picture is an essentially linear picture. Modes are field distributions inside a linear resonator. If one speaks of a "mode emitted by a laser" then one particular field distribution is selected (largely by linear mechanisms like the spectral gain distribution of the amplifying medium). But the selection involves also a non-linear mechanism (mode competition), and in particular, the amplitude of the selected mode is stabilized by the gain saturation, a nonlinearity that is usually described by a cubic field term. Disregarding these nonlinear effects the patterns compatible with the given resonator parameters can be imagined in a linear

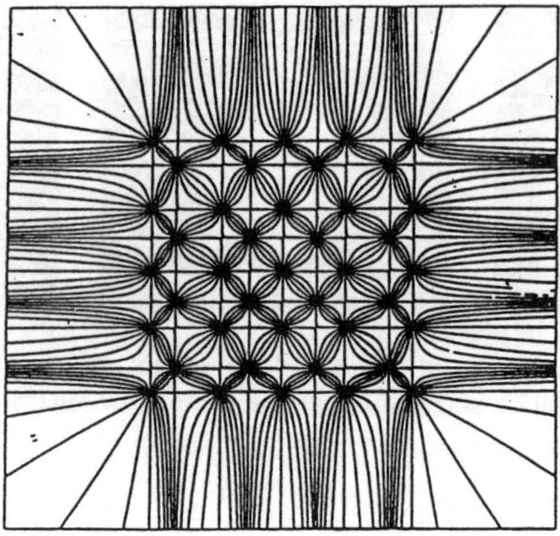

Fig. 2 Superposition of a TEM45- and a TEM54-mode with 90° phase difference produces a large vortex-"crystal". Note that near the crystal "surface" the lattice is deformed like in a real crystal , with "dangling bonds".

picture. For the selection or excitation of a mode or a super-position of modes at least a weak non-linearity is required. As most lasers have relatively weak gain, and thus operate in an only weakly nonlinear regime, the "linear" mode picture is generally qualitatively "good". The linear mode picture on the other hand fails completely for strong nonlinearities: in the linear regime a "vortex crystal" is just the (phase-locked) superposition of two TEMnm Gauss-Hermite modes, each of which fills the whole resonator cross section. The "crystal" is thus a space-filling "coherent" pattern. In the strongly nonlinear regime, however, vortices attain an identity as individual particles, which means that a picture which represents the essential properties of the "crystal" pattern in this regime is that of a periodic arrangement of individual "vortex-solitons", the vortices being loosely bound together by mutual interactions. Thus the transition from very weak to strong nonlinearity represents the transition from a "coherent" to an "incoherent" pattern. A process also known as "melting" (of a crystal). Fig. 6 shows such a transition (melting) of a vortex "crystal" in a laser. At relatively small pump (small nonlinearity) the pattern is ordered, periodic. Increasing the pump, "defects" in the crystal appear which are moving randomly. With further increased pump the defects become more numerous and move faster until the crystal is "molten" i.e. the field consists of independent "particles" moving randomly.

This "melting process" described when going from small to large non-linearity can also be characterized in a different way: a "coherent" space filling pattern in which the "particles" have no identity exists when the "rest energy" of the "particles" is small or at most comparable to the binding energy of two particles. This situation exists for very weak nonlinearity. For strong nonlinearity the rest energy of particles (energy of vortices [11]) is large compared to the binding energy of two particles. In this regime the particles have an "identity". One can speak of particles therefore meaningfully in case of strong nonlinearity only. For weak nonlinearities the picture of particles is useless and the field is best described in a wave picture; the wave filling all space. One wonders if these properties of optical fields might relate to the particle-wave duality of quantum mechanics [12]. We have described here the particle-wave duality for vortices. We mention that the transition from space-filling, "coherent" patterns to particles with increasing nonlinearity occurs equally for bright solitons [13] and most probably for all types of solitons.

Various analogies can be found between resonator optics and (super-) fluids. First of all lasers as well as superfluids exhibit quantized vortices. This means that the phase circulation around such a vortex is an integer number times 2π. Whereas the vortex in optics is a screw dislocation in the electromagnetic wave [14], it is a screw dislocation in the matter-wave of the superfluid [15]. Mathematically the analogies can be evidenced by considering a class A-laser, the laser with fast-relaxing material variables.

The description of pattern formation of lasers generalizes the plane wave mean field single mode laser equations by adding a non-local interaction (diffraction):

$$\frac{\partial E}{\partial t} = \kappa \left[-(1 + i\omega)\, E \; + \; P \; + \; i d \nabla^2 E \right]$$

$$\frac{\partial P}{\partial t} = -\gamma_\perp (P - ED) \tag{2.1}$$

$$\frac{\partial D}{\partial t} = -\gamma_\parallel \left[D - D_0 + \frac{EP' + E'P}{2} \right]$$

The optical field is linearly polarized, and the gain line is homogeneously broadened. It is also assumed that a single longitudinal mode family is excited.

The variables $E(\boldsymbol{r}_\perp, t)$ and $P(\boldsymbol{r}_\perp, t)$ are the slow envelopes in space and time of the electromagnetic field and polarization. $D(r_\perp, t)$ is the population inversion, which in the absence of stimulated emission is equal to its unsaturated value $D_0(r_\perp)$. κ is the relaxation rate of the optical field in the resonator (due to transmission of mirrors, also due to the linear losses of the optical elements in the resonator), $\gamma_\perp$ and $\gamma_\parallel$ are the decay rates of polarization and population inversion respectively. Eliminating the fast material variables in a physically meaningful way [16], [17] reduces (2.1) to a complex Ginzburg-Landau equation

$$\frac{\partial E}{\partial \tau} = (D_0 - 1)E + (d_{Re} + id_{Im})\nabla^2 E - (1 + i\omega)|E|^2 E \qquad (2.2)$$

with $p = (D_0 - 1)$ (threshold parameter), $d_{Im} = d$ (diffraction), $d_{Re} = 2d\omega\kappa/\gamma_\perp$ (diffusion) and $\tau = \kappa\dot\gamma^\perp/(\kappa + \gamma_\perp)$ (slow time). (2.2) has the non-linear Schrödinger Equation

$$\frac{\partial E}{\partial \tau} = id\nabla^2 E - i\omega|E|^2 E \qquad (2.3)$$

as the conservative limit. One can, by Madelung transformation: $A = \sqrt{\rho}c^{i\phi}$ convert (2.2) into the momentum- and mass- conservation equations, of hydrodynamic form.

$$\frac{\partial \rho}{\partial t} + \nabla(\rho v = 2(D_0(r) - 1)\rho - 2\rho^2$$

$$\qquad (2.4)$$

$$\frac{\partial v}{\partial t} + \nabla(v^2/2) = \nu\nabla^2 v - \nabla\beta(r) - \frac{\nabla p}{\rho} + \nabla\left[\frac{\nabla^2\rho^{1/2}}{2\rho^{1/2}}\right]$$

with the following correspondences:

$v = \nabla\phi$ is the velocity, equivalent to phase gradient;

$\rho = |A|^2$ is the photon density, equivalent to mass density;

$p = \rho^2/2$ is the internal pressure (compressibility);

$\nu = d_{Re}/(2d_{Im})$ is the viscosity, inversely proportional to the number of modes under the gain line of the laser;

$\beta(r) = c\dot r^2$ is the external potential (potential well of the laser due to resonator mirrors). The last term in the second equation (2.4) is called "quantum pressure" having no analogy in classical hydrodynamics.

In this fluid picture the circling vortices, which were above described as interference of transverse modes, can be pictured through the Magnus effect [9], Fig. 6. The vortex is embedded in a Gaussian fluid background. As intensity corresponds to mass density, there is a radial pressure gradient due to the compressibility. In this pressure gradient the vortex (corresponding to a bubble in the fluid) experiences a buoyancy force outward. However, it is not moving radially outward, but perpendicularly to the pressure gradient because it is not just a bubble but a rotating bubble. A rotating object in a flow moves perpendicular to the flow direction ("Magnus effect"). Consequently the vortex motion is on a circle around the optical axis. Generally optical "objects" move in intensity, or phase gradients, which we have demonstrated experimentally for bright solitons e.g. in [18]. The fluid analogy is even more vividly demonstrated in the "v. Karman Vortex Street", i.e. the period generation of vortices alternating in rotation sense, behind an obstacle in a flow. Such a demonstration was realized according to the laser fluid equivalence [4]. A flow is set up using a phase gradient, much like calculated in [16] by a tilt of one laser mirror.

In order to make the flow predominantly one-dimensional, a very astigmatic resonator geometry was chosen: in one transverse dimension the resonator is stable, while in the other it is plane. This is achieved by using cylindrical optics. The result is an "optical channel" with flow possible along the unstable direction. (Fig. 7 a). With one mirror tilted along the unstable direction it becomes a tilted channel. (Fig. 7 b) in which accelerated flow would take place if the "laser-fluid" (2.4) were not viscous. Due to the tilt of one mirror, the resonator length varies along the unstable direction, so that at different lateral positions along the unstable direction transverse modes of different index are resonant. (The frequency separation of the transverse modes is defined by the curvature of the cylindrical optics along the stable direction.) In the experiment the fundamental and the first order transverse mode (characterized by one node along the channel axis) were placed adjacently in a fashion so that flow goes from the first order mode towards the fundamental mode (Fig. 7). The obstacle in the flow is the zone between the two modes where light cannot exist due to the lack of a resonance condition (one can say the obstacle here is described by a complex number and takes the form of a dispersive rather than absorptive obstacle [19]).

Fig. 7 shows that the "flow" in this "channel" with "obstacle" corresponds to what we are used to in fluid flow: behind the obstacle vortices are created which alternatingly move upward and downward. They move laterally out of the "channel" because they correspond to "bubbles", (see the discussion of circling vortices above). An interferogram permits to visualize the handedness of the vortices. Fig.7 shows that the vortices following one another have alternating handedness just like in a real fluid. The Fresnel number in this arrangement corresponds well with the Reynolds number at which such a "vortex street" appears in real fluids. Fig. 7 is a numerical simulation, which shows how the vortices with alternating handedness detach consecutively from the dark line [19]. The phenomenon of turbulence occurs roughly at the Fresnel numbers of a few 1000 corresponding to the Reynolds numbers in flows for turbulence. To observe this "optical turbulence" it is simply necessary to use a large Fresnel number laser (without any particular provision for creating a "flow"). Fig. 7 shows such a "turbulent" field emitted by a photorefractive oscillator (which corresponds to a laser with very slow dynamics [20]). The field emitted by this laser has the appearance of a turbulent fluid. In optics one would call such a field a "speckle" field. Such speckle fields are known to contain vortices. Fig. 7 shows a section of the turbulent cross section in an interferogram. The large number of vortices is apparent. The dynamics is accompanied by a continuous pair-creation and pair-annihilation of vortices. We loosely use the word "turbulent" for this emission. This is probably not correct strictly because turbulent dynamics in the strict sense of the word is defined by a flow of energy from small to large wave numbers with a certain distribution [21]. One would have to verify such a distribution before calling this laser emission "turbulent". Thus it would be safer to call this laser emission "space-time chaotic". We would think that this space-time-dynamics of laser emission would range from regular-periodic at very small Fresnel number (see the circling vortices above), via "chaotic interaction" of a few modes at medium Fresnel numbers, to real turbulence for very large Fresnel numbers.

Vortices in lasers are associated with radial energy flow, a feature they share with many dissipative solitons. This radial energy flow results because the material inversion in a laser is not reduced at the vortex center as much as it is outside the vortex. At the center of the vortex there is no optical field, thus there is not stimulated emission reducing the inversion. Consequently material inversion diffuses radially away from the vortex core to the outer zones of the vortex where the optical field is stronger and the inversion can be radiated away; so to let the system get closer to thermodynamic equilibrium. In other words: a vortex "radiates" (like a radio transmitter). This radiation of inversion away from the vortex case becomes strikingly manifest where the radiations of two adjacent vortices collide. It was theoretically predicted [22] that this "collision" of energy flows would lead to a "shock-wave". Fig. 7 shows a numerical calculation of a vortex ensemble. The "shocks" between each two vortices are clearly visible. Fig.7 shows the experimental confirmation. Again a photorefractive oscillator was used to realize a laser [20], with a slow time scale for better observability. The shocks separating vortices are clearly recognizable.

There are conditions where ensembles of vortices arrange in a regular square crystal-like fashion. Fig. 6 a shows a calculated case (the existence of such vortex crystals was experimentally shown in [23]). Note that the shocks also present here. In Fig. 6 b, c it is shown that with increasing pump of the laser the vortex crystal develops "defects" and these defects become more numerous (and more mobile) as the pump increases further. Here the vortex crystal "melts" just like a regular crystal made up of atoms. The molten state corresponds to the irregular moving vortices of Fig. 7 and 7.

It is worthwhile noting the existence of the shocks in the "molten" state: the shocks "shield" the interaction of the vortices (vortices move in their mutual fields gradients), so that the dynamics of the vortex ensemble is drastically slowed compared to a vortex ensemble without shocks. This state of vortices accompanied by shocks has therefore been termed "vortex-glass" (as opposed to the "vortex-crystal" of Fig.6a) Whereas in the simple arrangements of vortices such as a TEM01 Gauss-Laguerre mode it can be said that vortices appearing obtain their stability through the boundaries, in vortex ensembles like the ones in Fig.6 the vortices are evidently entities which can move around without change or destruction. Thus they deserve being called solitons. Semi-regular motion of vortex solitons is shown experimentally in Fig. 7.

3   Class B-laser vortices

So far the lasers concerned were described by the field variable alone, because of the small decay times of the medium (inversion and polarisation) compared to the lifetime of the field in the optical resonator. In this class of lasers (termed class A-laser) the equivalence is with fluids and superfluids. Evidently there are lasers where not two of the three variables can be eliminated. In the case where decay times of field and inversion are comparable (termed class B-lasers), and the laser has to be described by two equations the equivalence is with chemically reacting fluids. A first evidence of such chemical (or "exciteable") phenomena occurred in a numerical calculation by M. Brambilla [24]. The four vortices in the field configuration of a laser belonging to the second order transverse mode family are found to be not stationary as in a class A-laser, but to move in time as shown in Fig. 7. The reason for this movement is shown in Fig. 7 schematically. As the decay of inversion in absence of stimulated emission is slow, the inversion around the vortex, where there is stimulated emission, is reduced to the steady state value (corresponding to the resonator loss) whereas at the vortex core the unsaturated inversion remains. In order to reduce this excess inversion at the vortex core, stimulated emission is required. As the vortex with its dark core is structurally stable (to destroy it means to break a wavefront which requires energy) emission at the location of the excess inversion is only possible by moving the vortex sideways, which the laser decides to do, in order to reach a state closer to thermal equilibrium. However, as soon as the vortex occupies a new location the build-up of inversion at its core starts anew. The vortex has to move again. A continuous motion of the vortex is the consequence for this laser with slow decaying inversion (class B-laser). We have termed this motion "restless" because it results due to the inability of the system to find a potential minimum in phase space: the vortex is not drown towards a potential minimum but pushed away from its present position wherever it is.

The motion of the vortices in Fig. 7 is confined by the ring-shaped potential around the optical axis. The motion is free in the angular direction and oscillating in the radial direction. The motion of a vortex in a 2-D homogeneous class B-laser field is calculated in Fig.8. Here the motion is unhindered in all directions. Thus at the beginning of the motion a symmetry breaking occurs which determines the initial motion direction. A few abrupt changes of the direction of motion are visible in the initial stage of the motion. These occur because the vortex interacts with the inversion wave it has emitted itself earlier (see section 1), at certain points, in such a way that a change of motion direction occur. After a transient phase the motion settles on a circle (for the parameters chosen). This motion on the circle is determined by the continuous interaction of the vortex with its own emission field. For other parameters the asymptotic motion takes the form of flower patterns, as it is often observed in chemical systems [25]. In [26] the mathematical equivalence of the class B-laser system with "excitable" chemical media is shown, making the similarity of phenomena

of class B-lasers and chemistry plausible. In a class B-laser vortex lattice
the vortices will equally be nonstationary. Calculations show that they can
also move on circles and those two types of lattice vibrations result. If the
circular motion of adjacent vortices is locked in phase a lattice vibration
in the form of an "acoustic phonon" Fig.8a) results. Locking out of phase
results in vibration of the "optical phonon" type (i.e. with macroscopic
dipole moment) Fig.8b). One may ask with what kind of wave structure
the "optical phonon" i.e. the macroscopic dipole moment would interact.
As the vortices move in intensity-gradients as well as in phase gradients one
would expect that the macroscopic dipole moment of such a vortex lattice
interacts with phase- as well as intensity-waves. One can even imagine
combined waves where the interaction due to the intensity gradient cancels
the one with the phase gradients. A first but only very limited verification of
this "restless" motion of vortices in class B-lasers has been given in [27]. No
other experimental investigations in the apparently wide field of dynamics
of vortices in class B-laser has been done so far.

4   Dissipation and formation of coherent structures

In [1] an open question in structure formation was addressed. It was known
from experimental experience that the formation of the shortest possible
pulsed from optical parametric oscillators required pumping the oscillator
precisely two times the threshold pump power. If the pump is lower than
this, the pulse structuring is limited by the insufficient pump, if the pump is
higher than this, irregular pulse break-up is observed. One can draw a formal analogy of such wave-mixing processes with a thermodynamic Carnot
process, and it turns out that the optimum pulse formation occurs at the
point where the dissipation corresponds to that of the Carnot process. The
irregular pulse break-up at higher pump is related to the formation principle of coherent highly dissipative structures if the dissipativity is below
that of the Carnot process.

Thus in the case of a 1-D system the chaos observed represents the coherent structures (e.g. vortices), spontaneously appearing in 2-D and 3-D
systems appearing irregularly and spontaneously in case of a bottleneck in
the microscopic dissipation. We mention that it appears as if these "large
fluctuations" (or coherent structures) are at the root of the generally observed 1/f noise of dissipating systems. It would appear that earthquakes,

stock market crashes etc. are to be equally understood as such spontaneous irregular formation of coherent structures due to lack of microscopic dissipation. In 3-D the structures appearing spontaneously irregularly in such cases are vortices i.e. "Skyrmions" [28]. These appear in pairs of vortex–antivortex. One wonders if there is a relation with vacuum noise, i.e. the spontaneous creation of particle pairs out of vacuum.

5   Phase fronts and solitons in degenerate wave mixing

Whereas in lasers the phase is a free variable, the phase in fields generated by degenerate wave mixing usually has a preferred value. The reason is the necessity to fulfill a phase relation between all fields involved in the mixing process for generating light. In degenerate cases such as degenerate 3-wave mixing (DOPO) or degenerate 4-wave mixing (D4WM) there are two stable phase values. For a system with a large cross section (large Fresnel number) this leads to phase domains domains in which the field has one of the two phase values. The domains are then separated by "domain walls". Along a cut through a domain wall, the field (which, due to its defined phase value is described as a real number) changes from positive to negative values (or vice versa). Where the field strength crosses zero, the intensity is zero. Thus the phase domains are surrounded by black lines. The field profile along the cut is that of a kink-soliton [28], so that the front separating phase domains can be thought of as the 2-D extension of a 1 D (kink-) soliton.

One finds that the motion of the domain boundaries is determined by the boundary curvature. The front velocity (perpendicular to the front) is inversely proportional to the radius of front curvature [29]. With resonator detuning the velocity also changes and crosses zero for a certain small detuning. The consequence is that phase domains expand for large detuning with a speed given by the curvature of their boundaries and by the detuning value. For small detuning they contract, leading to asymptotic disappearance of all domain boundaries. For an intermediate detuning range i.e. for a detuning range around the value of zero boundary velocity, domains contract or expand until they reach a small but finite diameter at which they stabilize. These stable small domains are phase solitons which arise finally as a consequence of the phase bistability [30].

Fig.8 shows an overview of the front dynamics. The detuning is normalized to the interval in which the homogenous intensity A2 solution goes from 1 to 0. For the detuning ranges marked the dynamics of the fronts is shown by the structure at an early stage ($t = 15$) and a late stage ($t = 150$). Near zero detuning ($0 < \Delta < 0.25$) structures existing initially disappear asymptotically. The homogeneous solution is the asymptotically stable solution because all domains contract. For larger detunings ($0.46 < \Delta < 0.82$) domains expand in time but the topology is preserved. The number of domains is constant throughout the evolution; whereas for large detunings $0.82 < \Delta < 1$ domains expand and new domains nucleate continually. The detuning range $0.29 < \Delta < 0.46$ is the range in which stable solitons exist.

The domain walls must not in a monotonic fashion connect the two stable field states, but are often doing so by approaching the homogeneous field value through damped oscillations when going away from the front ("oscillating tails" of the front [31]). Such oscillations are pronounced when the system is near to a modulational instability. The oscillating tails contain field gradients which produce interactions between fronts. Foremostly this is a mechanism by which solitons are stabilized. Secondarily such interaction can lead to bound states of solitons as shown in Fig.8. This figure shows "oscillating intensity tails" clearly and depicts how front interaction leads to positioning fronts one spatial oscillation period apart. With pronounced oscillating tails, higher order phase solitons can occur showing more than one dark ring (Fig.8c). Although the calculations discussed this far were done for degenerate parametric generation (DOPO) and a simplified model of this (real Swift-Hohenberg equation [32]), the experiments aimed at verifying the predictions were done in a (mathematically equivalent) degenerate 4-wave mixing system which shows characteristic time scales in the 100 ms range, convenient for recording the 2-D field dynamics by ordinary video equipment.

Fig. 9 shows the principle of the experimental arrangement: the crystal for 4-wave mixing is arranged in a linear self-imaging (diffraction-free) resonator. This resonator has transverse modes degenerate in frequency and is thus suited to be resonant for any field distribution, allowing free pattern formation. Two pump beams are irradiating the crystal at an angle of about 30° from the resonator axis. The two generated waves propagate in the same resonator, which forces their degeneracy. A diaphragm in the Fourier plane blocks all resonant tilted waves except the central disk (wave with zero tilt). The resonator length is in this case chosen slightly longer than 4f, for a small residual diffraction to enhance the "oscillating tails" of fronts and thereby to increase the soliton stability. The pump radiation

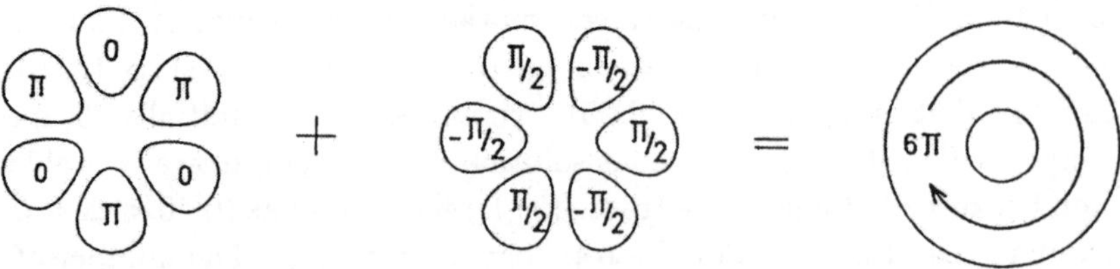

Fig. 3   Phase-locking of "Flower" Laguerre-Modes produces multiply charged vortices

comes from a single frequency Ar^+ laser at 514 mm. Pump intensities are around 0.5 W/cm^2. The field transmitted through the outcoupling mirror is simply directed to a CCD camera. Detuning is controlled by the resonator length.

The observations showed immediately domains separated by black lines as expected. Fig. 9 shows as an example a rather complicated domain boundary, and an interferogram, clearly showing that the phase on opposite sides of the boundary is different by π as predicted. The qualitative features predicted are also observed. For example, Fig. 9 shows the evolution of a domain boundary near zero detuning: as predicted the domain boundary shortens in time. For larger detuning the "elasticity" of the domain boundary is expected to be negative. This is illustrated in of Fig. 9. The domain walls lengthen in time until the whole area is filled by a labyrinthic pattern. Finally at very large detuning the topology is not preserved and new domains nucleate continually (Fig.9). The predicted phase solitons were also found experimentally. Fig. 9 shows four such solitons coexisting with domain walls. From such pictures as Fig. 9 it is not clear whether these small domains are really solitons. Therefore we tested the stability of these small domains experimentally and confirmed that they are stable solitons and not contracting or expanding domains [33]. Thus qualitatively we where able to confirm experimentally all the predictions based on the wave mixing model. Quantitative comparisons of model and experiment are presently being undertaken [34].

6 3-D Solitons

As the degenerate wave mixing produces a field with a fixed (though bistable) phase, it is described as a real-valued field. This being a computationally not so demanding problem, we looked for structures in 3-D space numerically. One finds indeed phase domains (Fig. 9) where the domains are now "volumes" i.e. three-dimensional. The 3-D domains are surrounded, completely analogously to the 2-D cases by a black zero field surface. Stable localized structures (3-D phase solitons) are equally found. The parameter range of their existence is, however, smaller then in 2-D (Fig. 9). Other conceivable localized structures such as tori are not found stable. One can probably, for a qualitative picture, conceive of the 3-D domains and solitons by referring to the hydrodynamics analogy (section 1). The domain wall (zero field surface) tends to contract ("surface tension"), thereby "compressing" the light "fluid" inside it. This "compression" is readily visible: in Figs. 9 and 9 the intensity enclosed by the domain walls is notably higher then in the surrounding. Recalling that in the hydrodynamics analogy light intensity corresponds to mass density, the "compression" is balanced by the surface tension to create a stable 3-D phase soliton. This simple balance works for convex boundaries. A torus has apparently too complex a shape for such a balance. It may be worthwhile mentioning that an extended (space filling) pattern is also found stable, Fig. 9. It is a diamond structure of intertwined domains of positive and negative field values [35].

Encouraged by the existence of structures and solitons in 3-D for the real field degenerate wave mixing processes, we searched for stable 3-D structures also in the more computationally demanding case of nondegenerate wave mixing (where the field is complex-valued; corresponding to a "very fast" laser). Naturally for complex-valued fields one can look for vortex structures. Fig. 9 shows stable structures found. The torus-like rings are in fact vortex rings ("smoke rings") which are stabilized due to an extra twist around the rings (smoke rings without the extra twist can contract and disappear). The stabilization of these structures is a most intricate mechanism: first the vortex rings have to be (structurally) stabilized by the extra twist. This, however creates a phase singularity in the center of the ring plane (it would seem to us that this latter effect was overlooked in [36]) which necessitates the existence of another vortex tube perpendicular to the ring plane. This can only be fulfilled by the two structures shown in Fig. 9. In Fig. 9a) each ring provides the phase singularity in the plane of the other ring. In Fig. 9b) the central phase singularity for all of the infinite number of vortex rings is provided by the straight vortex line. It follows for b) that all vortex rings have the same helicity, otherwise they would collapse and annihilate in pairs. It is possible in principle that the vortex rings in Fig. 9 a could move and touch each other, which would lead to a breaking of both rings to form a big ring, which would then disappear because in all probability it would have no "extra twist". Equally could the rings in Fig. 9b) move and touch the central vortex line, in which case the rings and the vortex line would break with the result of a single vortex line remaining.

Thus the structures are not, as one would presume, structurally stable, (the structural stability ceases to exist once the vortex lines touch), but only dynamically. The configuration shown represents a minimum of energy so that a potential prevents the touching of vortex tubes. As even the stabilization of these most basic localized structures ("solitons") seems a rather fragile one, one would assume that (perhaps with the exception of simple knots) more complicated localized structures would not exist.

7   Bright solitons in laser-like resonators

As opposed to the solitons discussed above which rely on the bistability of a phase structure, more commonly solitons are related with a bistability between different intensities of the field in a resonator. A laser resonator is not usually bistable. However, an additional saturable absorber in the resonator allows to create a bistable on/off-characteristic of the laser resonator. The first experiments of such type were carried out using a broadband dye laser with a very slow absorber (bacteriorhodopsin [37]) in order to produce dynamics with a time scale convenient for recording. A related experiment was described in [38] where the gain and loss elements were both implemented in the form of photorefractive media. An experiment where the gain was provided by two phase-conjugating mirrors was described in [39]. The dye laser experiments were extended in [18] to show the manipulation of solitons as necessary for applications. The laser resonator is schematically shown in Fig. 9. Again it uses a self-imaging resonator for transverse mode degeneracy. The gain element is located in the Fourier-conjugated plane of the absorber plane. Thus all emissions compete for gain. Fig. 9 shows the emission of this laser. As a function of pump power, the laser

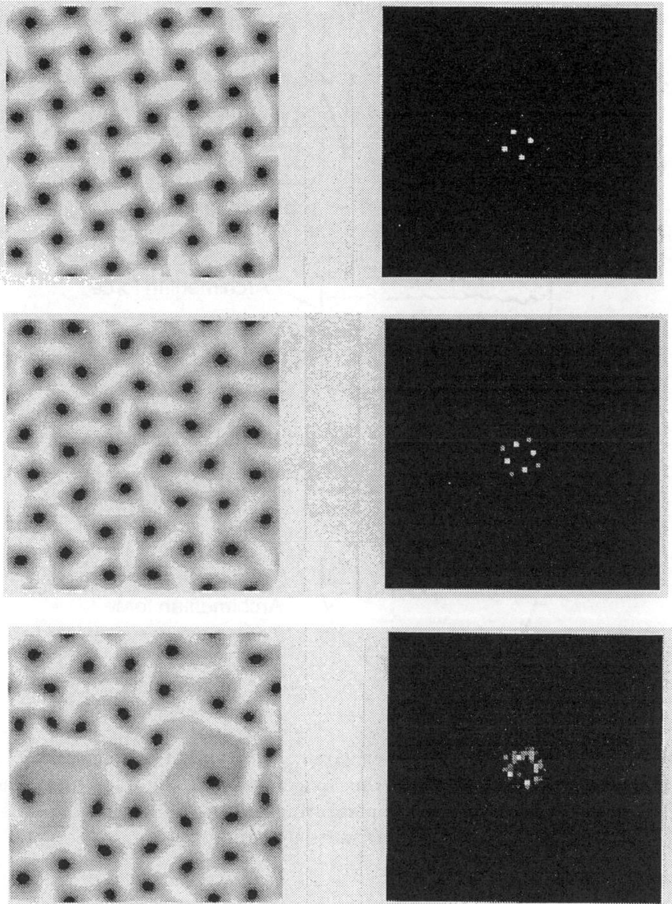

Fig. 4   With increasing pump a "vortex-crystal" develops "defects" until the crystal is molten. New field left, far field right

output power varies in a bistable manner. Outside the bistability range, the laser emits in an irregular shape, while inside the bistability range the emission occurs in a small circular spot or a spatial soliton. For inducing emission in a particular place the absorber is locally bleached by an additional (He-Ne-)laser. Fig. 9 shows that in such a way a spatial soliton can be "written" anywhere in the laser cross section. Fig. 9 proves also the bistability of this soliton: it can exist at a certain location or not.

Fig. 9 illustrates the competition between two solitons (which is a con-

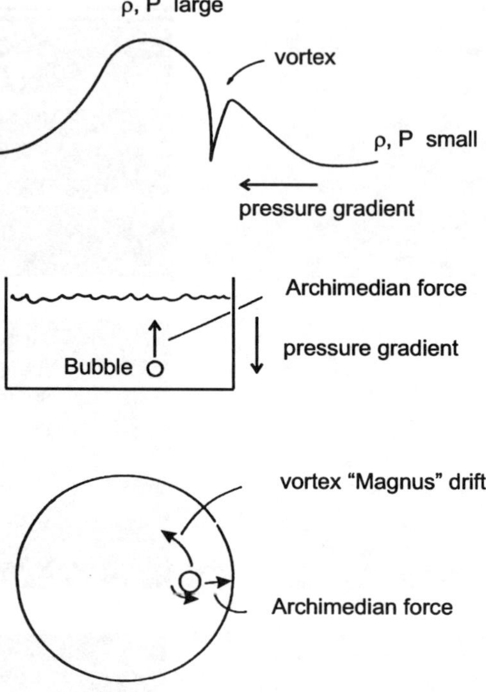

Fig. 5   Hydrodynamic interpretation of circling vortex: at the center of the field the intensity (density, pressure) is high, the vortex (a void, bubble) experiences a force outward (buoyancy). As the vortex is a rotating object, its motion is perpendicular to the force (Magnus effect) resulting in azimuthal motion.

sequence of arranging the gain element in the Fourier plane). Initially one soliton exists (a). A second soliton is written by the He-Ne-laser (b). (c) through (e) show that the second soliton persists and the first one extinguishes. What determines which of the solitons wins? If the bleaching of the absorber by the additional (He-Ne-)laser is stronger than the equilibrium bleaching of the first soliton, then the "new" soliton "wins". If on the other hand the bleaching by the "writing" beam is weaker than the bleaching by the first soliton, then the first soliton survives and the attempt to write a new soliton fails. The soliton arises in this case as usual as a balance between a nonlinear and a linear effect. For emitting a maximum power it is most favourable that the absorber is deeply saturated i.e. the beam in the absorber plane be as small as possible (nonlinear effect). This corresponds to widening the beam in the plane of the gain medium. There the finite size of the pump beam acts like a (linear) beam stop. It is clear that a balance arises, which determines the soliton size. Fig. 9 shows the motion of a soliton under the influence of a phase gradient. We recall that a phase gradient corresponds in the fluid picture to a flow velocity, which can "advect" objects such as solitons. Experimentally, a phase gradient was created by a small tilt of one of the two resonator mirrors. As is seen, this induces the soliton to drift across the aperture. Equally, localisation (or trapping in a certain location) can be accomplished using suitable phase gradients. In Fig. 9 the resonator length was slightly changed with respect to the self-imaging length of 4f. This produces a phase gradient directed radially inward everywhere, (a "phase trough" or, in nonlinear parlance, a "sink"). As is evident, the soliton is pulled from any side towards the optical axis, where it remains "trapped".

Periodic motion is also possible in phase gradients. Fig. 9 shows the situation where a phase gradient moves a soliton towards the edge of the field. The bleaching laser is continuously bleaching at the location marked by the arrow. Under the influence of the phase gradient, the soliton created drifts to the boundary, where it extinguishes. A new soliton appears then at the bleaching location and the process repeats indefinitely. In these experiments the laser was a longitudinal multimode dye laser. Such a system is independent of resonator tuning because of the very wide gain spectrum of the medium. The situation is different for materials with very narrow gain spectrum. Here essentially only one longitudinal mode with its transverse mode family can be amplified and its resonator tuning with respect to the gain line "tilts" the wave front of the generated light wave when the resonator is shortened. Fig. 9a explains the effect. Suppose a resonator consists of two plane mirrors, one at the origin of the wavevector k and the other one dashed. As the width of the gain spectrum is small, radiation of essentially fixed wavelength λ (wavenumber k) can only be generated.

If the second mirror (dashed) has a distance $l = n\lambda/2$ from the other resonator mirror, then the wavevector will be precisely normal to the mirror surfaces. Shortening the resonator so that $(n-1)\lambda/2 < l < n\lambda/2$, the wavevector must tilt to fit into the resonator (which is assumed to be extending infinitely laterally, so that the tilt does not introduce loss). In this way emission occurs in rings in the far field. The ring diameter, correspond-

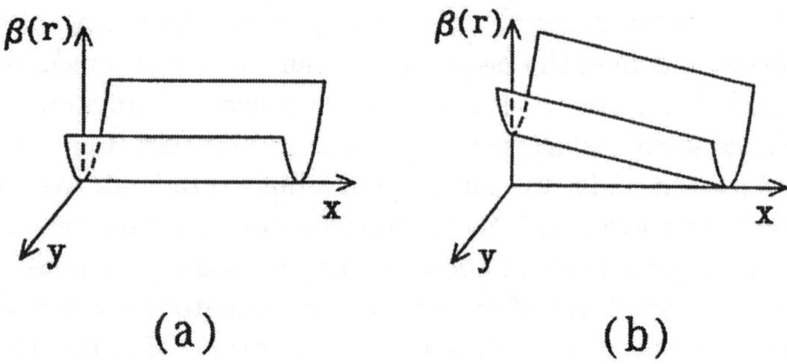

Fig. 6   The "potential" confining the light of resonator with cylindrical optics a) and with one mirror tilted b)

ing to the wavefront tilt depends then on resonator tuning. Vice versa, if a gain medium with a narrow gain spectrum is inside a resonator of fixed length, its emission will be in rings in the far field, a ring corresponding to a value of n, n+1, n +2, etc.. Rings thus correspond to longitudinal modes in a plane mirror resonator. Fig. 9 shows such multilonguitudinal wave emission of a narrow gain line laser (realized here by a photorefractive gain medium inside a self-imaging 4f-resonator). The near field shows the typical "turbulent" emission in "speckle" structure and the rings corresponding to the different longitudinal modes are clearly visible in the far field. It is worth recalling that each ring corresponds to a different tilt of the emitted wavefront i.e. to a different phase gradient ($\equiv$ flow velocity). Also, due to the degeneracy of the resonator one ring corresponds to all transverse modes of a given longitudinal mode. As concerns spatial solitons under this condition of a narrow gain line, Fig. 9 outlines the essential phenomena. A tuned transverse familiy of a longitudinal mode corresponds to emission in the central spot of the far field. It has no wave front tilt, thus no motion. The corresponding soliton in the near field is stationary. An adjacent longitudinal mode corresponds to the first ring in the far field, and the small but finite spread of possible emission wavevectors, due to the finite gain line width, means that an elliptical section of a ring is occupied by the emission. In the near field this corresponds to an elliptical soliton. As the generated wave is tilted, this soliton moves, into the direction of its long axis. While solitons on the central disk in the far field can exist anywhere

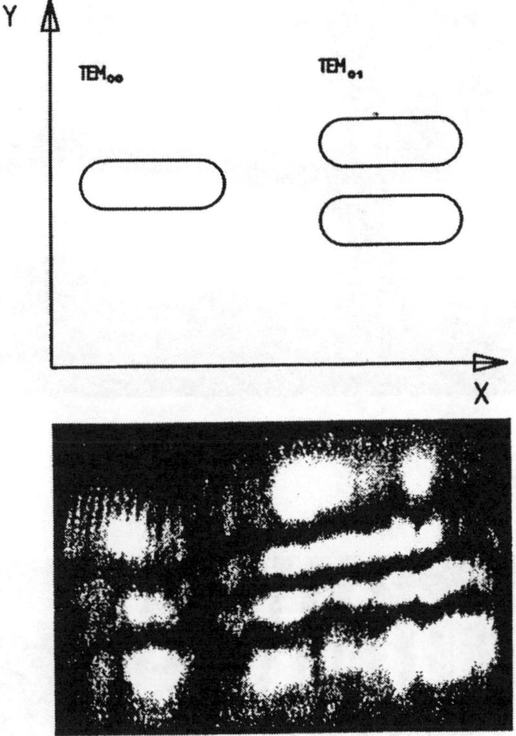

Fig. 7　Modes in a tilted channel (a) 00-,01-Mode schematic (b) 02-,03-Mode experimental. The mode orders are given by the dark node lines.

in the near field plane, and only one can exist because of the competition for gain, solitons on a ring can occupy different sections of their ring. This implies:

    a) Solitons can exist anywhere in the near field plane

    b) Soliton motion can be in any direction

    c) Several solitons can exist simultaneously, occupying different sections of the same ring.

    d) Solitons **compete in velocity space**

    e) Due to the discrete nature of the rings the magnitudes of the velocity of motion of the solitons is quantized.

    f) While stationary solitons will break symmetry only concerning their location, moving solitons have to break the direction of motion symmetry

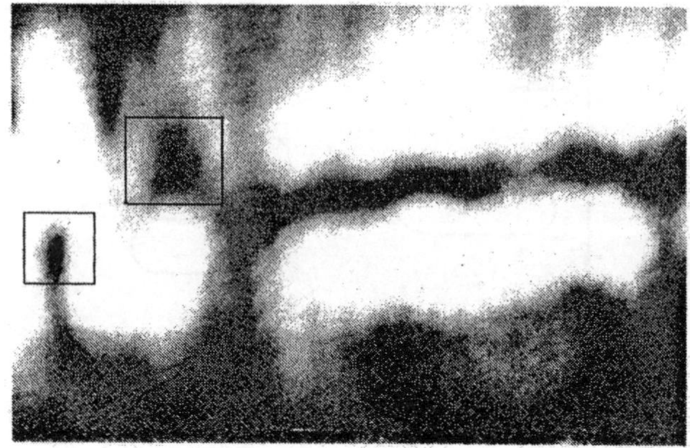

Fig. 8   Snapshot of the "vortex street". Vortices are marked by black square frames. Description see text

Fig. 9   Interferogram corresponding to Fig. 7. The direction of the interference fringe "forks" shows, that adjacent vortices have opposite rotation sense as in a fluid.

in addition. Two moving solitons can coexist in the same place (in the near field). Should they, however, no matter how far they are separated in the near field, have the same vectorial velocity, then they will compete and one can only survive. Simultaneous emission in adjacent rings as shown in Fig. 9 produces an "inchworm"-soliton, with the envelope moving, while the interference fringes are stationary.

The observations [40] were done in the resonator geometry sketched in Fig. 9. The active element was photorefractive $BaTiO_3$ generating gain by two-wave mixing when pumped by a single mode Ar^+ laser. The absorber is again in the Fourier-conjugated plane and the resonator is self-imaging in a 4f geometry to make it transversely equal to a short plane mirror resonator.

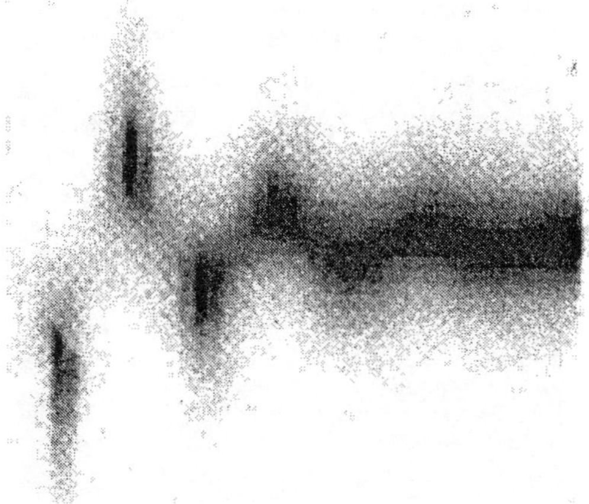

Fig. 10   Simulation of the "vortex street" by fields of the empty cylindrical resonator; indicating good correspondence with experiment Fig. 7.

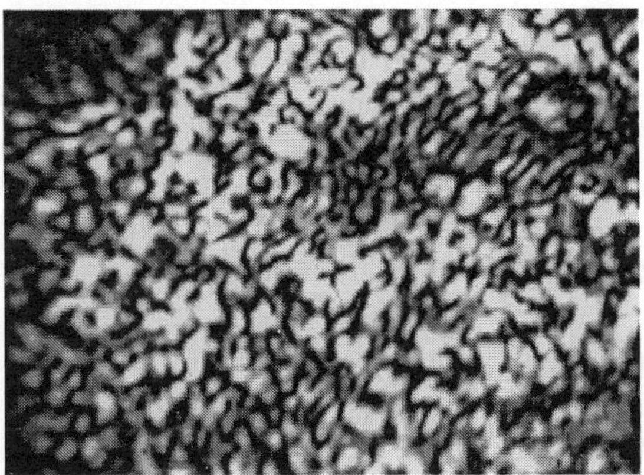

Fig. 11   Emission at large Fresnel number. The field is a dynamical "speckle" field, in which vortices are continually created and annihilated in pairs. The appearance is that of a turbulent fluid.

Fig. 9 shows photos of the stationary and moving solitons. Fig. 9a shows (in the near field) a stationary soliton and two solitons moving into different directions. The far field reveals that the solitons 2 and 3 are emitted in different rings. Thus the velocity of "2" is ∼twice the velocity of "3", while "1" is stationary.. The "inchworm"-soliton corresponding to emission in

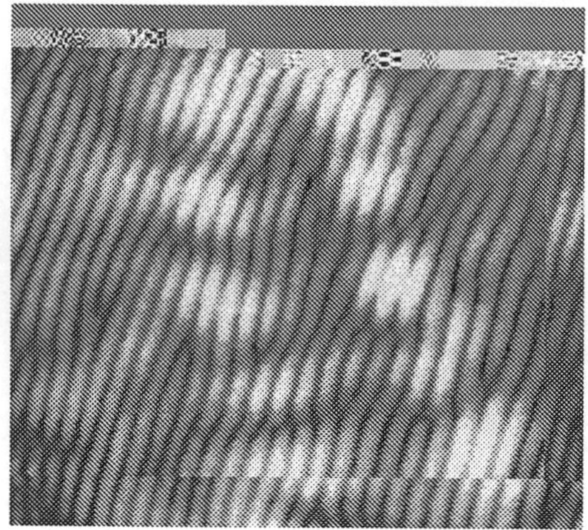

Fig. 12   Interferogram of optical turbulent emission of a (slow) laser. Each "fork" in the fringe pattern indicates a vortex. Vortices of opposite handedness correspond to forks opening upward or downward respectively.

adjacent rings is shown in Fig. 9b. For comparison Fig. 9 shows a numerical simulation of moving solitons. The model is a mean field model which can describe only one longitudinal mode family. It is chosen here to correspond to emission on the first ring. Four moving solitons in the near and far field are shown. One may note that the appearance and the "imperfections" are not any better in the simulation than in the experimental recordings, demonstrating that the simulation is quite realistic [40].

8   Coexisting stationary solitons

If one wants to use such solitons for information processing, the competition in velocity space may not be useful. In particular will one most likely want to use several stationary solitons which such competition would prevent. The resonator arrangement used to produce large ensembles of stationary solitons must therefore be like Fig. 9 to avoid competition. The experiments were conducted in the following way: even with the maximum pump power, the laser could not overcome the loss of the unsaturated absorber. To start the emission the absorber was bleached by an additional green laser. The laser would then start to emit, saturating the absorber completely, so that the system behaved like a laser without absorber (i.e. emit a radiation pattern showing vortices and the shocks separating them; see section 1).

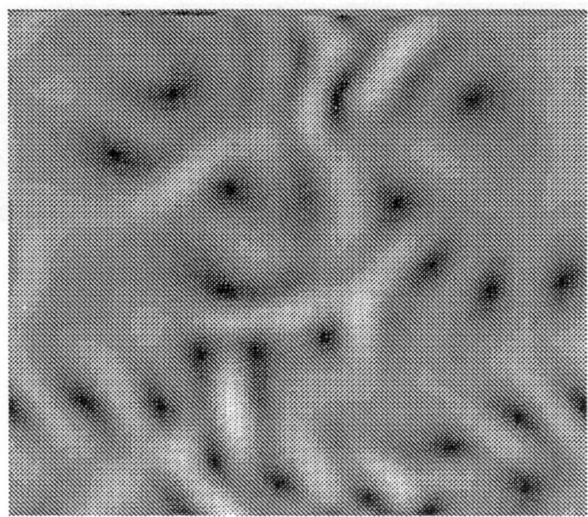

Fig. 13   Numerical calculation of "shocks" associated with vortices ("vortex glass") for a laser.

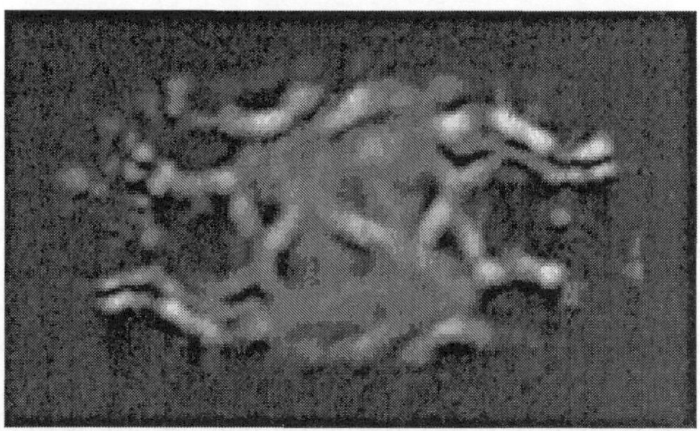

Fig. 14   Experimental observation of "shocks" associated with vortices recorded from a photorefractive oscillator.

By then reducing the pump power the absorber becomes less saturated and hence nonlinear. Sufficient reduction of pump power results then in the formation of large collections of (up to 50) solitons. Reducing the pump power further then reduces the number of solitons. Fig. 10 gives a sequence

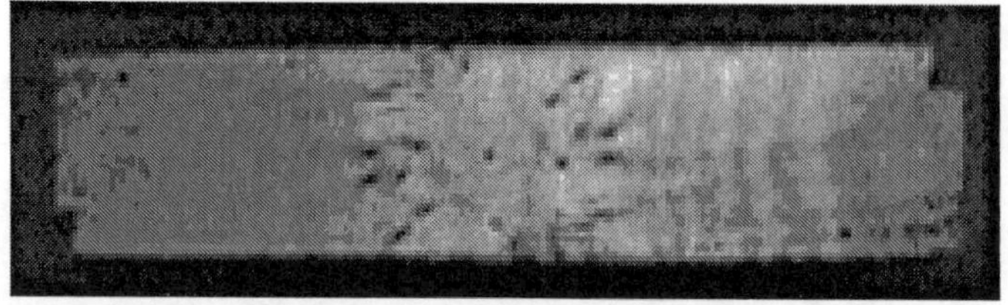

Fig. 15   Vortices being expelled from a "source" at the optical axis and moving outward towards the boundary. Recorded from a photorefractive oscillator.

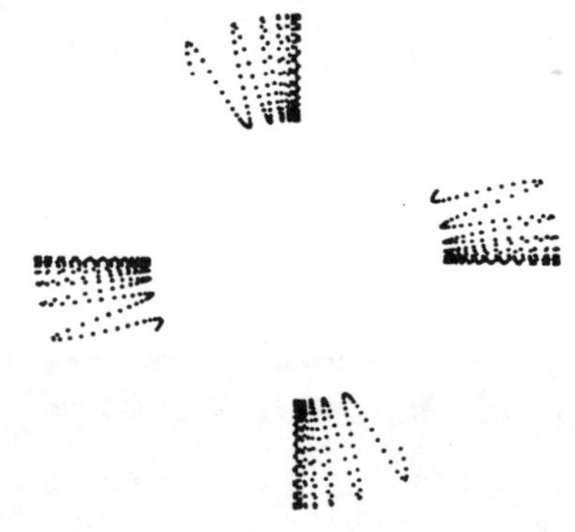

Fig. 16   Calculated dynamics of four vortices in the field of a class B-laser. The vortices oscillate around their angular trajectory.

of calculations showing in a simulation the transition from "laser" emission (with vortices, shocks) to spatial 2-D solitons.

The shock structures transform into bright stripes (actually 1-D solitons) when reducing the laser pump power. After these have formed, a slight increase of pump power converts the 1-D solitary structures into 2-D solitons i.e. bright spots. The 2-D solitons cannot be reached by monotonically decreasing the pump, the 1-D solitary structures have to be created first [41]. This is a consequence of 2-D solitons having a larger diffraction loss (i.e. a higher pump) than 1-D solitons. Fig. 10 shows collections of solitons thus created experimentally. The number of solitons is a monotonic function of pump power. The observations showed a peculiar mechanism to maintain the number of solitons at constant pump power: due to the

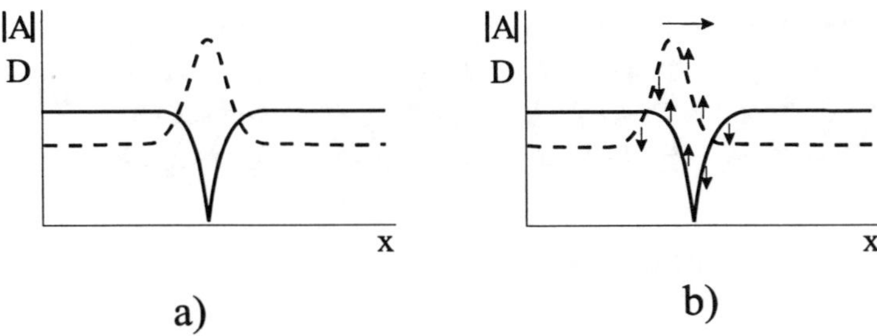

Fig. 17   (a) Field (solid) and inversion (dashed) for a vortex in a class B-laser. (b) When the vortex is moved laterally the excess inversion at the original location of the vortex can be reduced.

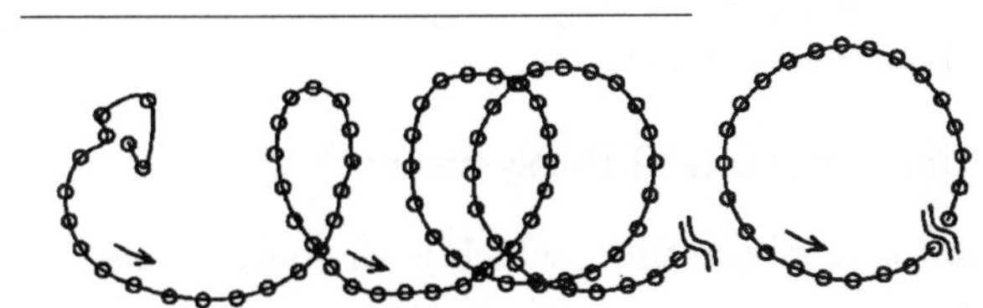

Fig. 18   Calculation of the motion of an initially stationary vortex for a plane resonator class B-laser

nonuniform pump field the solitons show a slow drift outward, where they eventually extinguish at the boundaries. To replace the lost solitons, new solitons are continually generated by splitting of solitons in the central area of the laser cross section (Fig. 10). This splitting ("self replication"(!)) is probably related to the periodic nucleation of solitons described above (Fig.9). Fig. 10 shows a first order soliton with a vortex at its center as numerically calculated. We have not been able to observe experimentally other than fundamental solitons.

9   Solitons in semiconductor microresonators

In order to be useful in technical applications it is mandatory that solitons can be switched and manipulated at the highest possible speed. There are

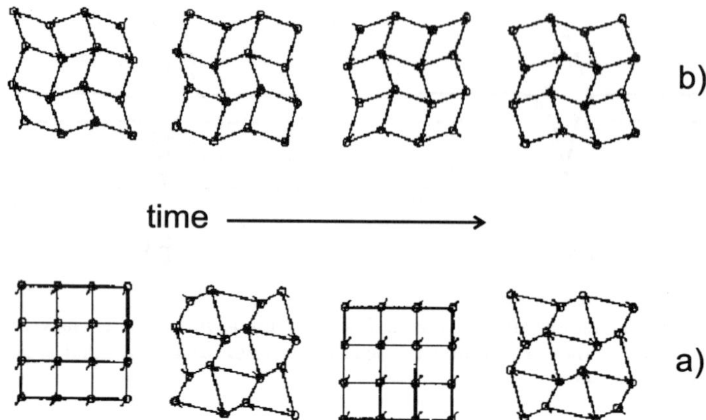

Fig. 19 The "restless" motion of the vortices in a "vortex crystal" of a class B-laser can synchronise with two different phases to form motion according to (a) an "acoustic phonon" and (b) an "optical phonon".

two limiting time constants of the systems.

- i) the resonator response time, which is the decay time of the field in the linear resonator
- ii) the nonlinear material relaxation time after an excitation

Material relaxation times can be smaller than 1 ps, so it is necessary first to minimize the resonator response time. Given that a certain finesse is required to make a passive resonator bistable as required for existence of solitons the resonator field decay time is simply proportional to the resonator length. Semiconductor materials are absorptive enough to provide substantial interaction already for propagation lengths of the order of an optical wavelength. Thus one can in principle use fundamental order resonators, i.e. resonators as short as one half optical wavelength. Such resonator length together with mirror reflectivities of 99.7 % yield field decay times of 200 fs to 1 ps. This being a definite limit for the speed of any system using a resonator, it remains to chose a material with a relaxation time not shorter than the resonator response time, because an unnecessarily fast material requires unnecessarily high light intensities to make it sizeably nonlinear.

It results that semiconductor materials, which can have relaxation times down to a few 10 fs are the best match for the microresonators. Thus

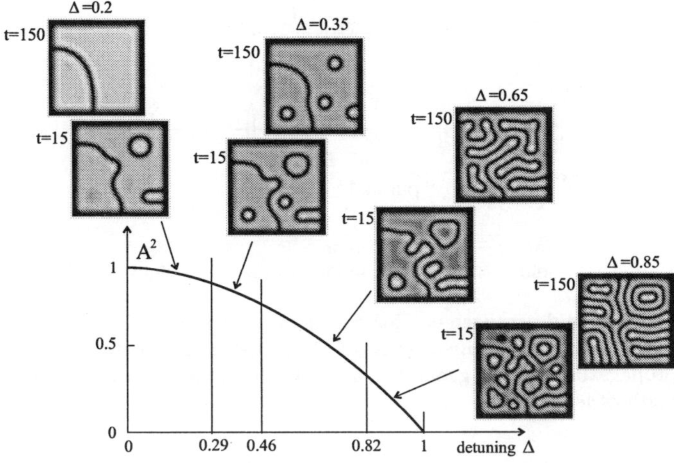

Fig. 20   The different types of dynamics of a degenerate OPO in different ranges of resonator tuning. The solid curve gives the stationary homogeneous intensity. Details see text.

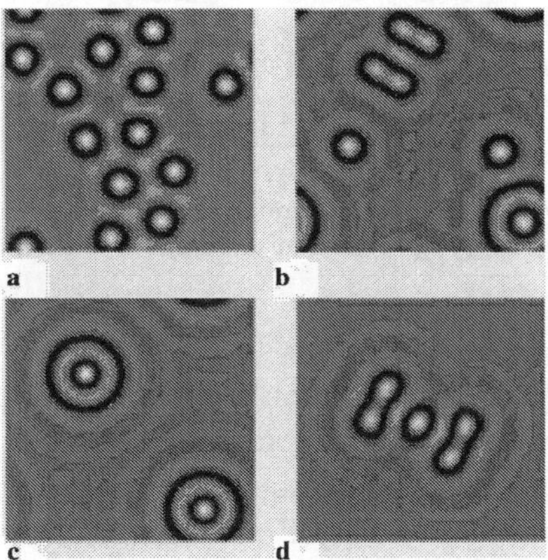

Fig. 21   Numerically calculated soliton structures in DOPO. (a) collection of fundamental solitons, (c) higher order solitons, (b), (c) bound states ("molecules") of solitons

for experiments on (and devices based on) microresonator spatial solitons, the structures typically used for VCSELs (vertical cavity surface emitting lasers) are adequate and so are all growth techniques developed to maturity for the purpose of VCSEL production. A typical semiconductor resonator

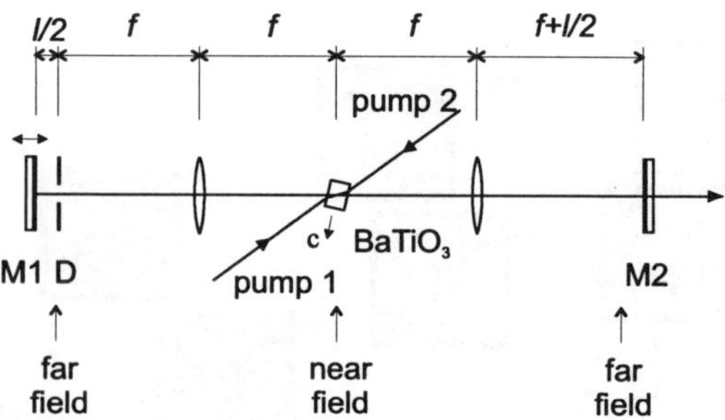

Fig. 22   Experimental arrangement for degenerate 4-wave mixing. The nonlinear element is a BaTiO₃ crystal pumped by two 514 nm fields. Degeneracy of the emission is forced by propagating the two generated waves in the same linear self-imaging resonator (f: focal length of lenses).

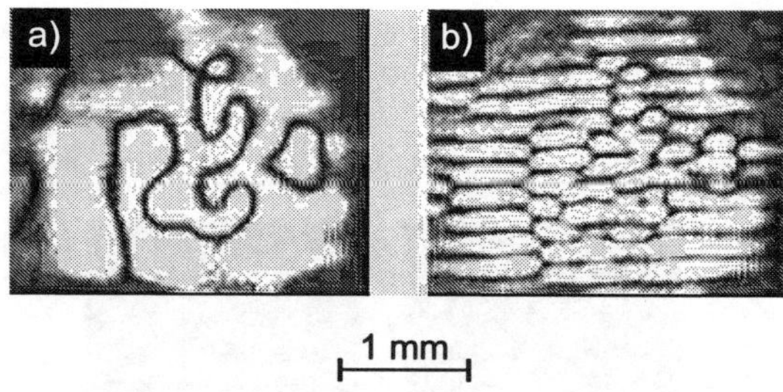

Fig. 23   A complicated-shaped domain boundary. (a) The boundary shows even self-crossing. The interferogram (b) proves that emission in adjacent domains is 180° out of phase.

structure useful for the experiments is shown in Fig. 10(a). It consists of a substrate onto which a Bragg mirror is grown. Spacers of inactive material are added to place the resonance wavelength at a desired value. Onto the Bragg mirror a number of quantum well layers are grown which constitute the active material. On top of the QW-layers is another Bragg mirror grown.

The optical length of the QW material is between $\frac{1}{2}\lambda$ and 2λ. The number of wells is between 5 and 20 and the mirror reflectivities are around 99.8%, yielding a finesse of several 100. The thickness of the whole structure is a few micron and the lateral extension is typically 5 cm, the standard size of a GaAs wafer, resulting in a huge nominal Fresnel number. But it is to be taken into account that the layer thickness as it is produced

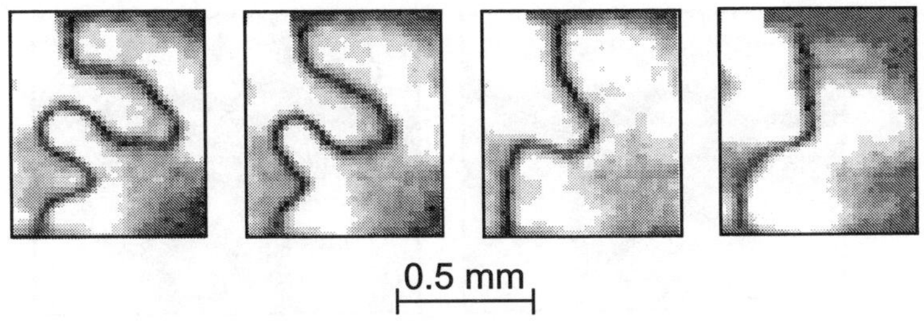

Fig. 24   A curved domain wall straightens and shrinks with time (small detuning, compare Fig. 8).

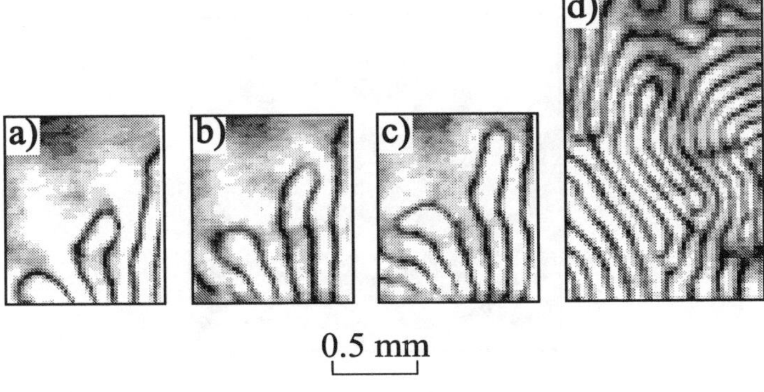

Fig. 25   Expanding domains (large detuning, compare Fig. 8).

in the growth process is not uniform across the 5 cm diameter. There is always a radial variation of thickness of the layers, such that the resonator wavelength varies in the most homogeneous structures still as much as 0.3 nm/mm. Given that a certain detuning of the laser field frequency from the resonance frequency is a condition for the existence of solitons, it follows that areas of only a few mm radial extent can be used for a given laser wavelength.

On the other hand is the radial variation of resonance frequency an advantage in that it permits to work on one wafer for different wavelengths with respect of the semiconductor QW band edge. It may be mentioned that the whole structure i.e. Bragg-mirrors, quantum wells, spacers, and substrate, use only 3 components (Ga, As, Al). GaAs has its band edge at 850 nm, while the band edge of AlAs lies in the green. The composite

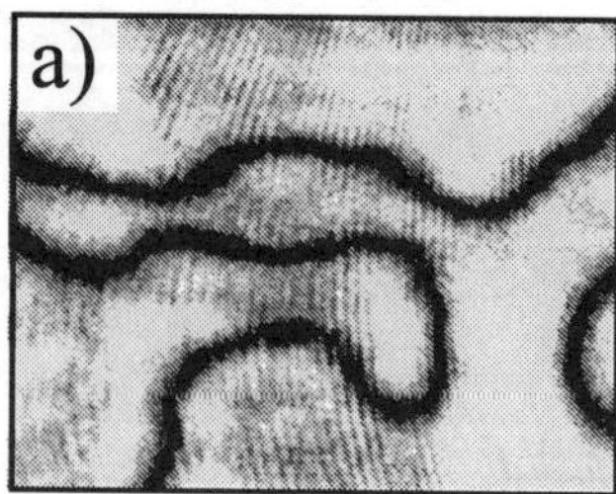
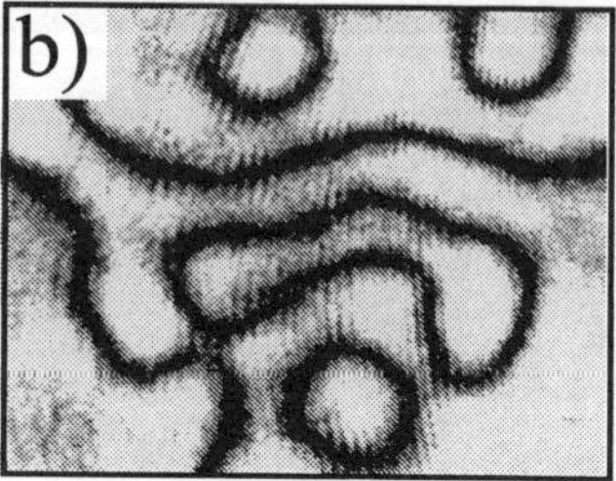

Fig. 26   Nucleation of new domains (large detuning, compare Fig. 8). (b) is later in time than (a).

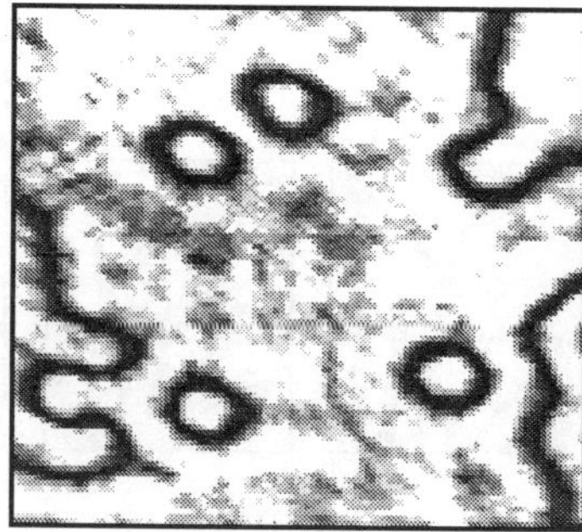

Fig. 27   Several solitons coexisting with non-stationary domains (small detuning compare Fig. 8).

$Ga_xAl_{1-x}As$ has a band edge wavelength depending on the composition "x" varying between those of GaAs (850 nm) and AlAs (green). Thus by varying "x" the material can be a dielectric without losses (such as needed for the mirrors and the spacers) or an absorbing (active) material (as needed for the quantum wells): in the case of GaAs and a resonance near the band edge, the substrate is opaque. Observations can then only be done in reflection. Fig. 10b shows the bistable characteristic of such a structure measured in reflection. At small input intensities the resonator reflects according to the reflectivity of the first mirror. At a certain input the resonator becomes suddenly transmitting (reduced reflection), which persists if the input is decreased below the switch-on intensity. At a certain intensity (switch-off intensity) the resonator then switches back to the high reflectivity. The easiest way to picture this bistability is to assume a purely reactive response

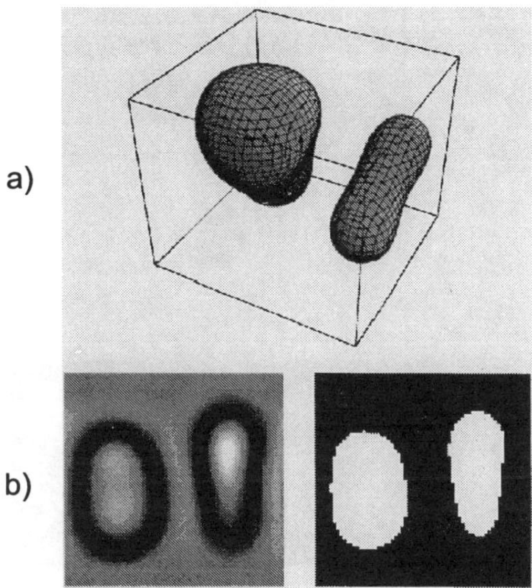

Fig. 28   Nonstationary 3-D domains calculated for a DOPO. Cross section shows that the phase inside and outside of the domain wall is of opposite sign.

of the active material. Initially the light is out of resonance. At sufficiently high input the intensity-dependent refraction of the active material in the resonator shifts the resonator resonance in coincidence with the laser light frequency. Whereupon the field in the resonator becomes large (and consequently so does the resonator transmission). Decreasing then the input intensity leads not directly to a switchback. The high intensity inside the resonator keeps the resonator tuned so that the switchback occurs only at substantially smaller intensity. Thus a pronounced bistability loop exists.

This is the plane wave characteristic, i.e. a homogeneous field in the resonator is assumed. Spatial solitons exist usually near the switching intensities. A plausible picture for this is the following: if the intensity is not high enough to bring the whole resonator into resonance, the system may decide (loosely speaking) that at least it will be able to bring certain small part of the cross-section in resonance by concentrating the available field in certain small areas (solitons). Such freedom to arrange relatively arbitrarily the internal spatial state is given to the system only at high nonlinearities [13]. One can thus say that solitons will always develop which "anticipate" the plane wave state to be switched to. Fig. 10c shows that

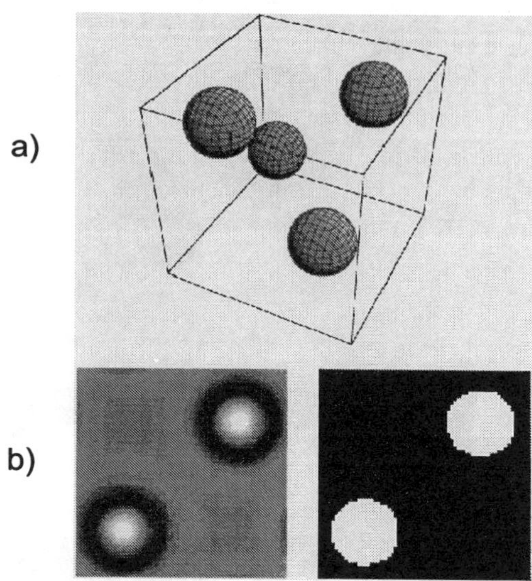

Fig. 29   A collection of phase-solitons in 3-D calculated for a pulsed DOPO.

the resonator can be switched on and off by pulses increasing or decreasing the illuminating intensity. The experimental arrangement used to search for and investigate spatial solitons in semiconductor resonators is shown in Fig. 10. The main laser beam illuminates the semiconductor resonator sample in an area of 30–100 m diameter with intensities around kW/cm^2. A part of the laser beam is split away from the main beam and recombined with the main beam after sharp focusing. The focused beam diameter on the sample is around 5μm diameter. This latter, or switching-beam can be opened for a few nanoseconds by an electro-optic modulator. The phase of the switching beam relative to the main beam phase can be adjusted by a piezo translator so that the pulse is negative (destructive interference for switching off) or positive (constructive interference for switching on). As the switching beam is very narrow, it can be directed to various points in the illuminated area to write (or erase) a soliton there.

The observations are done in reflection because of the opaque substrate. Snapshots with 10 nanosecond resolution are made by a CCD camera combined with an electrooptic modulator used as a fast shutter. In addition a fast photodiode can be imaged onto selected points of the illuminated area of the resonator sample to follow locally the reflected intensity in time. The

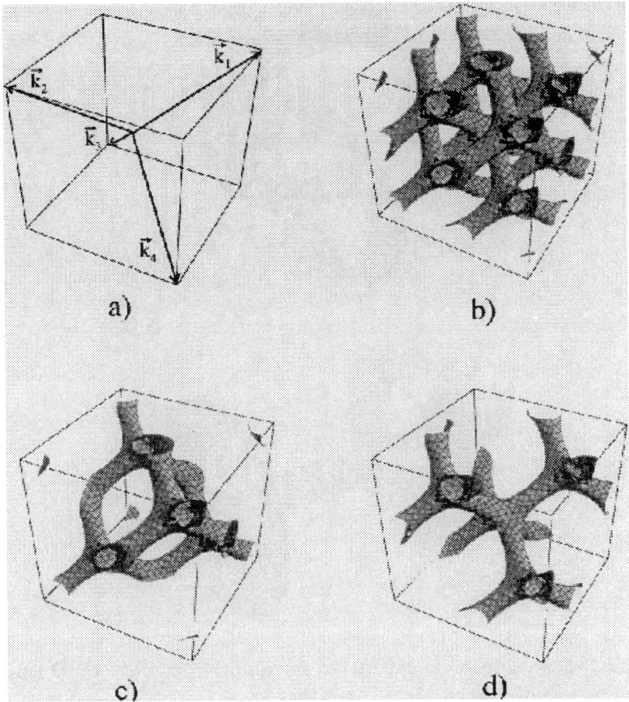

Fig. 30   A stable non-localized field structure calculated for the DOPO. The tetrahedral tree structure originates from a combination of wave vectors as shown in (a). (b), (c) give the 90% intensity contours for the 3-D domains of opposite phase. The total structure consists of both domains intertwined, (d).

measurements and observations were predominantly done within a time of a few microseconds during which time the illumination light was admitted to the sample through a mechanical chopper and/or an acoustooptic modulator. The illumination and thus the observations were repeated every millisecond. In this way pertubations due to heating effects of the sample were limited, and it was possible to follow the dynamics of the pattern/soliton formation. On the other hand, the intensities used (kw/cm^2) were well below a damage threshold of the material, so that continuous experiments are also possible and were also done.

The initial search for spatial solitons was done in the "dispersive limit" i.e. for laser frequencies below the band edge. As one might have expected, hexagonal patterns were observed here [42], (Fig. 10) confirming a historic

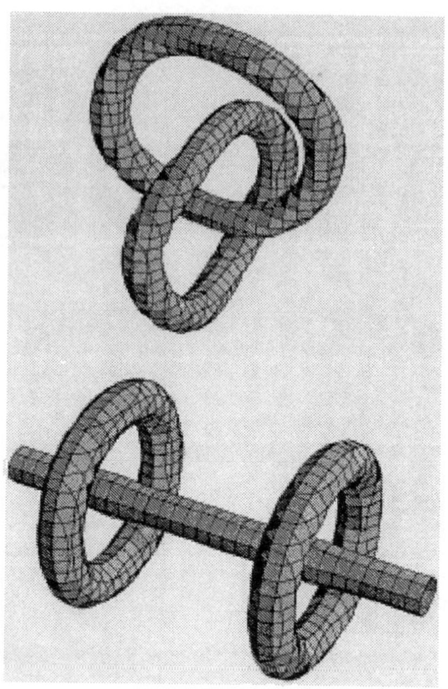

Fig. 31   Stable 3-D structures as calculated for a nondegenerate OPO (mathematically equivalent: a "very fast" laser). Details see text.

theoretical prediction [43].

Closer to the band edge bright and dark solitons were found. Fig. 10 gives single bright and dark solitons and also collections thereof. In an arrangement with an elongated illuminating beam (Fig. 10) we find that these solitons are really independent of one another, i.e. they can exist or not, irrespective of the existence of other solitons.

To demonstrate the switching on and off of solitons a series of measurements is shown in Fig. 10. (Note that the observations are done in reflection). (a) shows the switching on of a soliton. A pulse of 20 ns duration of the 5 μm wide switching beam is applied in phase with the background field (as can be seen from the initial increase of the local intensity). The switching threshold is then reached and a soliton is formed, as apparent from the reduced reflectivity corresponding to high transmission. The soliton evidently exists after the end of the switching pulse. It is extinguished only by the final decrease of the background light. (b) shows a "failed switch", proving that a certain intensity value is necessary to create a soli-

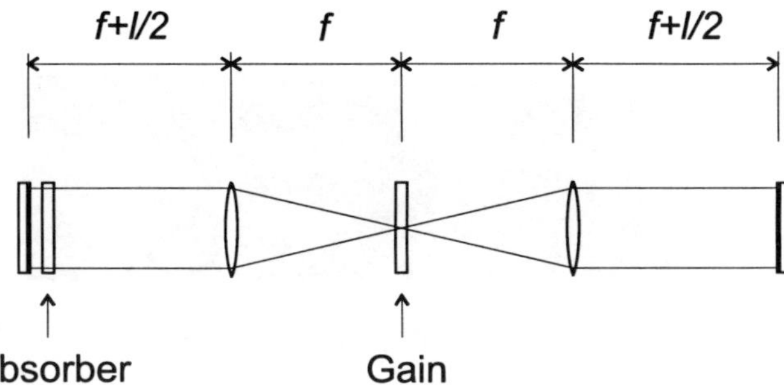

Fig. 32 Self-imaging resonator containing a gain element and a saturable absorber (bacteriorhodopsin) in Fourier-conjugated planes

ton (c) shows switching off of a solition. A soliton is initially spontaneously created. The narrow, pulsed switching beam is aimed at the location of the soliton. Its phase is opposite to the background field phase (as is visible by the reduced light reflected, directly after application of the switching light. The existing soliton is then destroyed as is apparent from increased reflection (opposite to case (a)) (d) shows analogously to (b) that also for switching a soliton off a certain minimum switching intensity is needed.

A rough estimate is possible of the energy necessary to switch a soliton on or off. The power necessary to sustain a bright soliton is roughly 1 mW. Such power has to be applied roughly for the response time of the medium (which is of the order of 1 ns for the sample used) so that the switching energy is of the order of 1 pJ. This value will not change appreciably if other faster semiconductor material is used: faster material means smaller nonlinearity, requiring higher intensity to switch. The switching energy will always be the energy necessary to generate the necessary charge carriers.

One way is possible, however, to reduce the sustaining power, and with it presumably the switching energy: a semiconductor can be "pumped". This means that carriers can be generated by other means than the soliton sustaining light. For this purpose obviously light of shorter wavelength can be used, as well as electrical pumping. Although electrical pumping has already allowed to observe soliton structures in VCSELs above transparency, it has its drawbacks: in large area VCSELs the current distribution is quite inhomogeneous, (in fact most such devices show pronounced current filamentation) and carriers have a small penetration depth into the quantum wells, meaning that the number of quantum wells can only be 1 to 3 in VCSELs resulting in small nonlinearities. (Nonetheless, structures with

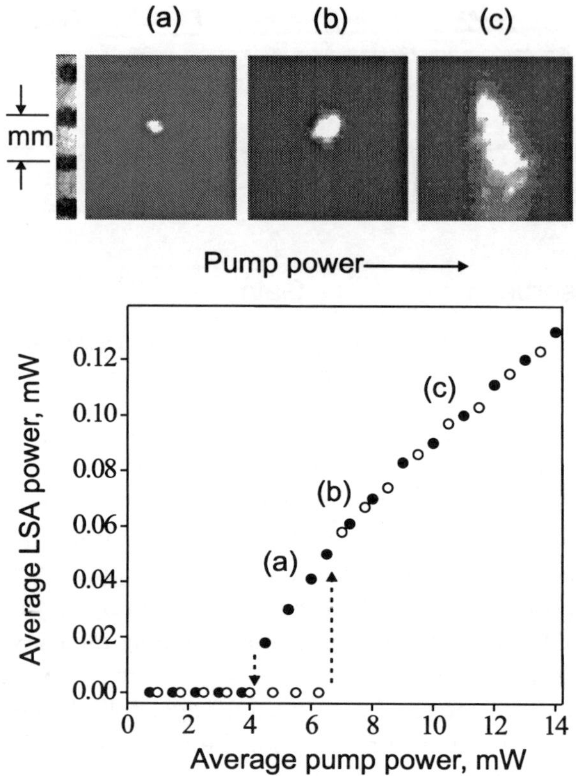

Fig. 33   Emission of the laser Fig. 9 for different pump powers. The emission is bistable. Within the bistability range the emission is in the form of a spatial bright soliton.

the appearance of solitons were recently observed [44] in such large area VCSELs).

We chose therefore optical pumping. We pump the semiconductor structure which has the band edge at 850 nm, where we work with the soliton sustaining- and switching- light, at a wavelength of 810 nm. As opposed to electrical pumping it is no problem pumping all of the 18 quantum wells, and the pump profile is relatively homogeneous (Gaussian beam). Fig. 10a shows measured switch on- and switch-off intensities for a soliton as a function of pump intensities (for the unphysical crossing of the "on"- and "off"-curve see figure caption). It is apparent that the intensity necessary to sustain a soliton is reduced at the maximum pump intensity available

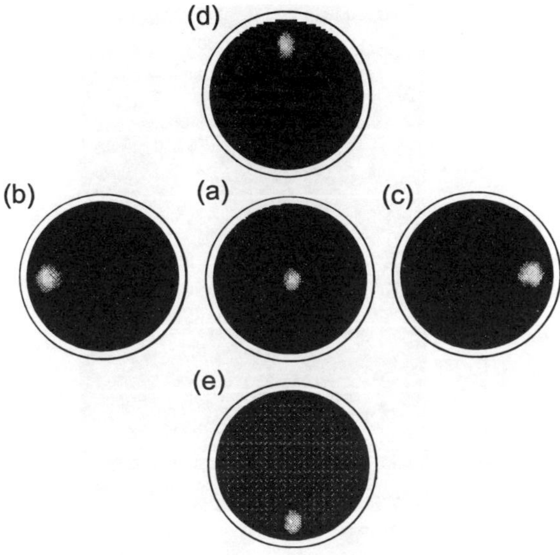

Fig. 34 Local bleaching of the BR-absorber allows to "write" a soliton in arbitrary locations.

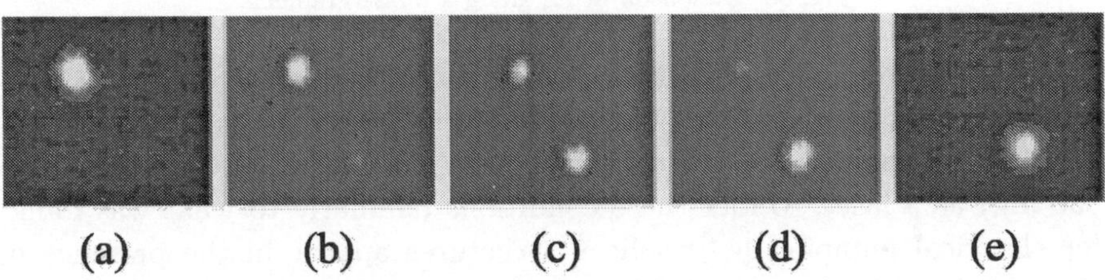

Fig. 35 Competition between two solitons. Details see text.

from the pump laser diode more than a factor of 10 compared to the unpumped material. We believe that sustaining power as small as 10μW will be attainable with increased pumping. Fig. 10b shows a calculation for comparison. One can see that the bistability (and the solitons) cease to exist at some pump level. This results because near the material transparency (equal populations in the valence and conduction band) all nonlinearities go to zero. Above transparency the nonlinearities grow again with pumping level, although with opposite sign (defocusing $\leftrightarrow$ focusing). This renders

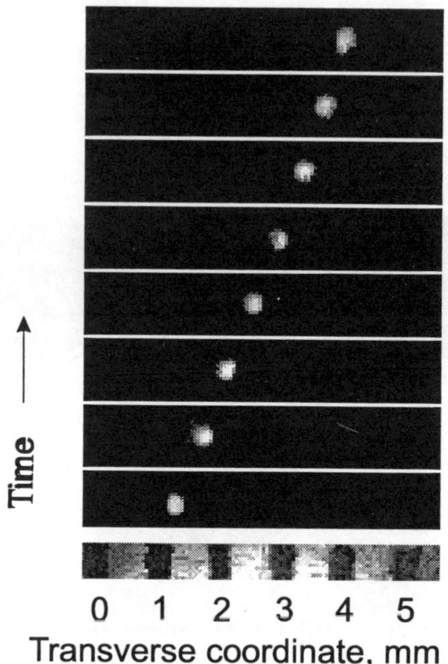

Fig. 36   A soliton drifts along a phase gradient

the conditions above transparency particularly favourable for bright solitons (focusing nonlinearity). In our pumping experiments we have been able to reach enough population inversion so that the structure emitted (at 850 nm) as a laser. Under these conditions (similarly to what was found for electrical pumping [44]) soliton structures appear in the presence of the additional background field. Fig. 10 shows collections of solitons observed for optical pumping above laser threshold. These soliton clusters are vividly reminiscent of the very first observations of spatial solitons in Kerr-resonators [45]. We have also been able to supply the soliton sustaining light from a low power diode laser, as it is important for applications [46].

10   Patterns and solitons

At the beginning of the search for semiconductor resonator solitons we observed hexagonal patterns such as Fig. 10. One surprising result was that individual bright (in reflection) spots could be switched off by an optical

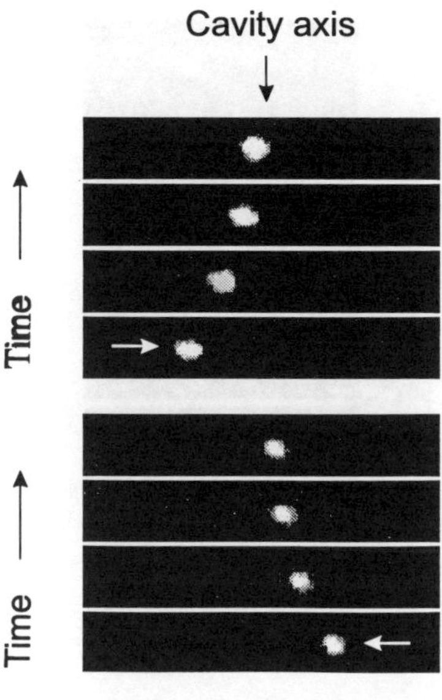

Fig. 37   A soliton is drawn from all sides into a phase trough (and trapped at the trough center)

address beam [42]. The relation of these switchable spots with solitons was unclear. It was termed "localised switching". Fig. 10 shows that individual spots and also "triples" of spots can be switched, completely independently of the rest of the structure.

A later investigation revealed the relation of "patterns" (space filling structures) and solitons. Fig. 10 shows results of model calculations clarifying this. Fig. 10 shows the plane wave steady state characteristic together with patterns observed at different input intensities. At small input intensities hexagons of bright spots appear, which with increasing intensity convert, by way of stripe patterns, to hexagons of dark spots. The hexagon period increases with intensity as visible in Fig. 10 (a)–(d), and at the highest intensity the hexagon solution coexists with solutions where individual dark spots are missing (as in the experiment) as shown in Fig. 10. Thus the following interpretation results:

(1) The increase of the pattern period with intensity is the result of the non-linear resonance effect in the following way: the pattern period at small intensity is given by the resonator detuning: for a resonator shorter

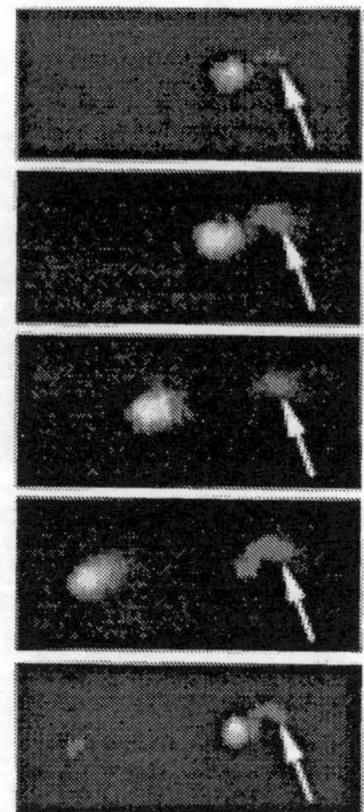

Fig. 38   Periodic nucleation, and extinguishing at the boundary, of a soliton.

than an integer number of wavelengths the resonant wave cannot propagate exactly perpendicular by the mirror plane, but is inclined with respect to the normal on the mirror plane (tilted wave solution [47]). In fact the structure Fig. 10 a is composed of 6 phase-locked, tilted waves. Thus the appearance of this pattern is largely a linear phenomenon, dictated by the boundary conditions (resonator detuning). Then the increased period at higher intensities is to be interpreted as a reduction of the tilt angle and thus as a reduction of the resonator detuning. The latter is obviously caused by the intensity-dependent index of refraction of the active material. The pattern period from (a) to (d) is increased by almost a factor of 2. This means that the "nonlinear detuning" is comparable to the external (linear) detuning. This signifies a strongly nonlinear regime of the active material. (2) The fact that the periodic and defected solutions coexist at the high intensities shows that due to the stronger nonlinearity the system obtains a freedom to arrange its internal structure (presumably to minimize some thermodynamic potential) rather freely. Thus many defected solutions coexist, meaning that individual of the dark spots can be switched off, as observed experimentally. We found that almost arbitrary arrangements of

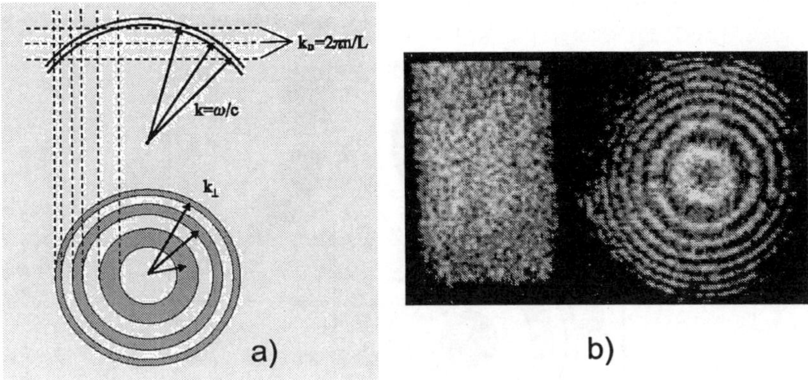

Fig. 39   Shortening of the length of a plane resonator results in tilt of the emitted wave, corresponding to rings in the far field, (a). Emission in the near and far field of a photorefractive oscillator (b). Details see text.

dark spots coexist including solutions with isolated solitons [48].

The key feature—the freedom to arrange the spatial state of the resonator provided by the strong nonlinearity—is the same as permitting the formation of solitons. Thus the individual dark spots of the pattern of Fig. 10d and Fig. 10 have to be interpreted as solitons. The space-filling pattern Fig. 10a, however, does not consist of solitons. It is "coherent" in the sense that the bright spots have no individual character. One spot cannot be switched without destroying the entire pattern. This pattern is "linear" in the sense that it results from boundary conditions. As opposed, the structures Fig. 10d and Fig. 10 are "incoherent" and "nonlinear". Though the individual building blocks (solitons) are largely defined in their shape and size by external system parameters, their spatial arrangement is largely free. Thus these latter patterns are really "crystals of solitons".

We see that increasing nonlinearity gives the system increasing freedom. At small intensities a rigid pattern forms. At medium intensities the patterns disintegrates into individual free solitons. And one would extrapolate that at very large intensities the freedom of the system would be even more complete and the structure would become turbulent in space and time.

One may draw a speculative general conclusion from this observation of transition from a "coherent" to an "incoherent" pattern: at small nonlinearity there is rigid structure, while at very high nonlinearity the system is totally free i.e. space-time turbulent. At intermediate nonlinearity rigid

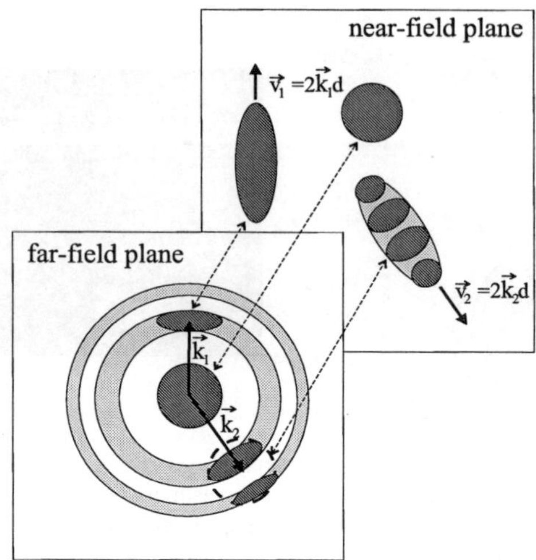

Fig. 40   Spatial solitons with wave front tilt (schematic). Emission in the central disk produces a stationary soliton. Emission in rings produces moving solitons. Details see text.

building blocks exist with properties defined by the external system parameters, but the system has a large freedom to arrange these building blocks spatially. This mean a large number of coexisting spatial steady states, which defines a memory capacity.

It follows that intermediate nonlinearities are the range in which useful optical processing schemes could be realised. It also follows, given that life and information processing are somehow the same, that biological structures operate in the intermediate nonlinearity regime.

References

E. Gaizauskas, K. Staliunas Opt. Comm. 114 (1995) 463

P. Bak et al. Phys.Rev.Lett. 59 (1987), 381

V.B. Taranenko, C.O. Weiss IEEE J. of sel. topics in QE 8 (2002) 488

Chr. Tamm Phys. Rev. A38 (1988) 5960

See e.g. C.O. Weiss Phys.Rep. 219 (1992) 311

C.O. Weiss et al. J.Mod.Optc. 37 (1990) 1825

Chr. Tamm In: Measures of Complexity and Chaos (Plenum Press, New York 1989) p. 456

M. Vaupel, C. O. Weiss Phys. Rev. A51 (1995) 4078

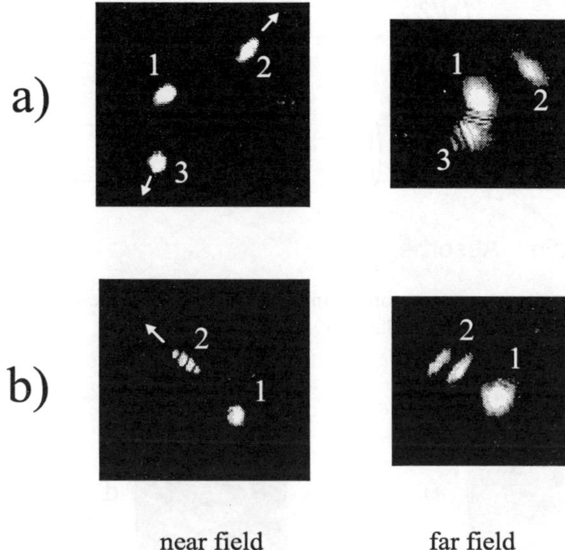

Fig. 41 Experimental recording of stationary and moving solitons of a photorefractive oscillator with BR-absorber. Details see text.

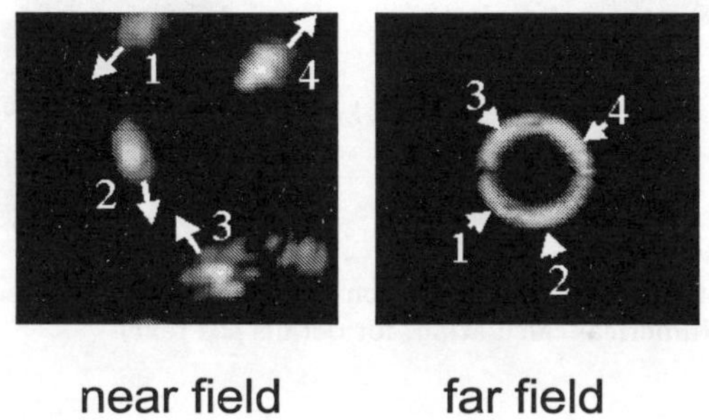

Fig. 42 Numerical calculation of moving solitons emitted in the first far field ring of a laser with nonlinear absorber, for comparison with Fig. 9.

Supplementary electronic material to: C.O. Weiss, M. Vaupel, K. Staliunas, G. Slekys, V. B. Taranenko Appl. Phys. B68 (1999) 151 at http://link.springer.de/journals/apb

N.R. Heckenberg et al. Phys.Rev. A 54 (1996) 2369; J.T. Malos et al. Opt.Comm. 128 (1996) 123

C. P. Smith, Y. Dihardja, C. O. Weiss, L. A. Lugiato, F. Prati, P. Vanotti Opt. Comm. 102 (1993) 505

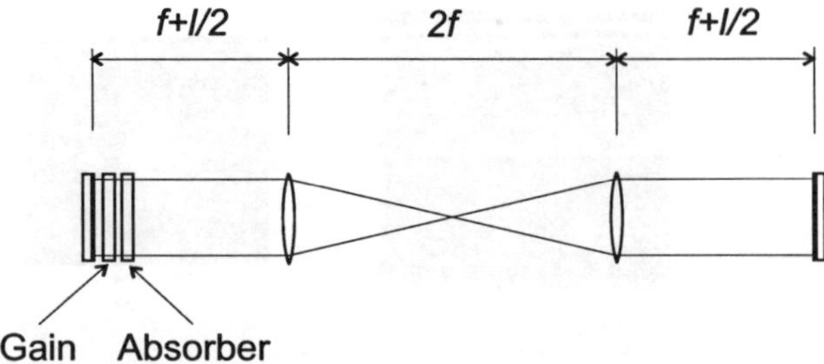

Fig. 43   Self-imaging resonator containing gain element and nonlinear absorber both in the near field. (f: focal length of lenses)

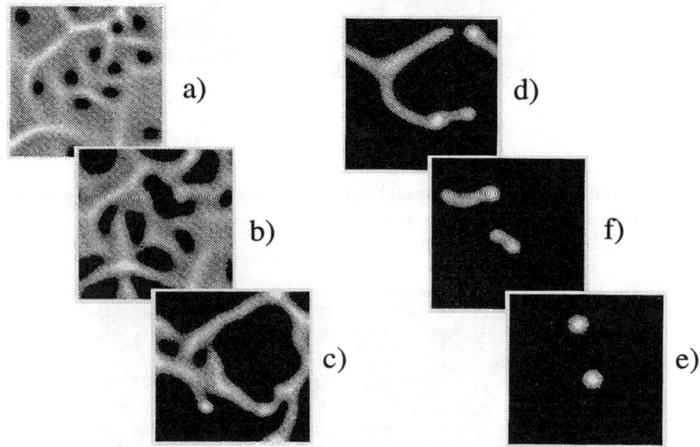

Fig. 44   Laser emission ("vortex glass") converts gradually to emission of solitons when reducing pump. (Numerical calculation, for details see text)

Stationary vortex clusters in trapped wave fields L.C. Crasovan et al. xxx.lanl.gov preprint physics/0202074

V.B. Taranenko et al. Phys.Rev. A 65 013812 (2002)

Optical Vortices M. Vasnetsov, K. Staliunas/Eds. (Horizons in World Physics 228) Nova Science Publishers, Commack, New York (2001)

M. Mathews et al. Phys.Rev.Lett. 83 (1999) 2498

K. Staliunas Phys. Rev. A48 (1993) 1573

L. Lega et al. Phys.Rev.Lett. 73 (1994) 2978

V. B. Taranenko, K. Staliunas, C. O. Weiss Phys. Rev. A56 (1997) 1582

M. Vaupel, K. Staliunas, C. O. Weiss Phys. Rev. A54 (1996) 880

M. F. H. Tarroja, K. Staliunas, G. Slekys, C. O. Weiss, L. Dambly Phys. Rev.

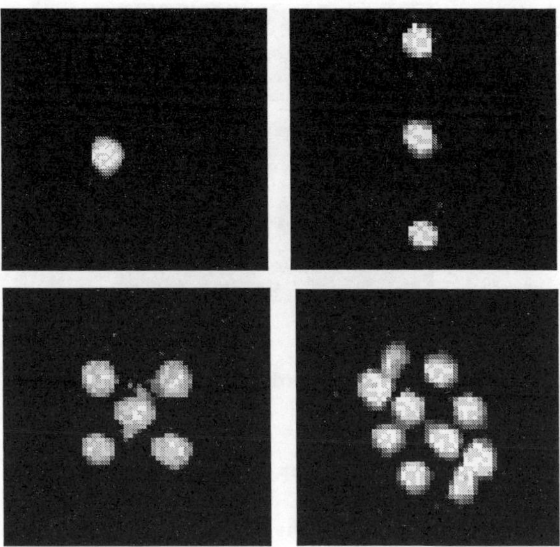

Fig. 45   Collections of stationary solitons recorded from a photorefraction oscillator with BR-absorber arranged as in Fig. 9.

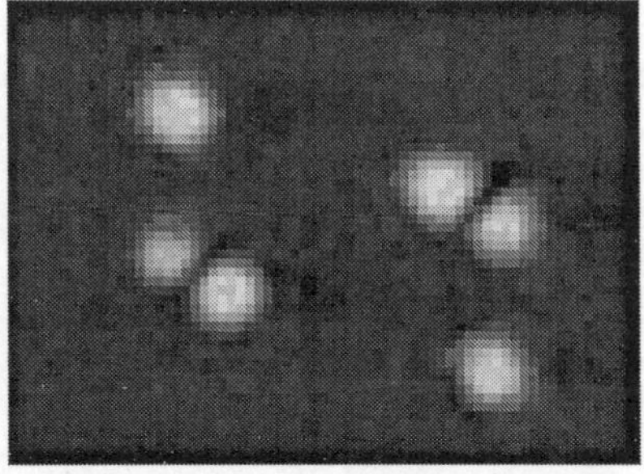

Fig. 46   Splitting of bright solitons in arrangement corresponding to Fig. 43/45

A51 (1995) 4140

U. Frisch Turbulence (The Legacy of A.N. Kolmogorov) Cambridge University Press 1995, pp. 296

I.S. Aaronson, L. Kramer, F. Plaza Phys. Rev. E 48 (1993) R/9

K. Staliunas, G. Slekys, C. O. Weiss Phys. Rev. Lett 79 (1997) 2658

C. O. Weiss, H. R. Telle, K. Staliunas, M. Brambilla Phys. Rev. A47 (1993) R1616

Z. S. Zygov Biofisica 31 (1986) 862 (in Russian)

K. Staliunas, C. O. Weiss J. Opt. Soc. Amer B12 (1955) 1142

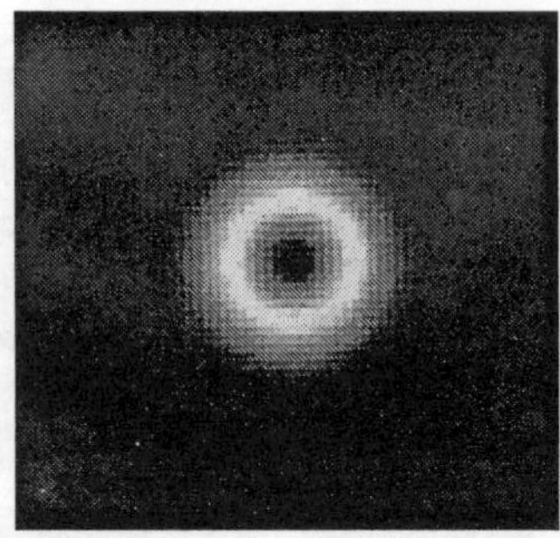

Fig. 47   Numerical calculation of a first order bright soliton (laser with non-linear absorber) with a vortex at the center.

G. Slekys, K. Staliunas, C.O. Weiss Opt.Comm. 119 (1995) 433

V.E. Zakharov, A.B. Shabat Sov. Phys. JEPT 37 (1973) 823

K. Staliunas, V. J. Sanchez-Morcillo Phys. Lett. A241 (1998) 28

V.B. Taranenko et al. Phys.Rev.Lett. 81 (1998) 2236

N. N. Rosanov, Progr. in Optics Vol. XXXV ed. by E. Wolf, Elsevier Science BV
       (1992)

K. Staliunas, V. J. Sanchez-Morcillo Phys. Rev. A57 (1998) 1454

V. B. Taranenko, M. Zander, P. Wobben, C. O. Weiss Appl. Phys. B69, (1999)
       337

Quantitative experiments presently conducted by A. Esteban Dept.Optica Univ.
       Valencia, Spain

K. Staliunas, Phys. Rev. Lett 81 (1998) 81

L. Faddeev, A.J. Niemi Nature 387 (1997) 58

V. Yu. Bazhenov, V. B. Taranenko, M. V. Vasnetsov Proc. SPIE 1840 (1992) 183

M. Saffmann et al. Opt. Lett. 19 (1994) 518

B. Fischer et al. Appl.Phys.Lett. 58 (1991) 2729

K. Staliunas et al. Phys.Rev.A 57 (1998) 599

G. Slekys et al. Opt.Comm. 149 (1998) 113

V. B. Taranenko, I. Ganne, R. Kuszelewicz, C. O. Weiss Phys. Rev. A61 (2000)
       063818

G. S. McDonald, W. J. Firth J. Mod. Opt. 37 (1990) 37

see report by Tredicce et al. in: www.pianos-int.org

M. Kreuzer et al. Mol.Cryst.Liq.Cryst. 207 (1991) 219

V.B. Taranenko et al., preprint xxx.lanl.gov nlin.PS/0204048

P.V. Jacobsen et al. Phys.Rev. A 45 (1992) 129

V. B. Taranenko at al. Phys. Rev. A 65 (2002) 013812

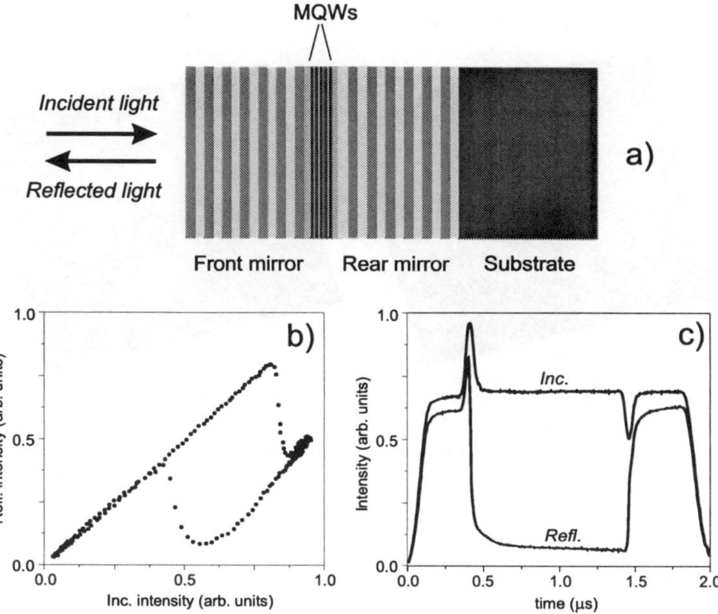

Fig. 48   Structure of a nonlinear semiconductor microresonator, bistability loop in reflection, and switching on and off of this resonator.

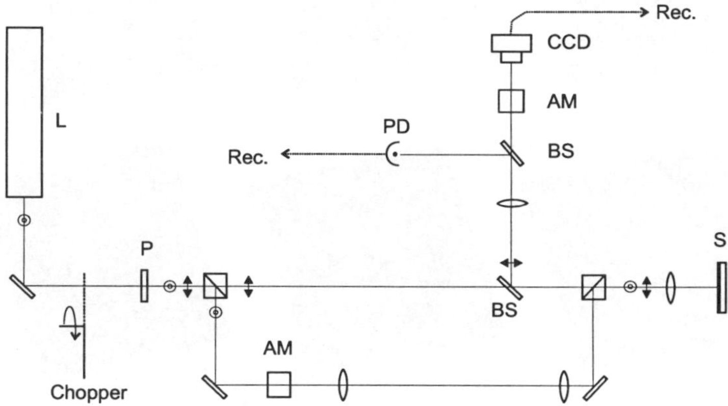

Fig. 49   Optical arrangement for the observations of spatial solitons in semiconductor microresonator. L: TiSa-Laser, P: phase plate, AM: intensity modulator, S: semiconductor resonator sample, CCD: camera

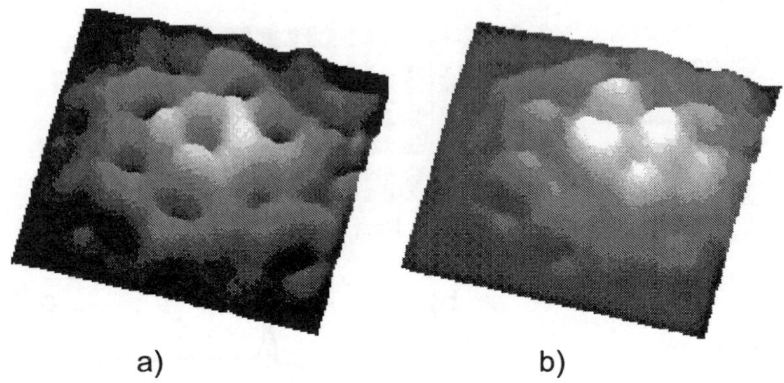

Fig. 50   Hexagonal patterns spontaneously formed in the semiconductor microresonator for the defocusing/dispersive case and for different resonator tunings/illumination intensities.

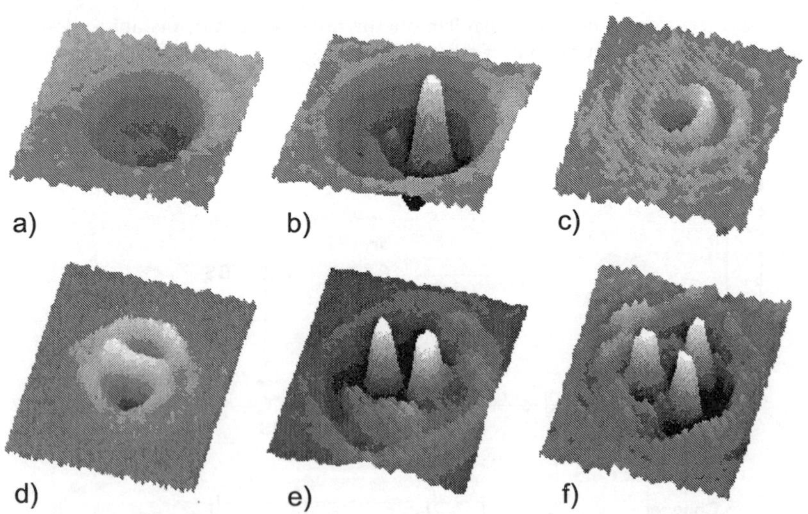

Fig. 51   Solitons in semiconductor microresonator (a) switched area; (b) dark soliton on switched background; (c) bright soliton on unswitched background; (d) 2 bright solitons; (e), (f) 2,3 dark solitons.

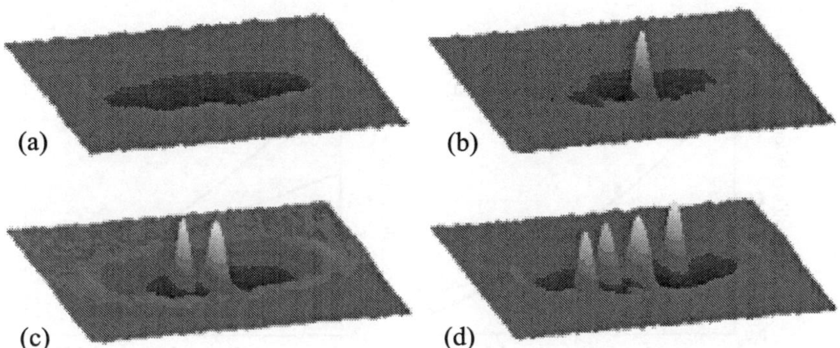

Fig. 52 Solitons in an illuminated area of elongated shape (a) switched domain, (b), (c), (d): 1, 2, 4 solitons in the switched domain showing the independence of the individual solitons.

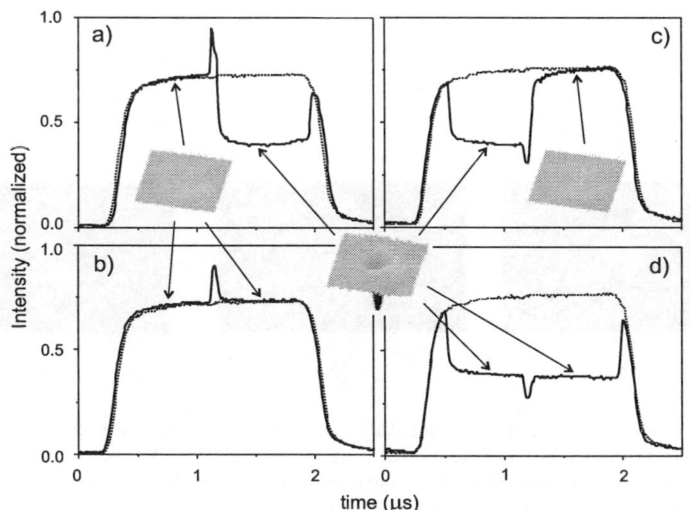

Fig. 53 Coherent switching on and off a bright soliton. Details see text

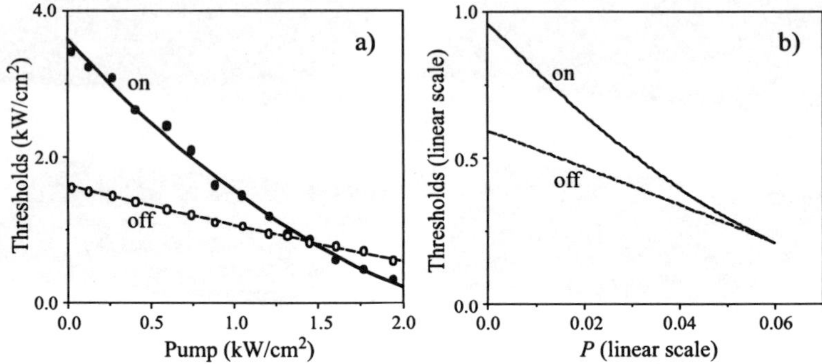

Fig. 54   Thresholds for switching a bright soliton on and off as a function of pump intensity. The unphysical crossing of the "on" and "off" curves is an artefact of the measurement method and should be disregarded (a). Experimental measurement (b) switching thresholds calculated numerically for comparison with (a).

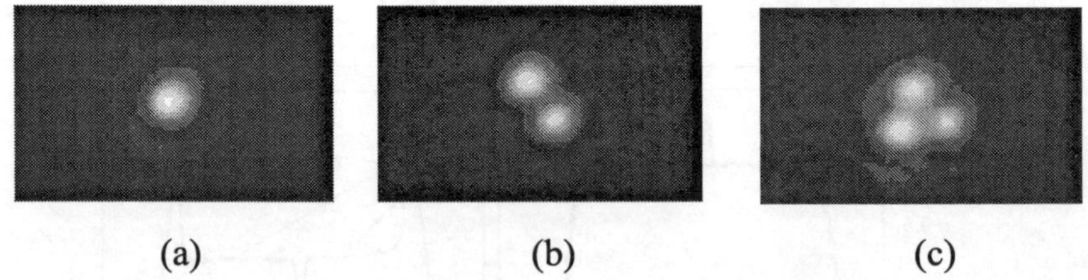

(a)   (b)   (c)

Fig. 55   Soliton structures experimentally observed when pumping the semiconductor above laser threshold, but in presence of soliton sustaining background light. The three-fold structure of (c) proves that these patterns are not "modes", since in "modes" adjacent bright spots must have p phase difference.

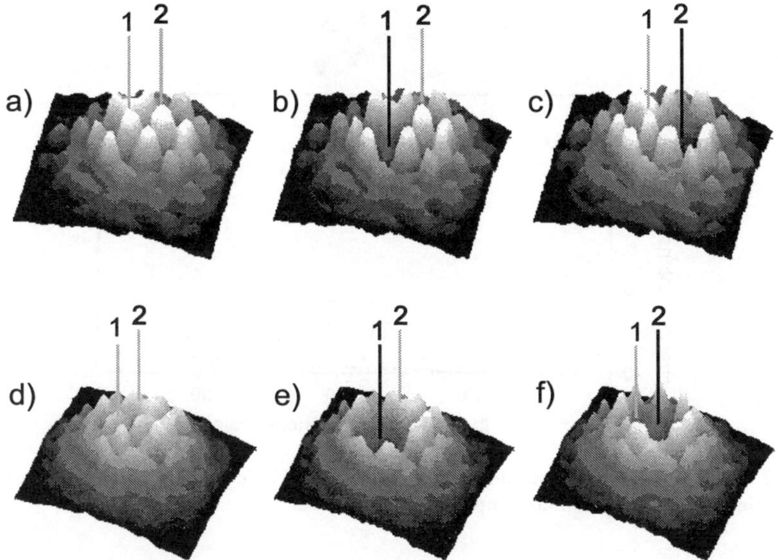

Fig. 56   Hexagonal patterns in which individual spots can be switched independently of the rest of the pattern. (a), (b), (c) different single solitons are switched, (d), (e), (f) different "triples" of solitons are switched.

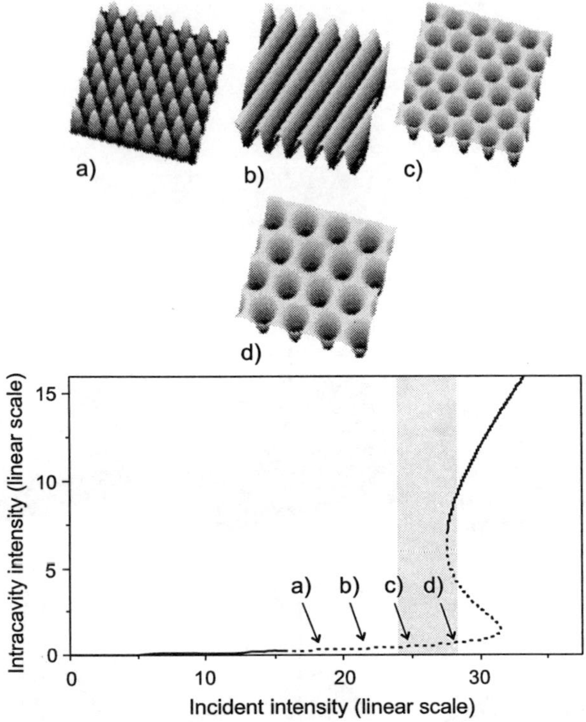

Fig. 57   The steady state bistable plane wave characteristic together with "in-resonator" intensities for various resonator input field intensities

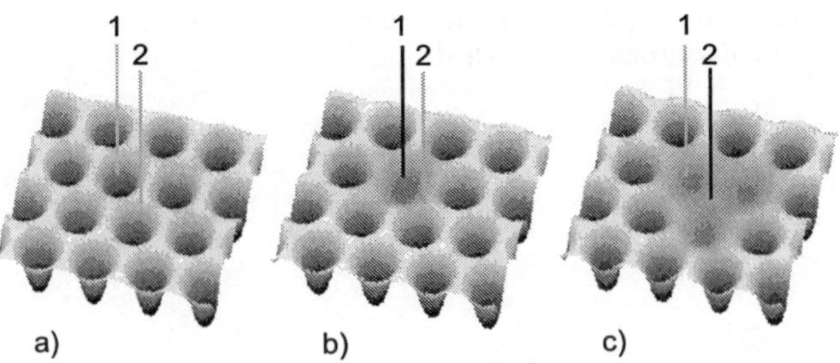

Fig. 58   Model calculation reproduces the switching of Fig. 10. The shown field structures are stable

Chapter 9

Nonlinear Waves and (Interesting) Applications

Mark J. Ablowitz, Toshihiko Hirooka, and Ziad H. Musslimani

Department of Applied Mathematics, University of Colorado at Boulder, Boulder, Colorado 80309-0526, USA

Abstract.   This paper consists of three self contained chapters. They are a summary of lectures given by MJA in the conference: Dynamic Summer in Canberra, Australia, January, 2002. The three chapters are:
1. Waves Everywhere
2. Nonlinear Waves in High Bit-Rate Communications
3. Discrete solitons

1   Waves Everywhere

This chapter begins with a discussion of John Scott Russell's remarkable discovery of the "Great Wave of Translation", which many scientists now often refer to as a solitary wave or soliton. Many years after Russell's discovery, Boussinesq and Korteweg-deVries showed that solitary waves could be obtained from the water wave equations. In recent years researchers have been able to effectively study a class of nonlinear evolution equations which admit multi-soliton solutions and a linearization procedure termed the Inverse Scattering Transform (IST). Some of the types of equations for which the IST procedure can be applied are discussed.

1.1   *Introduction*

Waves are of interest to anyone who looks at the sea or even a pond of water. Many of us have sat beside beach watching waves roll in and eventually break as they become "too heavy" to support the relatively smooth hump of water they once were. John Scott Russell (1808-1882), a Scottish Naval architect, was intrigued by waves, and was especially concerned about

369

whether there were some special wave forms that could help him improve the design of boats and ships. In 1834 he made a remarkable discovery. He described it in his "Report on Waves" in 1844 [1]

"I was observing the motion of a boat which was rapidly drawn along a narrow channel by a pair of horses, when the boat suddenly stopped–not so the mass of water which it had put into motion; it accumulated round the prow of the vessel in a state of violent agitation, then suddenly leaving it behind, rolled forward with great velocity, assuming the form of a large solitary elevation, a rounded, smooth and well-defined heap of water, which continued its course along the channel apparently without change of form or dimuntion of speed. I followed it on horseback, and overtook it still rolling on at a rate of some eight or nine miles an hour, preserving its original figure ... Its height gradually diminished, and after a chase of one or two miles I lost it in the windings of the channel. Such in the month of August 1834, was my first chance interview with that singular and beautiful phenomenon which I have called the Wave of Translation."

In his paper, Russell did experiments on these solitary waves, the main one which he called the Great Wave of Translation. He also called for mathematicians to study this phenomena:

"...it now remained for the mathematician to predict the discovery after it had happened...".

Unfortunately it took many years for this to occur. In 1871 Boussinesq [2] and later in 1895 Korteweg and deVries [3] showed that solitary waves indeed resulted from the known equations of water waves. The work of Korteweg and deVries was devoted to the detailed approximations involving long waves and small amplitude which led them to the equation:

$$\frac{\eta_t}{\sqrt{gh}} + \eta_x + \frac{3}{2h}\eta\eta_x + h^2\eta_{xxx} = 0, \tag{1.1}$$

where η is the elevation of the wave, h is the height of the undisturbed fluid, and g is gravity. Equation (1.1) can be put in dimensionless form by using: $t' = \sqrt{gh}t/6h, x' = (x - \sqrt{gh}t)/h, \eta = 2hu/3$ to find (after dropping primes for convenience),

$$u_t + 6uu_x + u_{xxx} = 0. \tag{1.2}$$

Equation (1.1) or (1.2) is called the Kortweg deVries (KdV) equation It is straight forward to verify that KdV has the special solitary wave solution,

$$u(x,t) = 2\kappa^2 \text{sech}^2 \kappa(x - 4\kappa^2 t - x_0).\tag{1.3}$$

Between 1895 and 1960 virtually all applications of the KdV equation and solitary waves were in the area of water waves. It found application in coastal engineering and from a theoretical point of view held the fascination of mathematical hydrodynamicists. With the development of more sophisticated singular perturbation methods, notably the method of multiple scales, the KdV equation was shown to arise in other important fields; e.g. plasma physics [4], internal waves [5], and as an approximation to lattice dynamics [6]. Indeed the work of Fermi, Pasta, Ulam in lattice dynamics motivated Kruskal and Zabusky to computationally study the KdV equation [7]. From the computations they observed the remarkable fact that solitary waves interacted elastically. Namely their amplitude and speed were asymptotically unchanged from before to after an interaction. Based on this observation they renamed these solitary waves to be solitons. The term soliton has become widely used in science today. Most researchers now use the term to denote any localized, stable structure (i.e a solitary wave). When the interaction of solitary waves possess the remarkable "elastic property" this usually indicates that there is considerably more mathematical structure that underlies the equations of motion. This was found to be the case for the Korteweg-deVries equation; in 1967 Gardner et al [8] showed that the KdV equation could be linearized by using certain inverse scattering methods, developed earlier by Gel'fand and Levitan [9].

It turned out that the KdV equation was no fluke. In 1972 Zakharov and Shabat [10] showed that inverse scattering could be applied to linearize the nonlinear Schrödinger equation

$$iu_t + u_{xx} \pm 2|u|^2 u = 0,\tag{1.4}$$

which had already begun to become a well known nonlinear equation (cf. [11; 12]).

In the 1970's the method was significantly generalized and shown to apply to certain classes of interesting nonlinear evolution equations including the modified KdV equation

$$u_t + 6u^2 u_x + u_{xxx} = 0,\tag{1.5}$$

and the sine-Gordon equation

$$u_{xt} = \sin u. \tag{1.6}$$

Central to whether one can apply inverse scattering at all is to show that the nonlinear evolution equation can be written as the compatibility of two (or possibly more) linear operators. One of the linear operators usually plays the role of the scattering problem. From a conceptual point of view the technique was shown to be analogous to the method of Fourier transforms but now applied to certain classes of nonlinear problems where the direct and inverse Fourier transform are replaced by direct and inverse scattering. The solution contains "N-soliton" solutions, which are connected to the discrete spectrum of the scattering problem plus a radiation tail which is due to the continuous spectrum. The method was termed the Inverse Scattering Transform (IST) (cf. [13; 14]). Amongst other results it was shown that the long time asymptotic structure could be obtained (cf. [14]) and this uncovered that there was a deep connection to Painlevé equations (cf. [14; 15]).

Importantly the IST method was applied to other classes of physically important nonlinear evolution equation. In 1976 Ablowitz and Ladik showed that the method extended naturally to semi-difference nonlinear evolution equation. The following integrable semi-discrete NLS equation is one important case:

$$iu_{n,t} + (u_{n+1} + u_{n-1} - 2u_n)/h^2 \pm |u_n|^2 u_n = 0, \tag{1.7}$$

where h is the grid size. Clearly Eq. (1.7) reduces to the continuous NLS equation as h tends to 0 (cf. [14] for a review). It was also found that there were classes of nonlinear partial difference equation to which the IST method applied, including examples of integrable doubly-discrete NLS equation Interestingly, the IST method also applies to certain singular integral evolution equation such as the Benjamin-Ono equation

$$u_t + 6u^2 u_x + (Hu_{xx}) = 0, \tag{1.8}$$

where

$$Hu = \frac{1}{\pi} \fint \frac{u(x', t)}{x' - x} dx',$$

($\fint$ denotes Cauchy Principal value integral; see [15] for a review of this work). Indeed equations such as the the KdV equation and Benjamin-

Ono equation and another equation that "interpolates" between these two, the so called intermediate long wave equation, all describe long stratified internal waves with small amplitude, in different parameter regimes. These equations all have N-soliton solutions, but the Benjamin-Ono equation has algebraic solitons (of "Lorentzian" form).

In 1980 Osborne and Burch [16] reported on their oceanic observations near Indonesia of "giant" internal soliton waves some 300 feet in amplitude and 5000 feet in length. They also showed that the KdV equation was a good approximation to the dynamics. These solitons are formed because of the lunar tides. Subsequently many researchers have reported on similar observations in different ocean locations. Thus solitons are "real, living" everyday creatures!

It turns out that there are multidimensional nonlinear wave equations which too are solvable by IST. The paradigm equation two space-one time (2+1) dimensional equation is the so-called Kadomtsev-Petviashvili equation [17]

$$u_{xt} + (3u^2)_{xx} + u_{xxxx} \pm 3u_{yy} = 0, \tag{1.9}$$

which is derived under similar circumstances as the KdV equation but with the additional proviso that the waves are slowly varying in the transverse (i.e. y) direction. This equation also applies to water waves [18], possesses N-line soliton solutions and, for a choice of sign in (1.9), has lump type solutions (cf. [14; 15]). In [15] other 2+1 equations are also discussed including an integrable 2+1 NLS type equation, often called the Davey-Stewartson equation.

There is also a four dimensional system that fits into the IST framework, called the self dual Yang-Mills (SDYM) system. The SDYM system can be obtained as the compatibility of two linear operators each of which contains two independent variables. Therefore it is natural to consider the SDYM system as an equation in "2+2" dimensions. It is not natural to consider SDYM as an evolution equation The compatibility of SDYM shows that there is significant freedom in the resulting SDYM system. Namely one may choose the underlying group containing the dependent variables in a virtually arbitrary manner. With this enormous flexibility, researchers showed that SDYM could be reduced/related to virtually all the 1+1 (cf. [15]) and 2+1 dimensional [19] PDE's mentioned above plus interesting nonlinear ODE's, including all six classical

Painlevé transcendents [20] and other important nonlinear ODE's which contain unusual singularity structure, such as the Chazy equation (see [15; 21]).

The first two of the Painlevé equation and the Chazy equation are given by:

$$d^2w/dz^2 + 6w^2 + z = 0, \tag{1.10a}$$

$$d^2w/dz^2 - zw - 2w^3 - \alpha = 0, \tag{1.10b}$$

$$d^3w/dz^3 - 2wd^2w/dz^2 + 3(dw/dz)^2 = 0, \tag{1.10c}$$

where α is an arbitrary constant, and z is a complex variable.

Eqs. (1.10a–1.10b) have solutions which are meromorphic in the z-plane; the location of their poles (double and simple respectively) depends on initial conditions. The solution of (1.10c) is analytic in the z-plane apart from a natural boundary of singularities which consists of a circle whose radius depends on initial conditions. In cases (1.10a–1.10b) the solutions can be obtained by a modification of the IST procedure, sometimes referred to as the inverse monodromy transform (cf. [15]). The solutions of (1.10c) can be obtained by an implicit change of variables (cf. [15; 22]).

In the study of integrable systems, the SDYM system plays the role of a master "integrable" system from which most, if not all, integrable systems can be obtained by reduction.

At this time, there are no known physically significant or even reasonably simple nonlinear evolution equation in 3+1 dimensions to which the IST procedure applies. This is a major open problem in the study of the Inverse Scattering Transform.

1.2   *Fully Discrete Waves*

It is extremely interesting that nonlinear waves which are "fully" discrete exist. One well known case is the so called "game of life" developed by J. Conway [23]. In its simplest form one considers 2 discrete spatial dimensions plus discrete "time" with the dependent variable $x(i,j,t)$ taking values in the finite field $GF(2)$; i.e. x has two states 0,1. Usually such finite state systems are referred to as cellular automata (CA). In the game of life one associates 0 with a "dead cell" and 1 with a "live cell". The rule of evolution depends on the 8 element neighborhood N of $x(i,j,t)$; i.e. $x(i+l,j+k,t)$ where l, k can be $0,1,-1$ but not both 0.

The rule is:

A. $x(i,j,t+1) = 1$ if

i. $x(i,j,t) = 0$ and there are exactly 3 living cells in N (i.e we have a "birth" from N)

or

ii. $x(i,j,t) = 1$ and there are 2 or 3 living cells in N (i.e a "living cell remains living")

B. $x(i,j,t+1) = 0$ if $x(i,j,t) = 0$ or 1 and there are less than 2 or more than 3 living cells in N (i.e. "death due to loneliness or overpopulation")

This simple rule is shown to contain a number of localized structures including states which are stationary, have periodic oscillation, and states which move with internal oscillation–i.e solitary waves!

Let us turn our attention to one dimensional cellular automata. The study of one dimensional cellular automata is covered extensively in the book by S. Wolfram [24]. Explicit 1-d rules take the form:

$$x(i,t+1) = F[x(i-r,t),...,x(i,t),...,x(i+r,t)], \qquad (1.11)$$

where the "radius" r is given as are the values $x(i+j,t)$ for $j = -r,...r$. Despite the interesting and varied nature of such 1-d explicit cellular automata, none of them appear to yield a soliton type solution. However if we consider a "filter", or implicit, cellular automata then it turns out that there is a soliton CA. More precisely, consider the 1-d "filter" automata

$$x(i,t+1) = F[x(i-r,t+1),...,x(i-1,t),x(i,t),...,x(i+r,t)], \quad (1.12)$$

supplemented with the additional proviso that we assume all values sufficiently far to the left are zero (i.e we assume "compact support to the left) and "sweep" left to right to obtain new values.

One such filter automata, termed the parity rule filter automata, was proposed in [25]. It is:

$$\begin{aligned}
x(i,t+1) &= 1 \ \text{ if } S_i \text{ is even and nonzero,} \\
&= 0 \ \text{ if } S_i \text{ is odd or zero,} \qquad (1.13)
\end{aligned}$$

where $S_i = \sum_{j=0}^{r} x(i+j,t) + \sum_{j=1}^{r} x(i-j,t+1)$

This rule is found to have a wide class of particle-like solutions. Some of these particles are found to interact elastically, hence possess the "strict soliton property". In addition the rule can be rewritten in another way to analytically show "stability" and this particle-soliton property [26], and cf. [15]. However the rule is not reversible! For example the initial states of all zeros or the state of one 1 surrounded by all zeroes both evolve to zero–hence there is no unique predecessor.

Subsequently in [27] it was shown that this rule could be modified to be reversible . The idea is to write the above rule as a difference equation:

$$x(i, t+1) = S_i + \delta(x(i,t))P_i - 1, \tag{1.14}$$

where $\delta(x(i,t))$ is 1 if $x(i,t)$ =0,mod2, 0 otherwise, and $P_i = \Pi^r_{j=1}\delta(x(i+j,t))\delta(x(i-j,t+1))$

Indeed the reversible rule is found to be:

$$x(i, t+1) = S_i + P_i - 1, \tag{1.15}$$

which clearly is reversible. It possesses a symmetry which says "forward" evolution corresponds to a left to right sweep, while "backward" evolution corresponds to a right to left sweep.

The reversible rule has a large class of particle-like solutions including all previously obtained elastic "reversible" solitons obtained from the parity rule automata plus many others, including the "elementary particle" of 1 followed by r zeros. This rule can be generalized to be "k-state", put on a finite interval and the coefficients of the rule can be easily modified. For example, the term S_i can be taken to be $S_i = \sum^r_{j=0} a(i+j,t)x(i+j,t) + \sum^r_{j=1} a(i-j,t+1)x(i-j,t+1)$ with $a(i,t) = 1$, the others arbitrary. This allows us to connect the linear part of (1.15) with linear feedback shift registers, which have been heavily studied in the literature. This reversible rule can take data at a given time, transform it via this nonlinear feedback shift register (NLFSR), to another state, and then one can exactly reverse the scenario. This situation has been recently discussed in detail in [28].

1.3 *Conclusions*

In this first chapter, of this three part manuscript, a discussion of a class of nonlinear wave problems was presented. We began with the famous observations of John Scott Russell (1834-44) and mentioned the work of Korteweg and deVries (1895). The "modern" history of the KdV equation was described including the concept of a soliton and the method of solution via the inverse scattering transform. While solitons were originally defined to have the property of interacting with each other elastically, most scientists today use the term more broadly, and often call a localized, stable wave structure (i.e. a solitary wave) a soliton. In subsequent sections we shall use the term soliton in this latter, more general context.

We also mentioned that there were other classes of nonlinear equations beyond PDE's in 1+1 and 2+1 dimensions to which the inverse scattering transform could be applied. These include certain classes of: semi-difference, partial difference, singular integral evolution equations and nonlinear ODE's such as the Painlevé equations. It was mentioned that the self dual Yang-Mills system occupies a special place in the study of integrable systems. In a sense it is a master system from which many important integrable systems can be derived.

2 Nonlinear Waves in High Bit-Rate Communications

This chapter describes an unexpected encounter of applied mathematics (solitons and nonlinear waves) with a state-of-the-art technology (optical fiber communications). Propagation of optical pulses in fibers in the presence of dispersion and nonlinearity is governed by the nonlinear Schrödinger (NLS) equation – an integrable equation which is mentioned in chapter 1 and has been well studied in applied mathematics. The NLS equation has a special solution called a "soliton". A soliton is a stable localized pulse propagating in a dispersive and nonlinear media without change of shape. This motivates the use of solitons for an information carrier (soliton communications). The most recent transmission systems employ a technique called "dispersion management" to improve system performance. The idea and motivation of dispersion management is discussed, and the analytical tools to study the behavior of dispersion-managed pulses are developed. Based on a unified analytical framework, two different formats of dispersion-managed pulses: "dispersion-managed solitons" where nonlinearity balances dispersion and "quasi-linear pulses" where nonlinearity is managed, are studied.

2.1 *Introduction*

Optical fiber communications have made enormous progress in recent years. Long-haul high-speed transmission over transoceanic distances with total capacity in excess of tera (10^{12}) bits per second has already been achieved. Behind the success of these achievements, there have been significant technical advances such as the development of optical amplifiers and extensive fundamental studies of signal transmission in optical fibers. Indeed, in such a complex transmission system, signal deformation during propagation due to various properties of optical fibers such as dispersion and nonlinearity becomes one of the major limitations to transmission speed and distance. In order to design a transmission system and optimize system performance, it is crucial to understand the fundamental physical properties of optical fibers and investigate carefully their impact on signal transmission.

2.2   *Dispersion and Nonlinearity in Optical Fibers*

Among the physical properties of optical fibers, group-velocity dispersion (GVD) and nonlinearity are the main source of signal deformation. (Fiber loss, which used to be the major limitation, is now overcome by the use of optical amplifiers.)

GVD is the frequency dependence of the group velocity and originates from the frequency dependence of the refractive index of the fiber. In the presence of GVD, different spectral components of an optical pulse propagate at different group velocities and thus arrive at different times. This leads to pulse broadening, resulting in signal distortion.

Fiber nonlinearity responsible for signal deformation is the Kerr effect, where the refractive index changes in proportion to the intensity of the optical pulse. The Kerr effect brings about an intensity dependent phase shift and results in spectral broadening during propagation. In the presence of GVD and Kerr nonlinearity, the refractive index is expressed as

$$n(\omega, E) = n_0(\omega) + n_2|E|^2 \tag{2.1}$$

where ω and E represent the frequency and the electric field of the lightwave respectively, n_0 corresponds to the linear refractive index, and n_2 (the Kerr coefficient) has a value $\sim 10^{-22} \mathrm{m}^2/\mathrm{W}$. Although fiber nonlinearity is small, the nonlinear effects accumulate over a long distance and can have a significant impact because of low loss and high intensity of the lightwave over a small area of fiber cross section.

2.3 *Nonlinear Schrödinger Equation and Optical Solitons*

When one transmits information using lightwaves as a carrier, the information is usually modulated on the amplitude of the envelope $\mathcal{E}$ of the lightwave, which is a slowly varying function of the distance along the fiber Z and time τ. A modulated lightwave E is written as

$$E(Z,\tau) = \mathcal{E}(Z,\tau)\exp(ik_0 Z - i\omega_0\tau) + \text{c.c.}. \tag{2.2}$$

Hasegawa and Tappert [29] derived the equation which describes the evolution of the modulated amplitude $\mathcal{E}$ along the axis of the fiber z. This can be obtained most conveniently from the dispersion relation $k(\omega, E) = (\omega/c)n(\omega, E)$ where c is the speed of light. (A detailed approach to derive it from the Maxwell's equation by means of the multiple scale method can be found in [30].)

A Taylor series expansion of $k(\omega, E)$ around the carrier frequency $\omega = \omega_0$ yields

$$k - k_0 = k'(\omega_0)(\omega - \omega_0) + \frac{k''(\omega_0)}{2}(\omega - \omega_0)^2 + \frac{\omega_0 n_2}{c}|E|^2, \tag{2.3}$$

where the prime represents the derivative with respect to ω. By replacing $k - k_0$ and $\omega - \omega_0$ by the operators $i\partial/\partial Z$ and $-i\partial/\partial\tau$ respectively and letting them operate on $\mathcal{E}$, we have

$$i\left(\frac{\partial\mathcal{E}}{\partial Z} + k_0'\frac{\partial\mathcal{E}}{\partial\tau}\right) - \frac{k_0''}{2}\frac{\partial^2\mathcal{E}}{\partial\tau^2} + \nu|\mathcal{E}|^2\mathcal{E} = 0, \tag{2.4}$$

where $\nu = \omega n_2/cA_{\text{eff}}$ with A_{eff} being the (effective) area of the cross section of the fiber (the factor $1/A_{\text{eff}}$ is required to take into account the variation of the field intensity in the cross section). If we introduce a retarded time coordinate $\tau_{\text{ret}} = \tau - k_0'Z$ moving at the group velocity $1/k_0'$ and normalize the variables by the characteristic parameters (denoted by the subscript $*$) $t = \tau_{\text{ret}}/\tau_*$, $z = Z/Z_*$, $u = \mathcal{E}/\sqrt{P_*}$ such that $Z_* = 1/\nu P_* = \tau_*^2/|k_0''|$, the envelope equation can be written as

$$i\frac{\partial q}{\partial z} - \frac{1}{2}\text{sgn}(k_0'')\frac{\partial^2 q}{\partial t^2} + |q|^2 q = 0. \tag{2.5}$$

This is known as the nonlinear Schrödinger (NLS) equation and is special in the mathematical literature, since the NLS equation can be integrated by the inverse scattering transform (IST). The NLS equation has a special solution called a (bright) soliton for $k_0'' < 0$ [10]. A soliton is a stationary and stable pulse, where the stationarity originates from the balance between dispersion and nonlinearity, and the stability is attributed to the integrability of the equation (via IST), allowing special interaction properties. In a technical context, solitons preserve their shape during propagation in a fiber even in the presence of dispersion and nonlinearity. Thus using solitons as an information bit allows us to overcome signal distortion due to dispersion and nonlinearity. Interestingly, the NLS equation also frequently appears in a wide variety field of science such as water waves, plasma physics, and magnetics (cf. [14]).

2.4   *All Optical Soliton Transmission Systems*

It took seven years from the theoretical prediction of optical solitons by Hasegawa and Tappert in 1973 until they were observed experimentally in optical fibers [31]. Since then, as a result of technical advances in lightwave systems such as the development of high power semiconductor lasers, fibers with low loss, and optical fiber amplifiers, there have been a large number of theoretical and experimental contributions aimed toward the development of all optical soliton transmission systems. By "all optical" one means that no optical-to-electrical (O-E) or E-O conversion of signals is employed in processing signals such as amplification in the middle of transmission. O-E/E-O conversion requires complicated processes and becomes a major limitation to transmission speed. The development of optical amplifiers, in particular, made a remarkable contribution by increasing the transmission capacity by the use of wavelength division multiplexing (WDM); namely by allowing the transmission of multiple signals having different wavelengths .

One important theoretical discovery is that the averaged dynamics of optical pulses in the presence of periodic perturbations such as weak variation of GVD or non-adiabatic loss and lumped amplification can also be described within the framework of the NLS equation, as long as the periodicity is short enough. In the presence of GVD variation $d(z)$, loss Γ and periodic lumped amplifiers $G(z)$, the NLS equation is modified as

$$i\frac{\partial q}{\partial z} + \frac{d(z)}{2}\frac{\partial^2 q}{\partial t^2} + |q|^2 q = -i\Gamma q + iG(z)q. \tag{2.6}$$

By introducing a new amplitude u through $q(z,t) = A(z)u(z,t)$ where $A(z)$ satisfies $dA/dz = (-\Gamma + G(z))A$, we have the perturbed NLS equation of the form

$$i\frac{\partial u}{\partial z} + \frac{d(z)}{2}\frac{\partial^2 u}{\partial t^2} + g(z)|u|^2 u = 0, \tag{2.7}$$

where $g(z) = A^2(z)$. Hasegawa and Kodama [32] showed that the perturbed NLS equation with $d(z)$ and $g(z)$ assumed to be $O(1)$ can be reduced to the standard NLS equation in the lowest order by means of a Lie transformation, where the GVD and nonlinear coefficients are replaced by the average of $d(z)$ and $g(z)$ respectively over a period. Thus the perturbed NLS equation (2.7) still supports the NLS equation and soliton solutions at the leading order in a perturbation theory. They referred to this as the guiding center soliton. Hereafter we generally refer to the soliton solution of the NLS equation as a "classical" soliton, when it is either the solution of the original NLS equation or in the perturbed sense of the guiding center soliton.

Soliton based communications also suffers technical difficulties unique to solitons. One of the most serious problems is the phenomenon referred to as the Gordon-Haus effect [33]. Gordon-Haus effect is the fluctuation of the temporal position of the soliton pulse which originates from random modulation of the carrier frequency of the soliton caused by the interaction with amplifier noise. The Gordon-Haus effect crucially limits the available transmission speed and distance in soliton based systems.

The nonlinear interaction between two neighboring solitons in the same time series also leads to fluctuations of their temporal position, where two solitons attract or repel each other during propagation depending on their relative phase [34]. Moreover, in WDM transmission, solitons in different wavelengths also suffer from strong interaction [35; 36; 37]. The situation becomes much worse in the presence of loss and lumped amplification.

Because of these difficulties in soliton transmission, all optical transmission systems in an early stage did not adopt solitons but employed the non-return-to-zero (NRZ) format. In the NRZ format a continuous wave is transmitted over the total time slot of successive "1" digits, whereas in the RZ format including solitons an individual pulse is transmitted for each "1" digit regardless of the sequence. NRZ pulses were popular in those systems because their intensity is allowed to be lowered compared with RZ pulses having the same average power within one bit slot and thus they were considered to suffer less nonlinearity. It should be noted, however, that the peak intensity must be enhanced as the bit rate increases and the pulse width decreases accordingly, since the average power within one bit slot must be maintained at a certain level so that the signal is not buried by the amplifier noise. Thus NRZ pulses also suffer from nonlinearity especially for high bit-rate transmission.

The effects of dispersion and nonlinearity in NRZ systems can also be analyzed in the framework of the perturbed NLS equation (2.7) [38]. The NLS equation (2.5) or (2.7) is now widely recognized as a fundamental equation to describe the evolution of optical pulses regardless of the transmission format. Indeed commercial software packages developed recently to simulate optical signal transmission are, in principle, based on the numerical calculations of the NLS equation. It is not feasible to numerically simulate Maxwell's equations over distances over thousands of kilometers.

2.5 *Dispersion Management*

After numerous of attempts to overcome the limitations mentioned above, a significant breakthrough for both soliton and non-soliton systems has come about. It is referred to as dispersion management. In a dispersion-managed system, the fiber is made up of alternating sections of positive and negative GVD in such a manner as to create a transmission line with high local GVD and low average GVD.

Dispersion management was originally proposed as a way to suppress nonlinear effects in non-soliton transmission. Since the local GVD is made large, the pulse experiences large broadening. This helps suppress certain nonlinear effects because of reduced peak intensity associated with large pulse broadening. At the same time, since the average GVD is close to zero, the pulse recovers its original shape periodically. Such observations led to recent successful experimental demonstrations of dense WDM transmission over transoceanic distances [39]. In dispersion-managed systems, however, nonlinearity is still not entirely avoided, and indeed residual nonlinearity can lead to serious penalties due to nonlinear interactions between neighboring pulses which are strongly overlapped locally as a result of their large pulse width broadening. For this reason, such a system is commonly referred to as "quasi-linear". Quasi-linear systems and their penalties are analytically studied in Sections 8 and 9.

Dispersion management is also found to be a powerful scheme for soliton systems but in a different context. Suzuki et al. [40] proposed the employment of dispersion management in soliton transmission in order to overcome the Gordon-Haus effect. Their motivation was to reduce the Gordon-Haus effect by compensating for the GVD accumulation periodically by alternating the sign of GVD. Since the Gordon-Haus effect is proportional to the accumulated GVD, they expected the Gordon-Haus effect to be reduced by lowering the average GVD, and indeed demonstrated it experimentally.

It should be noted that an ideal soliton solution does not exist in a dispersion-managed fiber because of the large variation of GVD. However, Smith et al. [41] showed numerically that there exists a nonlinear stationary solution in a dispersion-managed system whose shape is close to Gaussian rather than hyperbolic secant for classical solitons. Whereas the pulse width changes locally because of the periodic variation of GVD, their original shape is recovered exactly at every period even in the presence of nonlinearity. For this reason, this nonlinear pulse, which is, on average, stationary, is referred to as a dispersion-managed (DM) soliton.

The DM soliton is found to have enhanced energy compared with the energy of the classical soliton in a fiber with constant GVD equal to the average GVD of the dispersion-managed line. The energy enhancement is a particularly important property, since it allows the suppression of the Gordon-Haus effect by reducing the average GVD without sacrificing the degradation of the signal-to-noise ratio (SNR). In classical solitons, on the other hand, when the GVD is reduced the peak power must be lowered accordingly in order to maintain the balance between GVD and nonlinearity, which leads to the degradation of the SNR. DM solitons and their properties are analytically studied in Section 7.

2.6 *Dispersion Managed Nonlinear Schrödinger Equation*

The evolution of optical pulses in dispersion managed systems is also governed by the perturbed NLS equation (2.7), where $d(z)$ is now given by a *large*, rapidly varying function whose sign is alternating periodically. Because of the large perturbation, the pulse dynamics in dispersion-managed systems is expected to be completely different from that of the NLS equation. In this section, we derive a reduced model which describes long scale pulse dynamics in strongly dispersion-managed systems from the perturbed NLS equation by introducing multiple scales and thus eliminating fast dynamics due to the large and periodically varying GVD [42]. The obtained equation, referred to as the dispersion managed NLS (DMNLS) equation, elucidates the role of nonlinearity in dispersion-managed transmission, independently of the transmission format (DM soliton or quasi-linear).

In order to model strong dispersion management, we decompose the GVD $d(z)$ into two parts: a path-average constant δ_a and the rapidly varying function Δ corresponding to local GVD:

$$d(z) = \delta_a + \frac{1}{z_a}\Delta(z/z_a), \tag{2.8}$$

where $z_a(\ll 1)$ is the map period. Note that Δ/z_a represents a large variation about the average due to strong dispersion management and thus the proportionality factor $1/z_a$ is required in front of $\Delta(z/z_a)$ so that both δ_a and Δ are quantities of order one. Since the perturbed NLS equation with $d(z)$ given by (2.8) contains both slowly and rapidly varying terms, it is convenient to introduce the fast and slow scales as $\zeta = z/z_a$ and z respectively. We also expand the field u in powers of z_a:

$$u(\zeta, z, t) = u^{(0)}(\zeta, z, t) + z_a u^{(1)}(\zeta, z, t) + \cdots . \tag{2.9}$$

The perturbed NLS equation is now broken into a series of equations corresponding to the different powers of z_a. At the leading order in the expansion $O(1/z_a)$, we have

$$i\frac{\partial u^{(0)}}{\partial \zeta} + \frac{\Delta(\zeta)}{2}\frac{\partial^2 u^{(0)}}{\partial t^2} = 0, \tag{2.10}$$

namely the evolution of the pulse is determined solely by the large variations of $d(z)$ about the average, and nonlinearity and residual dispersion represent only a small perturbation to the linear solution. Eq. (2.10) can be solved by the Fourier transform

$$\hat{u}^{(0)}(\zeta, z, \omega) = \mathcal{F}[u^{(0)}] = \int_{-\infty}^{\infty} u^{(0)}(\zeta, z, t)\exp(-i\omega t)dt,$$

and the solution is given by

$$\hat{u}^{(0)}(\zeta, z, \omega) = \hat{U}(z, \omega)\exp[-iC(\zeta)\omega^2/2], \quad C(\zeta) = \int_0^{\zeta}\Delta(\zeta')d\zeta', \tag{2.11}$$

where $\hat{U}(z, \omega)$ is the integration constant in terms of ζ and represents the slowly evolving amplitude of $\hat{u}^{(0)}$, whose exact form is determined from the higher order in the expansion.

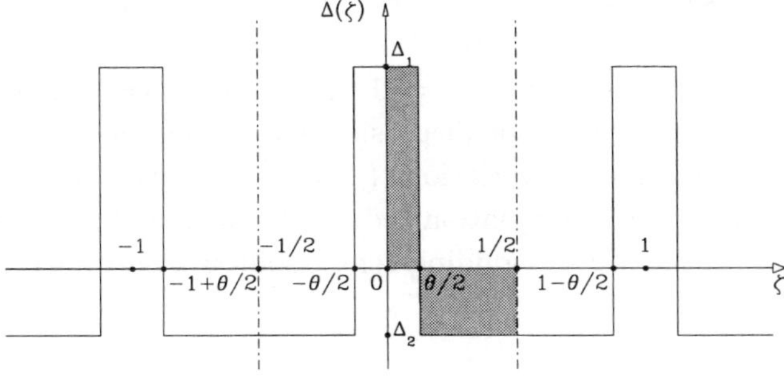

Fig. 1   Schematic diagram of a two-step dispersion map.

At the next order in the expansion $O(1)$, we have (in the frequency domain)

$$i\frac{\partial \hat{u}^{(1)}}{\partial \zeta} - \frac{\Delta(\zeta)}{2}\omega^2 \hat{u}^{(1)} = -P^{(1)}, \quad P^{(1)} = i\frac{\partial \hat{u}^{(0)}}{\partial z} - \frac{\delta_a}{2}\omega^2 \hat{u}^{(0)} + g(z)\mathcal{F}[|u^{(0)}|^2 u^{(0)}],$$

(2.12)

which is written in a more convenient form as

$$i\frac{\partial}{\partial \zeta}\left[\hat{u}^{(1)}\exp(iC\omega^2/2)\right] = -P^{(1)}\exp(iC\omega^2/2).$$

(2.13)

In order to remove secularities, namely to avoid resonant growth of $u^{(1)}$ so that the expansion of u in powers of z_a in (2.9) is remained to be well ordered, we require the following condition:

$$\int_0^1 P^{(1)}\exp(iC\omega^2/2)d\zeta = 0.$$

(2.14)

This condition yields the following equation for $\hat{U}$:

$$i\frac{\partial \hat{U}}{\partial z} - \frac{\delta_a}{2}\omega^2 \hat{U} + \int_{-\infty}^{\infty}\int_{-\infty}^{\infty} r(\omega_1\omega_2)\hat{U}(z,\omega+\omega_1)\hat{U}(z,\omega+\omega_2)\hat{U}^*(z,\omega+\omega_1+\omega_2)d\omega_1 d\omega_2 = 0,$$

(2.15)

where the kernel $r(x) = (1/(2\pi)^2)\int_0^1 g(\zeta)\exp(iC(\zeta)x)d\zeta$ represents the structure of GVD profile.

Let us consider two special cases. First, when there is no loss and amplification ($g(z) = 1$), the kernel $r(x)$ of a piecewise constant dispersion map (see Fig. 1) is given by $r(x) = (1/(2\pi)^2)\sin sx/sx$, where $s = [\theta\Delta_1 - (1 - \theta)\Delta_2]/4$ is a measure of dispersion map strength. Secondly, if the dispersion has no periodic variations ($\Delta = 0$ and thus $s = 0$), then $r(x) = 1/(2\pi)^2$ and the DMNLS equation reduces to the standard NLS equation, which was obtained via the guiding center soliton theory [32] when $d(z)$ is assumed $O(1)$.

2.7   *Dispersion Managed Solitons*

Special stationary solutions of the DMNLS equation are obtained by looking for solutions of the form $\hat{U}(z,\omega) = F(\omega)\exp(i\lambda^2 z/2)$ which yields the following nonlinear integral equation for $F(\omega)$:

$$(\lambda^2+\delta_a\omega^2)F(\omega) = 2\int_{-\infty}^{\infty}\int_{-\infty}^{\infty} r(\omega_1\omega_2)F(\omega+\omega_1)F(\omega+\omega_2)F^*(\omega+\omega_1+\omega_2)d\omega_1 d\omega_2.$$

$$(2.16)$$

A rapidly convergent procedure to solve this numerically is described in [43]. The obtained solution is the stationary DM soliton pulses. Some typical pulses are depicted in Fig. 2. We note that (2.16) contains a parameter λ which characterizes the mode, corresponding to the energy. The main features of the DM solitons for large s are a Gaussian like center with exponentially decaying and oscillating tails. For small s, the mode approaches the classical soliton profile. DM solitons have also been found to exist when the average GVD is zero or even positive, where no (bright) solitons exist for the standard NLS equation.

The fast scale evolution (i.e., with respect to ζ) can be reconstructed by multiplying $\exp(-iC\omega^2/2)$ with the slowly varying Fourier amplitude $\hat{U}$. When observed in this scale the pulse shape is changing periodically because of the local GVD variation, whereas the pulse propagates without changing its shape on a slow scale which is characterized by nonlinearity and the average dispersion. Thus DM soliton is considered to be a stationary pulse where nonlinearity balances the average dispersion.

2.8   *Quasi-Linear Pulses*

The evolution of pulses in the quasi-linear regime ($s \gg 1$) can also be studied based on the DMNLS equation [44]. To analyze the behavior of

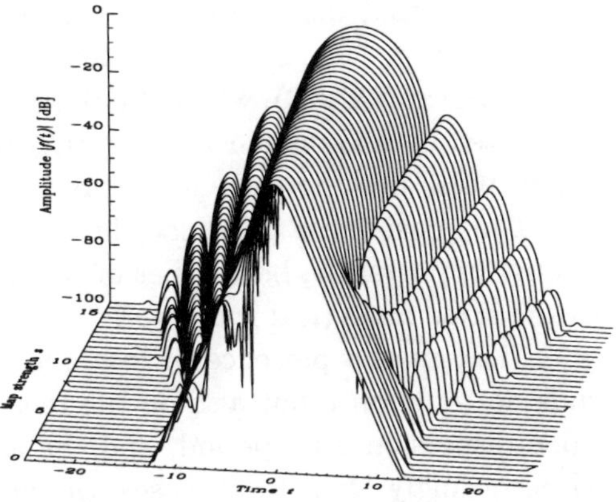

Fig. 2   The shape of the stationary pulses for $\delta_a = 1$, $\lambda = 1$ and various values of s.

quasi-linear pulses, we assume that $\hat{U}(z,\omega)$ depends only weakly on s and compute an asymptotic expansion of the nonlinear nonlocal term in (2.16) for $s \gg 1$.

When $g(z) = 1$ (i.e., in a lossless system), we have

$$i\frac{\partial \hat{U}}{\partial z} - \frac{\delta_a}{2}\omega^2 \hat{U} + \Psi[|\hat{U}|^2]\hat{U} = 0, \tag{2.17}$$

$$\Psi[|\hat{U}|^2] = \frac{1}{2\pi s}\left[(\log s - \gamma)|\hat{U}(z,\omega)|^2 - \int_{-\infty}^{\infty}|\hat{U}(z,\omega')|^2 f(\omega' - \omega)\right], \tag{2.18}$$

where $\gamma = 0.57722$ is Euler's constant and $f(\omega) = (1/\pi)\int_{-\infty}^{\infty}\log|t|\exp(i\omega t)dt$. Eq. (2.18) can be solved explicitly as

$$\hat{U}(z,\omega) = \hat{U}(0,\omega)\exp[-i\delta_a\omega^2 z/2 + i\Psi[|\hat{U}(0,\omega)|^2]z]. \tag{2.19}$$

From these results, we note the following important observations in terms of pulse dynamics in the quasi-linear regime. First, nonlinearity is mitigated by $O(\log s/s)$ and vanishes in the limit $s \to \infty$. In other words, the quasi-linear system is in the regime where nonlinearity is managed. Secondly, nonlinearity is responsible only for phase shift $\phi_{\mathrm{NL}}(z,\omega) = \Psi[|\hat{U}(0,\omega)|^2]z$ in frequency domain, and thus the spectral intensity $|\hat{U}(z,\omega)|$ is preserved during propagation, as opposed to the self-phase modulation in standard fibers. Finally we also note that the quasi-linear pulse can be viewed as degenerate limit of a DM soliton. The DM soliton satisfies the DMNLS equation with $\hat{U}(z,\omega) = F(\omega)\exp(i\lambda^2 z/2)$, where $F(\omega)$ is the solution of (2.17) and can be approximated by a Gaussian $F(\omega) \sim \alpha(\lambda)\exp(-\beta(\lambda)\omega^2/2)$. A quasi-linear Gaussian pulse has the same structure but is independent of the "eigenvalue" λ. The quasi-linear pulse can thus be looked at as a limit $\lambda \to 0$ in DM solitons. Importantly, both types of transmission format can be described with the same analytical framework.

When $g(z) \neq 1$ (i.e., in the presence of loss and amplification), the kernel $r(x)$ depends not only on s but also on the relative location of the amplifier within one dispersion map period, and the asymptotic analysis must be modified accordingly [45]. In all cases, the nonlinearity is again mitigated by $O(\log s/s)$. However, the spectral characteristics of the pulse evolution depends strongly on the amplifier locations. When the amplifiers are placed at the locations where the sign of GVD changes, the spectral intensity is found to be still preserved in the evolution. Otherwise, the spectral intensity varies as $O(1/s)$ and spectral compression or broadening is observed.

2.9   *Transmission Penalties in Quasi-Linear Systems*

We have seen in Section 8 that nonlinearity is mitigated by the factor $O(\log s/s)$ in the quasi-linear regime. This suggests the employment of strong dispersion management with large map strength s. Indeed dispersion management with large s is found to be effective in suppressing certain nonlinear interactions between pulses in different wavelength in WDM transmission (cf. [46]).

As s increases, however, different forms of nonlinear interactions take place and result in the main source of signal deformation [47]. In a strongly dispersion managed system, quasi-linear pulses in neighboring bit slots interact with each other because of their large overlap which is as a result of large pulse width broadening associated with high local GVD. This overlap induces nonlinear mixing (crosstalk) between pulses such as their frequency modulation and energy exchange, and leads to the fluctuation of the temporal position and amplitude of the pulses in "1"bits and the generation of "ghost" pulses in "0" bits [48; 49; 50]. The timing and amplitude jitter and the ghost pulse growth impose a major limitation to system performance in the quasi-linear regime with large s.

Let us analyze the growth of a ghost pulse q generated at the 0th bit slot by the interplay between the signals u_l, u_m, and u_n, where u_k, $k = l, m, n$ is the signal at the k-th bit [48]. Assuming that u in the perturbed NLS equation (2.7) is written as the sum of the signals and the ghost pulses $u = u_l + u_m + u_n + q$ and linearizing the equation, we have

$$i\frac{\partial q}{\partial z} + \frac{d(z)}{2}\frac{\partial^2 q}{\partial t^2} = -g(z) \sum_{l,m,n} u_l^* u_m u_n. \tag{2.20}$$

We note that the signals which contribute to the growth of ghost pulse at the 0th bit slot must satisfy the condition $-l + m + n = 0$; so the sum in (2.21) is taken for all the possible integers satisfying $l = m + n$ (e.g. $l = -1$, $m = -2$, $n = 1$, if there are "1" bits in -2, -1, and 1st slots). We solve (2.21) by the Fourier transform. After some manipulation [48], we find

$$\hat{q}(z,\omega)\exp(iC\omega^2/2) = P_0(\omega)z + (\text{periodic}), \tag{2.21}$$

where $P_0 = (1/z_a) \int_0^{z_a} i\mathcal{F}[\sum u_l^* u_m u_n]g(z)\exp(iC\omega^2/2)dz$. The first term on the right-hand side is the dominant cause of the resonant growth of the ghost pulse. The energy of the ghost pulse increases in proportion to z^2: $W(z) = (z^2/2\pi) \int_{-\infty}^{\infty} |P_0(\omega)|^2 d\omega$.

It should be noted that nonlinear pulse interactions in optical fibers are analogous to the so called quartet interactions which are responsible for energy changes between wavetrains in water waves, except for the interchange between time and distance. The condition $l = m + n$ necessary for the ghost pulse to grow resonantly corresponds to the wave number matching condition of the quartet interaction in water waves. The condition that the amplifier period equals the period of dispersion management corresponds to the frequency matching condition in quartet resonance.

2.10   *Conclusion*

We briefly reviewed the history and recent progress in the development of the mathematical theory of pulse propagation in optical fibers, including classical solitons, dispersion-managed solitons, and quasi-linear pulse transmission. Nonlinearity plays a crucial role in pulse dynamics of high-bit rate optical communication systems. The fundamental equation governing pulse propagation is the perturbed NLS equation. Solitons were originally proposed as a way to combat dispersion by the balance with nonlinearity, but today in dispersion-managed systems, dispersion is positively utilized to manage nonlinearity.

Because of its rich applications and underlying mathematics, engineers, physicists and applied mathematicians have contributed to the development of modeling and analysis of high-bit rate optical fiber communications. We hope this stimulates students and researchers with a wide variety of backgrounds to study this exciting and interesting scientific field.

3   Discrete Solitons

Localized, stable nonlinear waves, often referred to as solitons, are of broad interest in mathematics and physics. They are found in both continuous and discrete media. In this section a unified method is presented which is used to obtain soliton solutions to both discrete and continuous problems. In recent experiments discrete solitons were observed in an optical waveguide array. The fundamental governing system is the discrete nonlinear Schrödinger equation. A suitable modification of this system describes diffraction managed solitons. Mathematically speaking, in the continuous limit these solitons reduce to the dispersion managed solitons described in chapter 2 of this paper.

3.1   *Introduction*

In physics and applied mathematics certain fundamental equations arise frequently. One such set of equations are the nonlinear Schrödinger (NLS) systems-continuous and discrete. NLS equations in continuous media have been studied since the 1960's. In 1967[11] it was shown that the NLS equation (see eq. (2.5)) governs the slowly varying wave amplitude of a carrier wave in dispersive weakly nonlinear media. In 1972 Zakharov and Shabat[10] showed that the NLS equation was solvable by the inverse scattering transform (IST). As such they showed that it possesses soliton solutions which have special elastic interaction properties (see also Ref.[14]). Shortly thereafter, and as mentioned in ch. 2, Hasegawa and Tappert [29] showed that the NLS equation governs long distance pulse propagation of nonlinear waves in optical fibers. In 1974 Manakov [51] showed that a vector extension of the NLS equation was also solvable by IST.

Two well known discretizations of the NLS equation are:

$$i\frac{du_n(z)}{dz} + \frac{1}{h^2}(u_{n+1} + u_{n-1} - 2u_n) + |u_n|^2 u_n = 0 \, , \qquad (3.1)$$

$$i\frac{du_n(z)}{dz} + \frac{1}{h^2}(u_{n+1} + u_{n-1} - 2u_n) + |u_n|^2(u_{n+1} + u_{n-1})/2 = 0 \, , \quad (3.2)$$

Eq. (3.1) is often called the discrete nonlinear Schrödinger (DNLS) equation. It has been found to be a useful model in many applications (eg biophysics [56], discrete self trapping [57], optical waveguide array [64; 65]). On the other hand eq. (3.2), has been shown to be integrable by the inverse scattering transform method properly extended to discrete problems [75]. It too has been useful in the study of physical problems. Like the continuous problem there is a useful integrable vector extension of eq. (3.2) [84].

In this section we will concentrate on eq. (3.1) above. The physical context we shall describe herein is that of coupled optical waveguides which also is a convenient setting for laboratory experiments.

The first theoretical prediction of discrete solitons in an optical waveguide array was reported by Christodoulides and Joseph[58]. Later on, many theoretical studies of discrete solitons in a waveguide array reported switching, steering and other collision properties of these solitons[59; 60] (see also the review paper[61]). In all of the above cases, the localized modes are solutions of the discrete nonlinear Schrödinger (DNLS) equation which describes beam propagation in Kerr nonlinear media (according to coupled mode theory). Discrete bright and dark solitons have also been found in quadratic media[62]. In some cases, their properties differ from their Kerr counterparts[63].

It took almost a decade until self-trapping of light in discrete nonlinear waveguide array was first experimentally observed by Eisenberg et. al.[64; 65]. When a low intensity beam is injected into one or a few waveguides, the propagating field spreads over the adjacent waveguides hence experiencing discrete diffraction. However, at sufficiently high power, the beam self-traps to form a localized state (a soliton) in the center waveguides. Subsequently, many interesting properties of nonlinear lattices and discrete solitons were reported. For example the experimental observation of linear and nonlinear Bloch oscillations in: AlGaAs waveguides[66], polymer waveguides[67] and in an array of curved optical waveguides[68]. Discrete systems have unique properties that are absent in continuous media such as the possibility of producing *anomalous* diffraction[69]. Hence, self-focusing and defocusing processes can be achieved in the same medium (structure) and wavelength. This also leads to the possibility of observing discrete dark solitons in self-focusing Kerr media[70].

The recent experimental observations of discrete solitons[64] and diffraction management[69] have motivated further interests in discrete solitons in nonlinear lattices. This includes the newly proposed model of discrete diffraction managed nonlinear Schrödinger equation[71] whose width and peak amplitude vary periodically; optical spatial solitons in nonlinear photonic crystals[72] and the possibility of creating discrete solitons in Bose-Einstein condensation[73]. Also, very recently, it was shown, for the first time, that discrete solitons in two-dimensional networks of nonlinear waveguides can be used to realize intellegent functional operations such as blocking, routing, logic functions and time gating[74].

In this section, we introduce the Fourier transform method to analyze stationary solitons in nonlinear lattices. The essence of the method is to transform the DNLS equation governing the solitary wave into Fourier space, where the wave function is smooth, and then deal with a nonlinear nonlocal integral equation for which we employ a rapidly convergent numerical scheme to find solutions. A key advantage of the method is to transform a differential-delay equation into an integral equation for which computational methods are effective. Mathematically, the method also provides a foundation upon which an analytic theory describing solitons in nonlinear lattices can be constructed. We shall consider in this chapter two important models: the discrete nonlinear Schrödinger (DNLS) equation and the diffraction-managed discrete nonlinear Schrödinger (DM-DNLS) equation. Applying this method to the first model, shows that stationary solitons can readily be constructed. For the second model, we derive in the limit of strong diffraction, averaged equations governing the slow dynamics of the beam's amplitudes. Stationary solutions (in the form of bright-bright vector bound state) are obtained.

3.2   *Waveguide Array*

As mentioned above, an array of coupled optical waveguides is a setting that represents a convenient laboratory for experimental observations and theoretical predictions. Such system (see Fig. 3) is typically composed of

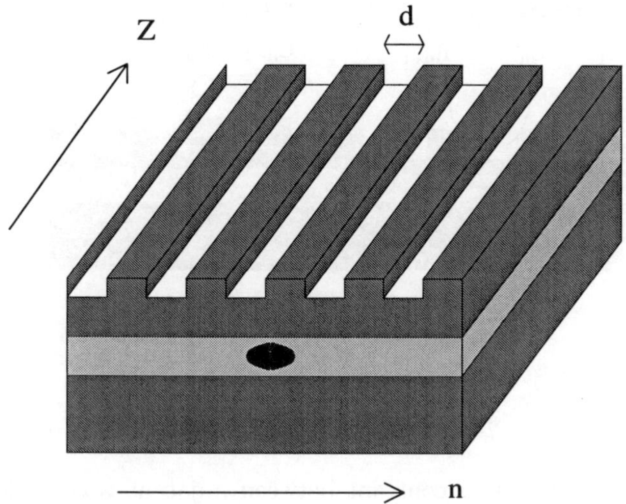

Fig. 3   AlGaAs waveguide structure. It is composed of three layers of AlGaAs material: a substrate with refractive index n_0, a core with higher index (n_1) and surface with index n_0. By etching the surface of the waveguide, one forms a periodic structure which is called a waveguide array.

three layers of AlGaAs material: a substrate with refractive index n_0, a core with higher index (n_1) and surface with index n_0. By etching the surface of the waveguide, one forms a periodic structure which is called a waveguide array. Self-traping of light in the "y" (i.e., vertical) direction is possible (even in the linear regime) by virtue of the principle of total internal reflection. On the other hand, the beam can diffract in the "x" direction unless it is balanced by nonlinearity. In the following we describe the propagation of light in such a periodic structure both in the linear and nonlinear regime.

3.3   *Linear Propagation: Anomalous Diffraction*

If the distance, d, between the waveguides is "large" then the propagating beams across each single waveguide do not "feel" each other. Therefore, the amplitude of each beam evolves independently. On the other hand when d is small then there is an overlap between modes in adjacent waveguides. As a result, the beam's amplitude is *not* constant anymore (see Fig. 4).

Assuming linear coupling between nearest neighbors, the dynamics of the beam's amplitude $E_n(z)$ at waveguide number n follows from coupled-

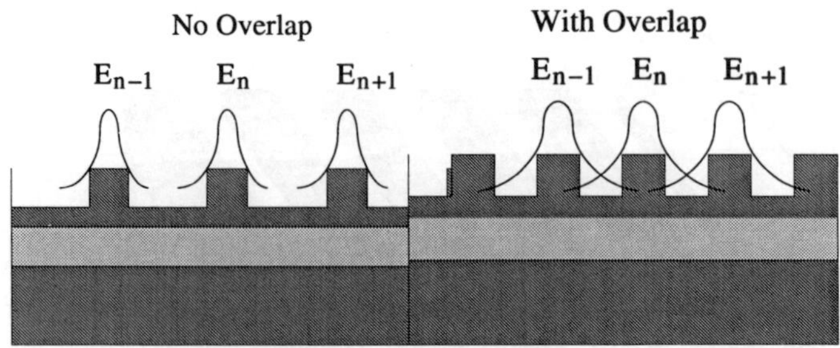

Fig. 4   Cross section of the waveguide array and mode overlap.

mode theory

$$i\frac{\partial E_n}{\partial z} + C(E_{n+1} + E_{n-1}) = 0 \,, \tag{3.3}$$

where C is the coupling constant between adjacent waveguides which is given by an overlap integral of the two modes in such waveguides; z is the propagation distance. To facilitate understanding, we first recall basic properties of discrete diffraction of a *linear* array. When a mode of the form

$$E_n(z) = A \exp[i(k_z z - n k_x d)] \,, \tag{3.4}$$

is inserted into Eq. (3.3) it yields the following diffraction relation

$$k_z = 2C \cos(k_x d) \,. \tag{3.5}$$

In close analogy to the definition of dispersion, discrete diffraction is given by $k_z'' = -2C d^2 \cos(k_x d)$. We consider $|k_x d| \leq \pi$. In that region, the diffraction is *normal* for wavenumbers k_x satisfying $-\pi/2 < k_x d \leq \pi/2$ ($k_z'' < 0$) and is *anomalous* in the range $\pi/2 < |k_x d| \leq \pi$. Moreover, contrary to the bulk case, diffraction can even vanish when $k_x d = \pm\pi/2$. In practice, the sign and value of the diffraction can be controlled and manipulated by launching light at a particular angle or equivalently by tilting the waveguide array. This in turn allows the possibility of achieving a "self-defocusing" (with *positive* Kerr coefficient) regime which leads to the formation of discrete dark solitons[70]. To understand more about diffraction management we consider three typical cases for which light enters the central waveguide array at different angles, say, $k_x d = 0, \pi/2$ and π. When $k_x d = 0$ then light tunnels between adjacent waveguides giving rise to discrete diffraction. The phase front in this case has a concave (negative) curvature. On the other hand, if $k_x d = \pi$, then diffusion of light still occurs but this time the phase front has convex (positive) curvature. Finally, at $k_x d = \pi/2$, diffraction vanishes (even though light can couple to different waveguides) and in the absence of any higher order diffraction the phase front looks almost flat (see Fig. 5).

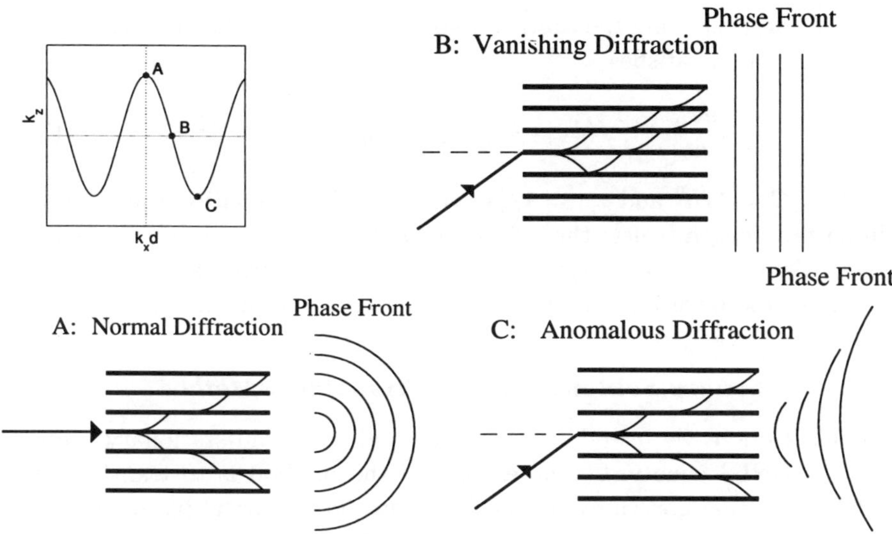

Fig. 5   Diffraction relation (top left) showing three typical examples of diffraction: (A) Normal in which the phase front is concave, (B) vanishing diffraction in which the phase front is flat, (C) anomalous diffraction with convex phase front.

3.4 *The Discrete Nonlinear Schrödinger Equation*

In the preceding section, we addressed the physical manifestation of propagation of light in a discrete linear medium. However, when the intensity of the incident beam is suffeciently high then the refractive index of the medium will depend on the intensity. The equation which governs the evolution of the electric field E_n, according to nonlinear coupled mode theory is given by

$$i\frac{\partial E_n}{\partial z} + C(E_{n+1} + E_{n-1}) + \kappa |E_n|^2 E_n = 0 \,, \qquad (3.6)$$

where C is the coupling constant between adjacent waveguides, κ is a constant describing the nonlinear index change, z is the propagation distance. First we put the NLS equation in a dimensionless form. To do so, we define

$$E_n = \sqrt{P_*}\,\phi_n \exp(2iCz)\,, \qquad z' = z/z_{\mathrm{nl}} \tag{3.7}$$

with P_* and z_{nl} being the characteristic power and z_{nl} the nonlinear length scale. Then ϕ_n satisfies

$$i\frac{d\phi_n}{dz} + \frac{1}{h^2}\left(\phi_{n+1} + \phi_{n-1} - 2\phi_n\right) + |\phi_n|^2 \phi_n = 0\,, \tag{3.8}$$

with $z_{\mathrm{nl}}C = 1/h^2$ and $z_{\mathrm{nl}} = 1/(\kappa P_*)$. In the NLS equation there are two important length scales: the diffraction and nonlinear length scales respectively defined by $L_D \sim 1/C$ and $z_{\mathrm{nl}} = 1/(\kappa P_*)$. Solitons which are self-confined and invariant structures are expected to form when $L_D \sim z_{\mathrm{nl}}$.

3.5   *Stationary Solitons: Fourier Transform Method*

In this section, we introduce a new method to obtain stationary solitons for the DNLS equation. The essence of the method is to transform the DNLS equation governing the solitary wave into Fourier space, where the wave function is smooth, and then deal with a nonlinear nonlocal integral equation for which we employ a rapidly convergent numerical scheme to find solutions. A key advantage of the method is to transform a differential-delay equation into an integral equation for which computational methods are effective (see also refs.[80; 42]). Mathematically, the method also provides a foundation upon which an analytic theory describing solitons in nonlinear lattices can be constructed.

3.6   *Stationary Solutions*

We look for a stationary solution to Eq. (3.8) in the form

$$\phi_n = F_n \exp(i\omega z)\,, \tag{3.9}$$

with F_n being real valued function and ω is a real eigenvalue. Then F_n satisfies

$$-\omega F_n + \frac{1}{h^2}\left(F_{n+1} + F_{n-1} - 2F_n\right) + F_n^3 = 0\,. \tag{3.10}$$

Equation (3.10) can be solved using Newton iteration scheme by which one gives initial value for F_0 and F_1 and then iterate. However, our aim here is to provide a different approach based on Fourier transform method. We introduce the discrete Fourier transform

$$\hat{F}(q) = \sum_{m=-\infty}^{+\infty} F_m e^{-iqmh} \,, \qquad F_m = \frac{h}{2\pi} \int_{-\pi/h}^{\pi/h} \hat{F}(q) e^{iqmh} dq \,. \qquad (3.11)$$

Then, the above equation, transforms into the following nonlinear integral equation

$$\hat{F}(q) = \frac{h^2}{4\pi^2 \Omega(q)} \int dq_1 \, dq_2 \hat{F}(q_1)\hat{F}(q_2)\hat{F}(q - q_1 - q_2) \equiv \mathcal{K}_\omega[\hat{F}(q)] \,, \quad (3.12)$$

where $\Omega(q) = \omega + 2(1 - \cos(hq))/h^2$. The important conclusion is that soliton can be viewed as a fixed point of an infinite dimensional nonlinear functional. To numerically find the fixed point, we employ a modified Neumann iteration scheme and write

$$\hat{F}_{n+1}(q) = \left(\frac{\alpha(\hat{F}_n)}{\beta(\hat{F}_n)}\right)^{3/2} \mathcal{K}_\omega[\hat{F}_n(q)] \,, \quad n \geq 0 \,, \qquad (3.13)$$

with α and β defined by

$$\alpha = \int \hat{F}_n^2(q) dq \,; \quad \beta = \int \hat{F}_n(q)\mathcal{K}_\omega[\hat{F}_n(q)] dq \qquad (3.14)$$

The factor 3/2 is chosen to make the right hand side of Eq. (3.13) of degree zero which yields convergence of the scheme[80; 42]. When F_m is real and even, it implies that $\hat{F}(q)$ is also real. Clearly when $\hat{F}_n(q) \to \hat{F}_s(q)$ as $n \to \infty$ then $\alpha/\beta \to 1$ and in turn $\hat{F}_s(q)$ will be the solution to Eq. (3.12). The factors α and β are introduced to stabilize an otherwise divergent simple Neumann iteration scheme. Note that when we apply the continuous Fourier transform on Eq. (3.10) (to find stationary solution), then the numerical scheme based on (3.13) does not converge which indicates that a *continuous* stationary solution to the DNLS may not exist. Fig. 6 shows a typical solution to (3.12) both in the Fourier domain (Fig. 6a) and in physical space (Fig. 6b) for different values of lattice spacing h. Importantly, with suitable modification of Eq. (3.13) this method yields new breathing localized modes: "discrete diffraction managed solitons"[71].

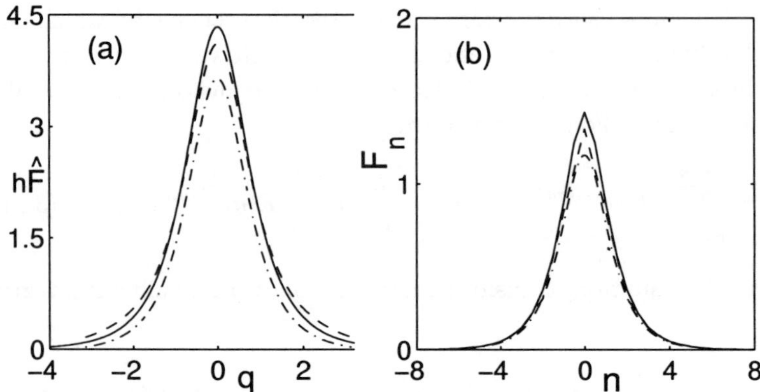

Fig. 6   Mode profiles obtained with $\omega_s = 1$ in Fourier space (a), for $h = 0.5$ (solid), $h = 1$ (dashed) and $h = 1$ (dashed-dotted) for the integrable case. (b) Soliton shape in physical space for $h = 0.5$ (solid), $h = 1$ (dashed) and for the integrable case at $h = 1$ (dashed-dotted).

3.7   *Nonlinear Diffraction Management*

We begin our analysis by considering again an infinite array of weakly coupled optical waveguides with equal separation d. The equation which governs the evolution of the electric field E_n, according to nonlinear coupled mode theory[58; 79; 64], is given by

$$\frac{\partial E_n}{\partial z} = iC(E_{n+1} + E_{n-1}) + ik_w E_n + i\kappa |E_n|^2 E_n , \qquad (3.15)$$

where C is the coupling constant between adjacent waveguides which is given by an overlap integral of the two modes of such waveguides; κ is a constant describing the nonlinear index change, z is the propagation distance and k_w is the propagation constant of the waveguides. Similar to[69], we use a cascade of different segments of waveguide, each piece being tilted by an angle zero and γ respectively. The actual physical angle γ (the waveguide tilt angle) is related to the wavenumber k_x by the relation[69] $\sin\gamma = k_x/k$ where $k = 2\pi n_0/\lambda_0$ ($\lambda_0 = 1.53\ \mu$m is the central wavelength in vacuum and $n_0 = 3.3$ is the linear refractive index). In this way, we generate a waveguide array with alternating diffraction. To model such a system, we define $E_n = \sqrt{P_*}\phi_n e^{i(k_w+2C)z}$ and $z' = z/z_*$ with P_* and z_* being the characteristic power and nonlinear length scale respectively. Substituting these quantities into Eq. (3.15) yields the following (dropping the prime) diffraction-managed discrete nonlinear Schrödinger (DM-DNLS) equation

$$\frac{d\phi_n}{dz} = i\frac{D(z/z_w)}{2h^2}\left(\phi_{n+1} + \phi_{n-1} - 2\phi_n\right) + i|\phi_n|^2\phi_n \qquad (3.16)$$

with $z_* = 1/(\kappa P_*)$ and $z_* C\cos(k_x d) = D(z/z_w)/(2h^2)$ where $D(z/z_w)$ is a piecewise constant periodic function that measures the local value of diffraction. Here, $z_w \equiv 2L/z_*$ with L being the actual length of each waveguide segment (see Fig. 7(a) for schematic representation). Eq. (3.16) describes the dynamical evolution of a laser beam in a Kerr medium with varying diffraction. When the "effective" nonlinearity balances the average diffraction then bright discrete solitons can form.

The coupling constants that correspond to the experimental data reported in[69] (for 2.5 μm waveguide separation and width) are found to be $C = 2.27\,\text{mm}^{-1}$[77], $\kappa = 3.6\,\text{m}^{-1}\text{W}^{-1}$. For typical power $P_* \approx 300$ W and waveguide length $L \approx 100\,\mu m$ we find $z_* \approx 1$ mm and $z_w \approx 0.2$. Hence, it is natural to construct an asymptotic theory based on small z_w. It is in this parameter regim that diffraction managed spatial solitons can be experimentally observed. To this end, we consider the case in which the diffraction takes the form

$$D(z/z_w) = \delta_a + \frac{1}{z_w}\Delta(z/z_w)\,, \qquad (3.17)$$

where δ_a is the average diffraction (taken to be positive) and $\Delta(z/z_w)$ is a periodic function. Since in this case, Eq. (3.16) contains both slowly and rapidly varying terms, we introduce new fast and slow scales as $\zeta = z/z_w$ and $Z = z$ respectively and expand ϕ_n in powers of z_w

$$\phi_n = \phi_n^{(0)}(\zeta, Z) + z_w\phi_n^{(1)}(\zeta, Z) + O(z_w^2)\,. \qquad (3.18)$$

Substituting Eqs. (3.18) and (3.17) into Eq. (3.16) we find that the leading order in $1/z_w$ and the order 1 equations are respectively given by

$$\mathcal{J}(\phi_n^{(0)}) = 0\,, \qquad \mathcal{J}(\phi_n^{(1)}) = -\mathcal{F}_n\,, \qquad (3.19)$$

where

$$\mathcal{J}(A_n) \equiv i\frac{\partial A_n}{\partial \zeta} + \frac{\Delta(\zeta)}{2h^2}\left(A_{n+1} + A_{n-1} - 2A_n\right)\,,$$

$$\mathcal{F}_n = i\frac{\partial \phi_n^{(0)}}{\partial Z} + \frac{\delta_a}{2h^2}\left(\phi_{n+1}^{(0)} + \phi_{n-1}^{(0)} - 2\phi_n^{(0)}\right) + |\phi_n^{(0)}|^2\phi_n^{(0)}\,.$$

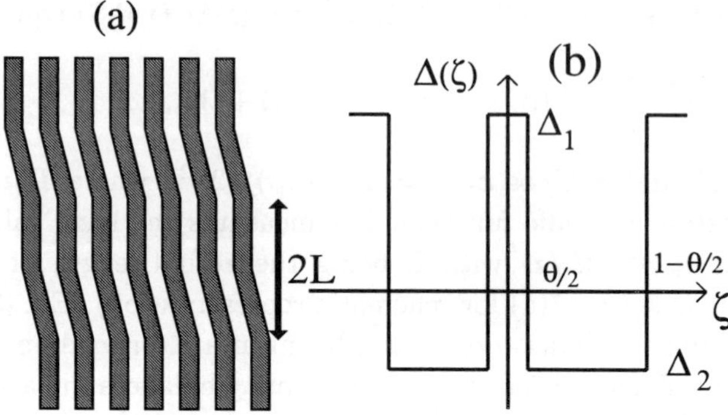

Fig. 7   Schematic presentation of the waveguide array (a) and of the diffraction map (b).

To solve at order $1/z_w$ (see Eq. (3.19)), we introduce the discrete Fourier transform

$$
\hat{\phi}_0(q, \zeta, Z) = \sum_{n=-\infty}^{+\infty} \phi_n^{(0)}(\zeta, Z) e^{-iqnh} ,
$$

$$
\phi_n^{(0)}(\zeta, Z) = \frac{h}{2\pi} \int_{-\pi/h}^{\pi/h} \hat{\phi}_0(q, \zeta, Z) e^{iqnh} dq .
$$

(3.20)

The solution is therefore given in the Fourier representation by

$$
\hat{\phi}_0(q, \zeta, Z) = \hat{\psi}(Z, q) \exp[-i\Omega(q)C(\zeta)] ,
$$

(3.21)

with $\Omega(q) = (1 - \cos(qh))/h^2$ and $C(\zeta) = \int_0^\zeta \Delta(\zeta')d\zeta'$. The amplitude $\hat{\psi}(Z, q)$ is an arbitrary function whose dynamical evolution will be determined by a secularity condition associated with Eq. (3.19). In other words, the condition of the orthogonality of $\mathcal{F}_n$ to all eigenfunctions Φ_n of the adjoint linear problem which, when written in the Fourier domain, takes the form

$$
\int_0^1 d\zeta \hat{\mathcal{F}}(q, \zeta, Z) \hat{\Phi}(q, \zeta, Z) = 0 ,
$$

(3.22)

where $\hat{\mathcal{F}}$, $\hat{\Phi}$ are the Fourier transform of $\mathcal{F}_n$, Φ_n respectively. Here, $\hat{\mathcal{J}}^\dagger \hat{\Phi} = 0$ with $\hat{\mathcal{J}}^\dagger$ being the adjoint operator to $\hat{\mathcal{J}} \equiv id/d\zeta - \Delta(\zeta)\Omega(q)$. Substituting (3.21) into $\hat{\mathcal{F}}$ and performing the integration in condition (3.22) yields the following nonlinear evolution equation for $\hat{\psi}(Z, q)$:

$$i\frac{d\hat{\psi}(Z, q)}{dZ} = \delta_a \Omega(q)\hat{\psi}(Z, q) - \mathcal{R}[\hat{\psi}(Z, q)] \,,$$

$$\mathcal{R} \equiv \int d\mathbf{q}\,\mathcal{K}(q, q_1, q_2)\hat{\psi}(q_1)\hat{\psi}(q_2)\hat{\psi}^*(q_1 + q_2 - q), \tag{3.23}$$

where $d\mathbf{q} \equiv dq_1 dq_2$ and the kernel $\mathcal{K}$ is defined by

$$\mathcal{K}(q, q_1, q_2) = \frac{h^2}{4\pi^2} \int_0^1 d\zeta \, \exp[iC(\zeta)\chi(q, q_1, q_2)] \,,$$

$$\chi = \frac{4}{h^2} \cos\left(\frac{h(q_1 + q_2)}{2}\right) \prod_{j=1}^{2} \sin\left(\frac{h(q_j - q)}{2}\right) \,.$$

Equation (3.23) governs the evolution in Fourier space of an optical beam in a coupled nonlinear waveguide array in the regime of strong diffraction. In the special case of two-step diffraction map shown in Fig. 7(b), i.e., when two waveguide segments are tilted by angle zero and γ alternatively, we have $\Delta(\zeta) = \Delta_1$ for $0 \leq |\zeta| < \theta/2$ and Δ_2 in the region $\theta/2 < |\zeta| < 1/2$ where θ is the fraction of the map with diffraction Δ_1. In this case, the kernel $\mathcal{K}$ takes the simple form $\mathcal{K} = h^2 \sin(s\chi)/(4\pi^2 s\chi)$ with $s = [\theta\Delta_1 - (1-\theta)\Delta_2]/4$. Importantly, these parameters can be related to the experiments reported in[69]. To achieve a waveguide configuration with alternate diffraction, we use two values of $k_x d = 0$ and $2\pi/3$ ($h = 1$) which corresponds to waveguide tilt angles $\gamma = 0$ and $3.43°$. In this case we find, for $\theta = 0.5$, $\Delta_1 = -\Delta_2 = 0.681$, $\delta_a = 1.135$ and $s = 0.17$. Different sets of parameters with a smaller angle γ are also realizable. Next, we look for a stationary solution for Eq. (3.23) (for the particular kernel given above) in the form $\hat{\psi}(Z, q) = \hat{\psi}_s(q) \exp(i\omega_s Z)$. Inserting this ansatz into (3.23) leads to

$$\hat{\psi}_s(q) = \frac{1}{\delta_a \Omega(q) + \omega_s}\mathcal{R}[\hat{\psi}_s(q)] \equiv \mathcal{M}[\hat{\psi}_s(q)] \,, \tag{3.24}$$

which implies that the mode $\hat{\psi}_s(q)$ is a fixed point of the nonlinear functional $\mathcal{M}$. To numerically find the fixed point, we employ a modified Neumann iteration scheme[80; 81; 82] and write Eq. (3.24) in the form

$$\hat{\psi}_s^{(m+1)}(q) = \left(\frac{\alpha(\hat{\psi}_s^{(m)})}{\beta(\hat{\psi}_s^{(m)})}\right)^{3/2} \mathcal{M}[\hat{\psi}_s^{(m)}(q)] \, , \quad m \geq 0 \, ,$$

$$\alpha = \int |\hat{\psi}_s^{(m)}(q)|^2 dq \, ; \quad \beta = \int \hat{\psi}_s^{(m)}(q)\mathcal{M}[\hat{\psi}_s^{(m)}(q)]dq \, .$$

The factors α and β are introduced to stabilize an otherwise divergent Neumann iteration scheme. This method is used to find stationary soliton solutions to the integral equation (3.23) which in turn provides an asymptotic description of the diffraction-managed DNLS Eq. (3.16). It should be also noted that we can obtain periodic diffraction-compensated soliton solutions directly from Eq. (3.16). The technique is similar to that originally proposed for finding periodic dispersion-managed solitons in communications problems[85; 86; 87]. The averaging procedure does not require that the map period, z_w, be small. To implement the method, initially we start with a guess, say $\phi_n^{(0)} = \text{sech}(nh)$ with $\mathcal{E}_0 = \sum_{n=-\infty}^{\infty} \text{sech}^2(nh)$. Over one period this initial ansatz will evolve to $\phi_n^{(0)'}$ which in general will have a chirp[88]. We then define an average: $\phi_n^{(0)''} = (\phi_n^{(0)} + \phi_n^{(0)'} e^{-i\Theta_n})/2$ where $\phi_n^{(0)'} = |\phi_n^{(0)'}| \exp(i\Theta_n)$ which has power $\mathcal{E}_0''$. Then $\phi_n^{(1)} = \phi_n^{(0)''} \sqrt{\mathcal{E}_0/\mathcal{E}_0''}$ is the new guess and in general the mth iteration takes the form

$$\phi_n^{(m+1)} = \sqrt{\mathcal{E}_0/\mathcal{E}_m''}\phi_n^{(m)''}, \quad \mathcal{E}_m'' = \sum_{n=-\infty}^{\infty} |\phi_n^{(m)''}|^2 \, . \tag{3.25}$$

In Fig. 8 the mode profiles associated with a stationary solution are depicted for two typical parameter values. The profiles are obtained by using both the integral equation approach as well as the averaged method. The evolution of these discrete diffraction managed solitons are illustrated in Figs. 9 and 10 for the same two set of parameter values as in Fig. 8 respectively. We note that initially the beam has zero chirp[88]. During propagation a chirp develops and the peak amplitude of the beam begins to decrease and, as a result, the beam becomes wider (due to conservation of power). A full recovery of the soliton's initial amplitude and width is achieved at the end of the map period. This breathing behavior is shown in Figs. 9 and 10 for both strongly and moderately confined beams respectively.

To establish the relation between the two approaches and to highlight the periodic nature of these new solitons, we calculate the nonlinear chirp by both the integral equation approach and the averaging method. It is

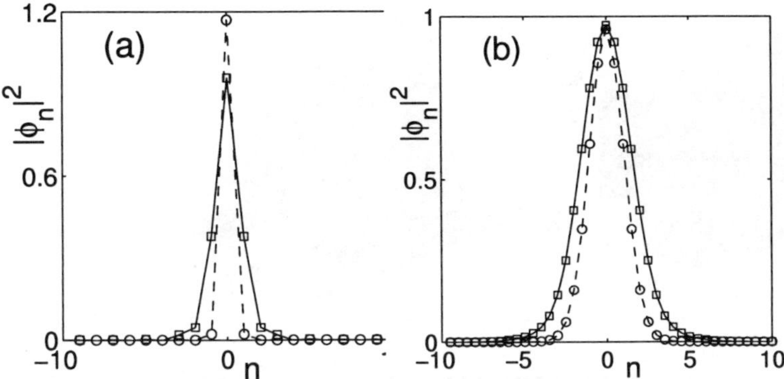

Fig. 8   Mode profile in physical space obtained from Eq. (3.24) (solid line) and from the average method (dashed line). Parameters are: (a) $\omega_s = 1$, $h = 1$, $s = 0.17$, $\Delta_1 = -\Delta_2 = 0.681$, $\delta_a = 1.135$ and $z_w = 0.2$; (b) $\omega_s = 1$, $h = 0.5$, $s = 1$, $\Delta_1 = -\Delta_2 = 4$, $\delta_a = 1$ and $z_w = 0.2$.

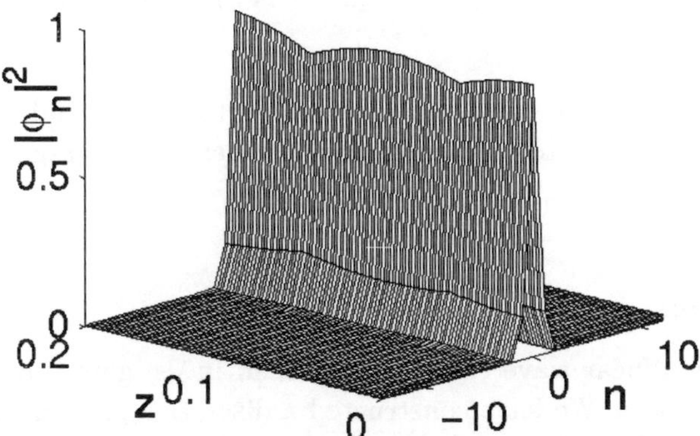

Fig. 9   Beam propagation over one period using Eq. (3.21) as initial condition obtained by a direct numerical simulation of Eq. (3.16). Parameters are the same as used in Fig. 8(a).

clear that for small values of the map period, z_w, the asymptotic analysis is in good agreement with the averaging method, as shown in Fig. 11.

We also mention briefly that the method of analysis associated with Eq. (3.16) can be modified to account for situations where the average diffraction is small, i.e., $\delta_a \ll 1$. In such a situation we write $\delta_a = \epsilon D_a$,

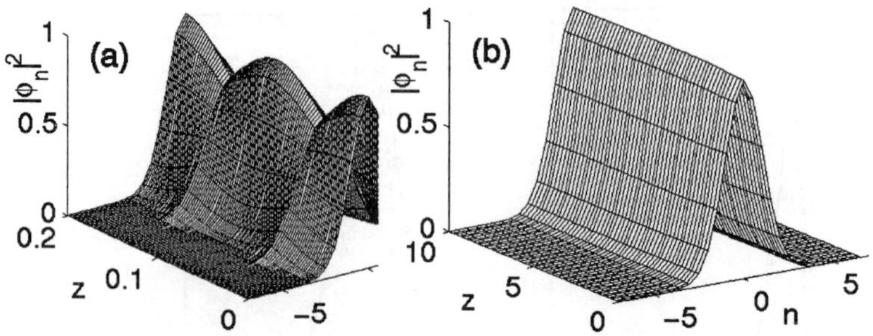

Fig. 10   Beam propagation over one period (a) and stationary evolution (b) obtained by a direct numerical simulation of Eq. (3.16) evaluated at end map period. Parameters are the same as used in Fig. 8(b).

$D = \epsilon D_a + \Delta(z)$, $z = \xi/\epsilon$, and $\phi_n = \sqrt{\epsilon}\Psi_n$. Then it is found that Ψ_n satisfies:

$$\frac{d\Psi_n}{d\xi} = i\frac{\mathcal{D}(\xi/\epsilon)}{2h^2}\left(\Psi_{n+1} + \Psi_{n-1} - 2\Psi_n\right) + i|\Psi_n|^2\Psi_n , \tag{3.26}$$

where $\mathcal{D}(\xi/\epsilon) = D_a + \frac{1}{\epsilon}\Delta(\xi/\epsilon)$. The model (3.26) is valid in parameter regimes which applies to physical situation where the average diffraction is small.

3.8   *Conclusion*

A discrete nonlinear wave equation has been investigated by the discrete Fourier transform. We have constructed a discrete equation governing the evolution of an optical beam in a waveguide array with varying diffraction. This equation has a novel type of discrete spatial soliton solution which breathes under propagation and, as a result, gains a nonlinear chirp. A full recovery of the soliton initial power is achieved at the end of each diffraction map. This open the possibility of fabricating a customized waveguide array which admits spechialized diffraction managed spatially confined solitary waves. A nonlocal integral equation governing the slow evolution of the soliton amplitude is derived and its stationary solutions are obtained.

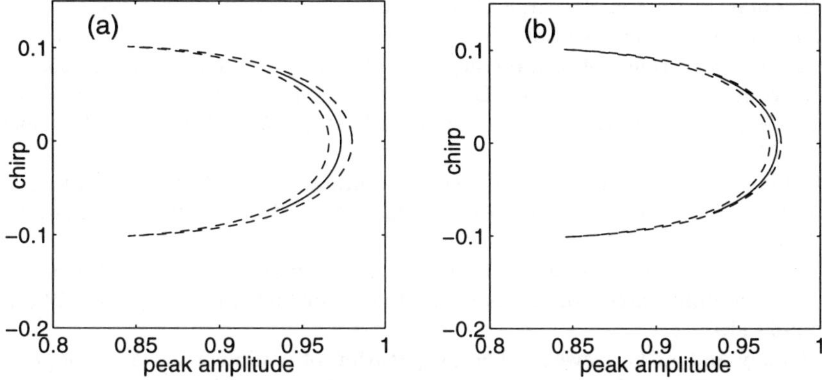

Fig. 11   Periodic evolution of the beam chirp versus maximum peak power. Solid line represents the leading order approximation, i.e., Eq. (3.21); dashed line represents numerical solution of the DM-DNLS. Parameters are the same as used in Fig. 8(b) with $z_w = 0.2$ (a) and 0.1 (b).

Acknowledgement

M.J.A. is partially supported by the Air Force Office of Scientific research, under grant number F49620-00-1-0031, NSF under grant numbers ECS-9800152, DMS-0070792.

References

J. Scott Russell, "Report on waves", Report of the 14th Meeting of the British Association for the Advancement of Science, Liverpool, pp. 417-496, 1844.

J. Boussinesq, "Theorie de l'intumescence liquid appelée onde solitaire ou de translation, se propagente dans un canal rectangulaire," *Compte Rendus Acad. Sci. Paris* **72**, 755-759, 1871.

D. J. Kortegweg and G. deVries, "On the change of form of long waves advancing in a rectngular canal, and on a new type of long stationary waves," *Philos. Mag. Ser.* 5, 39, 1039-1049, 1895.

C. S. Gardner and G. K. Morikawa, "Similarity in the asymptotic behavior of collision free hydromagnetic waves and water waves," Courant Inst. Math. Sci. Res. Rep. NYO-9082, New York University, New York (1960).

D. J. Benney, "Long nonlinear waves in fluid flows," *J. Math. Phys.*, (*Stud. on Appl. Math*), **45**, 52-63, 1966.

M. D. Kruskal, in Proc. IBM Scientific Computing Symposium on Large-Scale Problems in Physics, IBM Data Processing Division, White Plains, NY;

Thomas J. Watson Research Center, Yorktown Heights, NY (1965).

N. J. Zabusky and M. D. Kruskal, "Interaction of solitons in a collisioinless plasma and the recurrence of initial states," *Phys. Rev. Lett.* **15**, 240-243, 1965.

C. S. Gardner, J. M. Greene, M. D. Kruskal, and R. M. Miura, "Method for solving the Korteweg-de Vries equatoin," *Phys. Rev. Lett.* **19**, 1095-1097 (1967).

I. M. Gel'fand and B. M. Levitan, "On the determination of a differential equation from its spectral function," *Amer. Math. Sci. Transl.*, Ser 2, **1**, 259-309, 1955.

V. E. Zakharov and A. B. Shabat, "Exact theory of two-dimensional self-focusing and one-dimensional of waves in nonlinear media," *Sov. Phys. JETP* **34**, 62-69 (1972).

D. J. Benney and A. C. Newell, "The propagation of nonlinear wave envelopes," *J. Math. Phys.*, (*Stud. on Appl. Math*), **46**, 133-139, 1967.

V. E. Zakharov, "Stability of periodic waves of finite amplitude on the surface of a deep fluid," *Sov. Phys. J. Appl. Mech. Tech. Phys.* **4**, 190-194 (1968).

M. J. Ablowitz, D. J. Kaup, A. C. Newell and H. Segur, "The Inverse Scattering Transform – Fourier Analysis for Nonlinear Problems," *Stud. Appl. Math.*, **53**, 249-315 (1974).

M. J. Ablowitz and H. Segur, Solitons and the Inverse Scattering Transform, SIAM Studies in Applied Mathcmatics, 425 pages, SIAM, Philadelphia, PA, 1981.

M. J. Ablowitz and P. A. Clarkson, "Solitons, Nonlinear Evolution Equations and Inverse Scattering," London Mathematical Society Lecture Notes Series #149, 516 pages, Cambridge University Press, Cambridge, UK, 1991.

A. R. Osborne and T. L. Burch, "Internal solitons in the Andaman Sea," *Science* **258**, 451-460 (1980).

B. B. Kadomstsev and V. I. Petviashvili, "On the stability of solitary waves in weakly dispersing media," *Sov. Phys. Doklady* **15**, 539-541, 1970.

M. J. Ablowitz, and H. Segur, "On the Evolution of Packets of Water Waves," *J. Fluid Mech.*, **92**, 691-715 (1979).

M. J. Ablowitz, S. Chakravarty and L. A. Takhtajan, "A Self-Dual Yang-Mills Hierarchy and its Reductions to Integrable Systems in $1 + 1$ and $2 + 1$ Dimensions," *Commun. Math. Phys.*, **158**, 289-314 (1993).

L. J. Mason and N. M. J. Woodhouse, Integrability, Self-Duality, and Twistor Theory, Oxford University Press, LMS Monograph, New Series 15, Oxford, Oxford, 1996.

M. J. Ablowitz, S. Chakravarty and R. Halburd, "The Generalized Chazy Equation from the Self-duality Equations, *Stud. Appl. Math.* **103**, 75-88 (1999).

M. J. Ablowitz and A. S. Fokas, Complex Variables: Introduction and Applications, Cambridge University Press, Cambridge, UK, 1997.

E. R. Berlekamp, J. H. Conway, and R. K. Guy, Winning Ways, Academic Press, New York, 1982.

S. Wolfram, Theory and Applications of Cellular AUtomata, World Scientific,

Singapore, 1986.

J. H. H. Park, K. Steiglitz, and W. Thurston, "Soliton-like behavior in automata," *Physica* **19D**, 423-342 (1986).

T. S. Papatheodorou, M. J. Ablowitz and Y. G. Saridakis, "A Rule for Fast Computation and Analysis of SolitonAutomata, *Stud. in Appl. Math.* **79**, 173-184 (1988).

M. J. Ablowitz, J. M. Keiser and L. A. Takhtajan, "A Class of Stable Multistate Time-Reversible Cellular Automata with Rich Particle Content," *Phys. Rev. A* **44**, 6909-6912 (1991).

M. J. Ablowitz, "Nonlinear Feedback Shift Registers Via Nonlinear Evolution," *Stud. in Appl. Math* **107**, 127-136 (2001).

A. Hasegawa and F. Tappert, "Transmission of stationary nonlinear optical pulses in dispersive dielectric fibers. I. Anomalous dispersion," *Appl. Phys. Lett.* **23**, 142 (1973).

Y. Kodama and A. Hasegawa, "Nonlinear pulse propagation in a monomode dielectric guide," *IEEE J. Quantum Electron.* **23**, 510 (1987).

L. F. Mollenauer, R. H. Stolen, and J. P. Gordon, "Experimental-observation of picosecond pulse narrowing and solitons in optical fibers," *Phys. Rev. Lett.* **45**, 1095 (1980).

A. Hasegawa and Y. Kodama, "Guiding-center soliton," *Phys. Rev. Lett.* **66**, 161 (1991).

J. P. Gordon and H. A. Haus, "Random walk of coherently amplified solitons in optical fiber transmission," *Opt. Lett.* **11**, 665 (1986).

P. L. Chu and C. Desem, "Mutual interaction between solitons of unequal amplitudes in optical fibre," *Electron. Lett.* **21**, 1133 (1985).

L. F. Mollenauer, S. G. Evangelides, Jr. and J. P. Gordon, "Wavelength division multiplexing with solitons in ultra-long distance transmission using lumped amplifiers," *J. Lightwave Technol.* **9**, 362 (1991).

P. V. Mamyshev and L. F. Mollenauer, "Pseudo-phase-matched four-wave mixing in soliton wavelength-division multiplexing transmission," *Opt. Lett.* **21**, 396 (1996).

M. J. Ablowitz, G. Biondini, S. Chakravarty, R. B. Jenkins, and J. R. Sauer, "Four-wave mixing in wavelength-division-mulplexed soliton systems: damping and amplification," *Opt. Lett.* **20**, 1646 (1996).

Y. Kodama and S. Wabnitz, "Analytical theory of guiding-center nonreturn-to-zero and return-to-zero signal transmission in normally dispersive nonlinear-optical fibers," *Opt. Lett.* **20**, 2291 (1995).

N. S. Bergano, C. R. Davidson, M. Ma, A. Pillipetskii, S. G. Evangelides, H. D. Kidorf, J. M. Darcie, E. Golovchenko, K. Rottwitt, P. C. Corbett, R. Menges, M. A. Mills, B. Pedersen, D. Peckham, A. A. Abramov, and A. M. Vengsarkar, in *Digest of Optical Fiber Communication Conference* (Optical Society of America, Washington, D.C., 1998), postdeadline paper PD12.

M. Suzuki, I. Morita, N. Edagawa, S. Yamamoto, H. Taga, and S. Akiba, "Reduction of Gordon-Haus timing jitter by periodic dispersion compensation in soliton transmission," *Electron. Lett.* **31**, 2027 (1995).

N. J. Smith, F. M. Knox, N. J. Doran, K. J. Blow, and I. Bennion, "Enhanced power solitons in optical fibres with periodic dispersion management," *Electron. Lett.* **32**, 54 (1996).

M. J. Ablowitz and G. Biondini, "Multiscale pulse dynamics in communication systems with dispersion management," *Opt. Lett.* **23**, 1668 (1998).

M. J. Ablowitz, G. Biondini, and E. Olson, "On the evolution and interaction of dispersion-managed solitons," in *Massive WDM and TDM Soliton Transmission Systems*, Ed. A. Hasegawa (Kluwer, Dordrecht, 2000).

M. J. Ablowitz, T. Hirooka, and G. Biondini, "Quasi-linear optical pulses in strongly dispersion-managed transmission systems," *Opt. Lett.* **26**, 459 (2001).

M. J. Ablowitz and T. Hirooka, "Nonlinear effects in quasi-linear dispersion-managed pulse transmission," *IEEE Photon. Technol. Lett.* **13**, 1082 (2001).

R. Horne, "Collision induced timing jitter and four-wave mixingin wavelength division multiplexing soliton systems," Ph. D dissertation, Department of Applied Mathematics, University of Colorado at Boulder (2001).

P. V. Mamyshev and N. A. Mamysheva, "Pulse-overlapped dispersion-managed data transmission and intrachannel four-wave mixing," *Opt. Lett.* **24**, 1454 (1999).

M. J. Ablowitz and T. Hirooka, "Resonant nonlinear intra-channel interactions in strongly dispersion-managed transmission systems" *Opt. Lett.* **25**, 1750 (2000).

M. J. Ablowitz and T. Hirooka, "Intra-channel pulse interactions in dispersion-managed transmission systems: timing shifts," *Opt. Lett.* **26**, 1846 (2001).

M. J. Ablowitz and T. Hirooka, "Intra-channel pulse interactions in dispersion-managed transmission systems: energy transfer," *Opt. Lett.* **27**, 203 (2002).

S. V. Manakov, "On the theory of two-dimensional stationary self-focusing of electromagnetic waves," Sov. Phys. JETP **38**, 248 (1974).

D. Henning and G. P. Tsironis, Phys. Rep. **307**, 333 (1999); O. M. Braun and Y. S. Kivshar, Phys. Rep. **306**, 1 (1998); S. Flach and C. R. Willis, Phys. Rep. **295**, 181 (1998); F. Lederer and J. S. Aitchison, Les Houches Workshop on Optical Solitons, Eds. V. E. Zakharov and S. Wabnitz, Springer-Verlag (1999).

A. C. Scott and L. Macneil, Phys. Lett. **98A**, 87 (1983).

A. J. Sievers and S. Takeno, Phys. Rev. Lett. **61**, 970 (1988).

W. P. Su, J. R. Schieffer, and A. J. Heeger, Phys. Rev. Lett. **42**, 1698 (1979).

A. S. Davydov, J. Theor. Biol. **38**, 559 (1973).

P. Marquii, J. M. Bilbaut, and M. Remoissenet, Phys. Rev. E **51**, 6127 (1995).

D. N. Christodoulides and R. J. Joseph, Opt. Lett. **13**, 794 (1988).

Y. S. Kivshar, Opt. Lett. **18**, 1147 (1993); W. Krolikowski and Y. S. Kivshar, J. Opt. Soc. Am. B **13**, 876 (1996).

A. B. Aceves, C. De Angelis, S. Trillo, and S. Wabnitz, Opt. Lett. **19**, 332 (1994); A. B. Aceves, C. De Angelis, T. Peschel, R. Muschall, F. Lederer, S. Trillo, and S. Wabnitz, Phys. Rev. E **53**, 1172 (1996); A. B. Aceves, C. De Angelis, A. M. Rubenchik, and S. K. Turitsyn, Opt. Lett. **19**, 329 (1994).

F. Lederer, S. Darmanyan, and A. Kobyakov, *"Discrete Solitons"*, In: *Spatial Solitons*, Eds. S. Trillo and W. Torruellas (Springer-Verlag, Berlin, 2001), pp. 267-290.

T. Peschel, U. Peschel, and F. Lederer, Phys. Rev. E **57**, 1127 (1998);

S. Darmanyan, A. Kobyakov, and F. Lederer, Phys. Rev. E **57**, 2344 (1998).

H. Eisenberg, Y. Silberberg, R. Morandotti, A. Boyd, and J. Aitchison, Phys. Rev. Lett. **81**, 3383 (1998).

R. Morandotti, U. Peschel, J. Aitchison, H. Eisenberg, and Y. Silberberg, Phys. Rev. Lett. **83**, 2726 (1999).

R. Morandotti, U. Peschel, J. Aitchison, H. Eisenberg, and Y. Silberberg, Phys. Rev. Lett. **83**, 4756 (1999).

T. Pertsch, P. Dannberg, W. Elflein, A Bräuer, and F. Lederer, Phys. Rev. Lett. **83**, 4752 (1999).

G. Lenz, I. Talanina, and C. Martijn de Sterke, Phys. Rev. Lett. **83**, 963 (1999).

H. Eisenberg, Y. Silberberg, R. Morandotti, and J. Aitchison, Phys. Rev. Lett. **85**, 1863 (2000).

R. Morandotti, H. Eisenberg, Y. Silberberg, M. Sorel, and J. Aitchison, Phys. Rev. Lett. **86**, 3296 (2001).

M. J. Ablowitz and Z. H. Musslimani, Phys. Rev. Lett. **87** 254102, (2001).

S. F. Mingaleev, and Y. S. Kivshar, Phys. Rev. Lett. **86**, 5474 (2001); A. A. Sukhorukov and Y. S. Kivshar Phys. Rev. Lett. **87**, 083901 (2001); Phys. Rev. E (to be published).

A. Trombettoni and A. Smerzi, Phys. Rev. Lett. **86**, 2353 (2001).

D. N. Christodoulides and E. D. Eugenieva, Phys. Rev. Lett. **87**, 233901 (2001); Opt. Lett. **26**, 1876 (2001); E. D. Eugenieva, N. K. Efremidis, and D. N. Christodoulides, Opt. Lett. **26**, 1978 (2001); *Optics and Photonics News* **12**, 57, (2001).

M. J. Ablowitz and J. F. Ladik, J. Math. Phys. **17**, 1011 (1976).

M. Peyrard, and M. D. Kruskal, Physica D **14**, 88 (1984).

A. Yariv, Optical Electronics, Oxford University Press, Fifth Edition, (1997).

S. Darmanyan, A. Kobyakov, E. Schmidt, and F. Lederer, Phys. Rev. E **57**, 3520 (1998).

S. Somekh, E. Garmire, A. Yariv, H. L. Garvin, and R. G. Hunsperger, Appl. Phys. Lett. **22**, 46 (1973).

V. I. Petviashvili, Sov. J. Plasma Phys. **2**, 257 (1976).

M. J. Ablowitz and G. Biondini, Opt. Lett. **23**, 1668 (1998).

M. J. Ablowitz, Z. H. Musslimani, and G. Biondini, Phys. Rev. E **65**, 026602 (2002).

M. J. Ablowitz, Y. Ohta, and D. Trubatch, Phys. Lett. A **253**, 287 (1999).

M. J. Ablowitz, B. Prinari, and D. Trubatch, "Discrete and Continuous Nonlinear Schrödinger Systems", APPM report # 473, Dept. of Applied Mathematics, University of Colorado at Boulder (2001).

I. R. Gabitov and S. K. Turitsyn, Opt. Lett. **21**, 327 (1996).

S. K. Turitsyn, Sov. Phys. JETP Lett. **65**, 845 (1997).

J.H. B. Nijhof, W. Forysiak, and N. J . Doran, IEEE J. Sel. Top. on Quantum Elect. **6**, 330 (2000).

In close analogy to the definition of a chirp in dispersion managed solitons, here, the chirp of a discrete beam $\Phi_n(z) = |\Phi_n(z)| \exp[i\Theta_n(z)]$ is defined to be the coefficient of n^2 in the expansion of $\Theta_n(z)$ in a "Taylor series" around $n = 0$. In other words, if $\Theta_n(z) \approx c_0(z) + c_1(z)n + c_2(z)n^2 + \ldots$, then the chirp is given by $c_2(z)$. This definition is valid for moderately localized solitons and breaks down for strongly localized beams.

H. Feddersen, Lecture Notes in Physics, pp. 159-167 (Springer-Verlag, Berlin, 1991).

D. B. Duncan, J. C. Eilbeck, H. Feddersen, and J. A. D. Wattis, Physica D, **68** 1, (1993).

S. Flach and K. Kladko, Physica D **127**, 61 (1999).

S. Flach, Y. Zolotaryuk and K. Kladko, Phys. Rev. E **59**, 6105 (1999).

S. Aubry and T. Cretegny, Physica D **119**, 34 (1998).

Ch. Claude, Y. S. Kivshar, O. Kluth, and K. H. Spatschek, Phys. Rev. B **47**, 14228 (1993).

Chapter 10

Global Description of Patterns Far from Onset: A Case Study

N. Ercolani[1], R. Indik[1], A.C. Newell[1], and T. Passot[2]

[1] *Department of Mathematics, University of Arizona,
Tucson, AZ 85721*
[2] *CNRS, Observatoire de la Côte d'Azur, B.P. 4229,
06304 Nice Cedex 4, France*

Abstract.   The Cross-Newell phase diffusion equation $\tau(k)\Theta_T = -\nabla \cdot kB(k), k = \nabla\Omega, |k| = k$, and its regularization describe patterns and defects far from onset in large aspect ratio systems with translational and rotational symmetry. In this paper we show how director field solutions of this equation can be used to describe features of global patterns. The ideas are illustrated in the context of a nontrivial case study of high Prandtl number convection in a large aspect ratio, shallow, elliptical container with heated sidewalls, for which we also have the results of simulation and experiment.

1   Introduction

A macroscopic description of convection patterns far from the onset of the convective instability still presents a considerable challenge for both physicists and mathematicians. The path we have followed to tackle this problem is to study the phase diffusion equation that results from the elimination of the complex three-dimensional structure of the microscopic roll pattern and that focuses on the large-scale dynamics of the phase gradient, which is perpendicular to the rolls.

We recently established [1] rigorous proofs for the existence and form of solutions to the regularized phase diffusion equation in the limit where the inverse aspect ratio ε (ratio of the roll wavelength to the typical size

411

of the container) goes to zero. The basis of this work was first to remark that solving the phase diffusion equation amounts to solving a minimization problem for an energy E which is the sum of two terms, one associated with roll bending and the other with the mismatch between the local and the preferred wavenumber. Generally pattern forming systems are not gradient flows. However in the high Prandtl number limit the Oberbeck-Boussinesq equations behave as if they were [2]. A second point was to study a separate, although closely related, problem defined by equating the two components of the energy. This *self-dual reduction* leads to solutions which are also solutions of the original phase diffusion equation when the Gaussian curvature of the graph of the phase vanishes. For many of the observed structures in patterns, the curvature does vanish except in the neighborhood of certain points or curves. In general, however, the solutions of the self-dual equations are not solutions of this phase equation but they do provide, for each ε, an upper bound on the energy. The remarkable point is that, for a model form of the phase equation valid when the wavenumber is everywhere close to the selected one, the self-dual equation can be linearized and leads to explicit solutions whose limits as ε goes to zero, are well controlled. In special cases, a lower bound for the energy can be found, that equals the upper bound and thus provides an explicit minimizer (not necessarily unique).

One of the limitations of this work has to do with the fact that locally changing the phase gradient $\nabla\theta$ into $-\nabla\theta$ does not change the physical rolls. However, the manner in which the energy is defined associates a non-zero cost with such a reversal of the local wavenumber. Therefore, the solutions that have been found are possibly not the true minimizers if one defines the minimization problem in a space where the sign phase ambiguity is taken into account. Such a solution could for example be realized in terms of a two-sheeted Riemann surface, the branch cut being located on the line where this sign reversal occurs. The goal of this paper is to study the patterns obtained when the container has an elliptical lateral boundary which is heated in order to impose a constant phase on this boundary. Our theory of the phase diffusion equation leads, in the $\varepsilon \to 0$ limit, to rolls which are solutions of the eikonal equation $|\nabla\theta| = 1$. In this roll pattern, the rolls come in parallel to the boundary and meet at a grain boundary (see Figure 1; see also [3]). The main question is whether or not this solution is realized in experimental or numerical patterns. Several comparisons are made, first with one of the simplest toy models for infinite Prandtl number

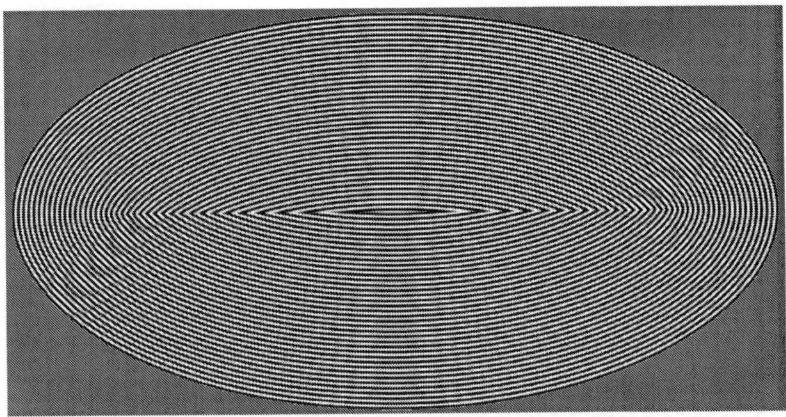

Fig. 1   The eikonal solution has rolls that are parallel to the elliptical boundary. This pattern propagates inward with fixed roll spacing. This leads to a phase grain boundary in the middle.

convection, the Swift-Hohenberg equation

$$W_t + (\nabla^2 + 1)^2 W - RW + W^3 = 0, \tag{1.1}$$

where W represents, for example, the temperature, and R represents the stress parameter deviation from onset. We also compare our predictions with numerical simulations of the Oberbeck-Boussinesq equations [4] and finally with experiments [5].

The most interesting feature to study is how the sign ambiguity in the wave vector of the pattern leads, in the numerical solutions of the model equation, to a solution with an energy possibly smaller than that of the eikonal solution. It turns out that the solution "prefers" to transform a grain boundary with sharp angles into pieces of straight rolls connected with dislocations.

In Section 2, the phase diffusion equation and its regularization are introduced, together with the energy and the linear stability properties. Section 3 briefly describes the main results of our previous work [1]. In Section 4, we introduce the self-dual equation and the exact phase grain boundary solution. Section 5 is a summary of the procedure to obtain weak solutions of the self-dual equation (steepest descent method) and the description of the eikonal solutions. In Section 6 we introduce the notion of twist associated with director fields (unsigned vectors) and explain in detail the limitations of our previous results. An argument is presented in Section 7 to predict the critical angle at which a phase grain boundary becomes unstable towards the formation of dislocations. Sections 8 and 9 respectively describe the numerical simulations of the Swift-Hohenberg equation and the comparisons with simulations of the Oberbeck-Boussinesq equation and laboratory experiments. Some concluding remarks are made in Section 10.

2   Background

Let us consider a straight roll solution of the Oberbeck-Boussinesq [6], or the Swift-Hohenberg (SH) equations, of the form

$$w(x, y, t) = F(\theta) = \Sigma A_n(k) \cos n\theta. \tag{2.1}$$

$F(\theta)$ is a 2π periodic function of a phase θ whose gradient k is the constant wavevector of the pattern. An important point, related to the question of twist discussed in Section 6, is to remark that the microscopic field is unchanged on the reversal of θ, $\theta \to -\theta$. Next, we seek nearby solutions in the form

$$w(x, y, t) = F(\frac{1}{\varepsilon}\Theta) + \varepsilon w_1 + \varepsilon^2 w_2 + \cdots \tag{2.2}$$

for which the local phase Θ varies over large spatial and temporal scales defined as $X = \varepsilon x, T = \varepsilon^2 t$. In particular, the wave vector $k = \nabla\theta = \nabla_X\Theta = (f, g)$, changes by order one over distances of $\frac{1}{\varepsilon}\lambda$, where λ is the wavelength and ε, $0 < \varepsilon \ll 1$ is the inverse aspect ratio of the system. For the modulated solution, the amplitudes A_n are still slaved to the local wavenumber $k = |\nabla\theta|$ via $A_n = A_n(k)$.

Because the original system is translationally invariant, the variation of w about $F(\theta)$ has a nontrivial null space and the non-homogeneous equation for the corrections $w_1, w_2, \cdots$ to $F(\theta)$ must satisfy solvability conditions. These express the fact that the right hand sides of these equations must lie in the range of the linear operator obtained by linearizing w about the exact solution $F(\theta)$. To dominant order in ε [7],

$$\tau(k)\Theta_T = -\nabla \cdot kB(k). \tag{2.3}$$

In equation (2.3), $\tau(k) > 0$, $kB(k)$ is a cubic like curve shown in figure 2 which is zero at the left and right ends of the region of existence of (2.1) and is zero at a point k_B between. The overall shape, one maximum, one

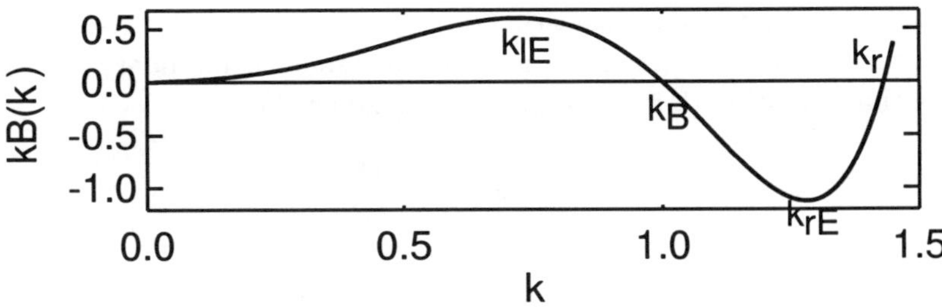

Fig. 2   Plot of a typical graph of $kB(k)$, There is a zero at k_B, k_{rE} labels the right Eckhaus instability boundary, while k_{lE} labels the left Eckhaus instability boundary.

interior zero, and one minimum of kB is universal for a broad class of microscopic systems. The spatial part of equation (2.3) becomes

$$\nabla \cdot \boldsymbol{k} B(k) = (B + \frac{f^2}{k} B')\Theta_{XX} + \frac{2fg}{k} B'\Theta_{XY} + (B + \frac{g^2}{k} B')\Theta_{YY}. \quad (2.4)$$

It is a quasilinear second order operator in Θ and is only negative definite when both B and $(kB)'$ are negative.

This occurs in the interval (k_B, k_{rE}) known as the Busse balloon. Outside this region, equation (2.3) is ill-posed. When the wavenumber k of the straight roll pattern lies to the left of k_B, the rolls lose their resistance to bending (this is the zig-zag instability). A regularization term is thus required in equation (2.3) to describe the resulting dynamics. Moreover, the stationary phase diffusion equation admits multivalued solutions [1] which can only relate to the physically relevant solutions if one adds a regularization coming from the contributions of the next orders of the asymptotic expansion. The dominant correction comes at order ε^3 and can be approximated by a linear term of the form $\varepsilon^2 \eta \nabla^4 \Theta$. After changing the definition of ε to absorb the constant η, the regularized equation reads

$$\tau(k)\Theta_T = -\nabla \cdot \boldsymbol{k} B(k) - \varepsilon^2 \nabla^4 \Theta. \quad (2.5)$$

The bi-Laplacian of the phase, $\varepsilon^2 \nabla^4 \Theta$, provides the necessary bending resistance to arrest and saturate the roll bending due to the zig-zag instability and leads to a stationary line defect which is called a phase grain boundary (PGB). Across the PGB, the pattern wavevector $\boldsymbol{k}$ changes abruptly (over distances of the roll wavelength).

Considering a compact domain Ω with smooth boundary, one easily finds that equation (2.5) is essentially a gradient flow

$$\tau(k)\Theta_T = -\frac{\delta F}{\delta\Theta} \tag{2.6}$$

with

$$F[\Theta] = \int_\Omega \left(\frac{1}{2}G^2(k) + \frac{1}{2}\varepsilon^2(\nabla^2\Theta)^2\right) d\mathbf{X} \tag{2.7}$$

and

$$G^2(k) = -2\int_{k_B}^{k} kB(k)\,dk. \tag{2.8}$$

In $F[\Theta]$, the G^2 term is zero when $k = k_B$, the preferred wavenumber. It represents the wavenumber mismatch energy density while the $\varepsilon^2(\nabla^2\Theta)^2$ term represents the bending energy density. The system would like to relax to the state where $k = k_B$ everywhere. In interesting circumstances, such as the problem posed in the introduction, this cannot be the case and patches of planform where $k = k_B$ are mediated by line and point defects.

Before analyzing the minimization problem for the energy (2.7), we emphasize the sense in which equation (2.5) is universal. First, we have assumed that there are no other slow modes than the one associated with the translational invariance of the system. In particular, we exclude cases (such as in small to moderate Prandtl number convection) where a mean flow, driven by roll curvature gradients, is present. Second, we have assumed that no *short scale* instability (e.g. a cross roll instability) can develop and trigger a modification to the local roll planform.

3   Description of Asymptotic Minimizers for RCN

For k close to k_B, G^2 is approximated to second order by $(k^2 - k_B^2)^2$. Choosing $k_B = 1$, the RCN (regularized Cross Newell) free energy divided by ε becomes

$$E^\varepsilon[\Theta] = \int_\Omega \left(\varepsilon(\nabla \cdot \mathbf{k})^2 + \frac{1}{\varepsilon}(k^2 - 1)^2\right) d\mathbf{X}. \tag{3.1}$$

Our goal is to study the asymptotic form, as $\varepsilon \to 0$, of the minimizers Θ^ε of $E^\varepsilon[\Theta]$ within the family of locally gradient director fields $\mathbf{k} = \nabla\Theta$ for which the phase Θ belongs to the class

$$\mathcal{A}^\varepsilon = \left\{ \Theta \text{in} H^2(\Omega);\ \Theta\Big|_\Omega = \alpha^\varepsilon(s),\ \frac{\partial\Theta}{\partial n}\Big|_\Omega = \beta^\varepsilon(s) \right\}, \qquad (3.2)$$

where $\alpha^\varepsilon(s)$ and $\beta^\varepsilon(s)$ are two sequences of smooth bounded functions of the boundary arclength s which are respectively uniformly convergent in ε.

The energy associated with the variational problem (3.1) has the form of the well-known Ginzburg-Landau functional [1]. The difference is that in the RCN variational problem this energy is restricted to vector fields which are gradient. We are able to say something about the asymptotic limit of these minimizers, in the class of $\mathcal{A}^\varepsilon$ coming from Dirichlet viscosity boundary conditions (these will be described in the next section). We summarize here what is known about this limit.

Theorem 3.1

(1) As $\varepsilon \to 0$, the minimizers $\Theta^\varepsilon \to \Theta^0$ in $H^1(\Omega)$ where Θ^0 solves the eikonal equation $|\nabla\Theta^0| = 1$. Defects are, therefore, supported on locally 1-dimensional sets which are a union of locally rectifiable curves. Let Σ denote this total defect locus.

(2) The asymptotic minimal energy is bounded:

$$\lim_{\varepsilon\to 0} \inf E(\Theta^\varepsilon) \le 1/3 \int_\Sigma |[\nabla\Theta^0]|^3 ds$$

where $[\nabla\Theta^0]$ is the jump in $\mathbf{k}^0$ across Σ and ds is the element of arclength.

(3) In the class of cases when Σ is a single straight line segment, the previous inequality becomes an equality; i.e., the upper bound is tight.

Details about the proofs of these results may be found in [1]. In that work we show that Θ^ε converges weakly to Θ^0; however, more recent results on these types of minimization problems [8; 9; 10] enable one to improve this to the strong convergence stated above. We will construct some explicit examples of eikonal solutions in section 5.

These one dimensional defect loci are what we have called *phase grain boundaries*. Our bound on the asymptotic minimum energy indicates that the bending term $\int_\Omega |\nabla\mathbf{k}|^2$ in this free energy will diverge as $1/\varepsilon$. This is a major increase in energetic cost compared to the $\log(1/\varepsilon)$ cost of the original Ginzburg-Landau problem whose defects are point vortices. We reiterate that this difference is due to the fact that, in comparison to Ginzburg-Landau, the minimization for RCN is done over the smaller class of irrotational vector fields; i.e., those k which are gradient.

4   Self-dual Test Functions

Although it is straightforward to conclude the existence of the minimizers Θ^ε of $E^\varepsilon[\Theta]$, it is far from straightforward to calculate them (they are solutions of a fourth order variational problem). Except in extremely special instances this has not, as far as we know, been done.

In earlier papers [1; 11; 12], we had proved that self-dual (or anti self-dual) solutions, namely phase functions $\Theta(X, Y)$ which satisfy

$$\varepsilon \nabla^2 \Theta = \pm \sqrt{G^2} \tag{4.1}$$

are solutions of the Euler-Lagrange equation for $F[\Theta]$,

$$\varepsilon^2 \nabla^4 \Theta + \nabla \cdot kB(k) = 0, \tag{4.2}$$

provided the Hessian (proportional to the Gaussian curvature of the graph of Θ) vanishes; i.e.,

$$\Theta_{XX}\, \Theta_{YY} - \Theta_{XY}^2 = 0. \tag{4.3}$$

The motivation for considering solutions of the self-dual equation come from the observation that, for many of the structures observed in convection patterns (PGB's, almost everywhere in the neighborhoods of dislocations and disclinations), the free energy is equipartitioned between the wavenumber mismatch and the bending energy components. In the weak bending limit we proved [11] that this equipartition was a fact in general; i.e., solutions of the associated self-dual equations were also solutions of the fourth order variational equations. Although this is not the case here, because of the Hessian obstruction, it turns out that the self-dual solutions are either a good approximation (in the small ε limit, the Hessian has its support at point defects [11] and along curved line defects) or, at worst, serve as good test functions for bounding the free energy in the $\varepsilon \to 0$ limit. This is the fundamental idea underlying our analysis. Because they solve a second order PDE, one can get precise estimates on the asymptotic behavior of self-dual solutions in the limit as $\varepsilon \to 0$. For $G^2 = (k^2 - 1)^2$, which is a good quantitative approximation near $k = 1$, and a good qualitative match for $k_{lE} < k < k_{rE}$, equation (4.1) becomes

$$\varepsilon \nabla^2 \Theta = \pm(1 - (\nabla\Theta)^2). \tag{4.4}$$

The transformation

$$\Theta = \varepsilon \ln \Psi \tag{4.5}$$

linearizes (4.4) as

$$\varepsilon^2 \nabla^2 \Psi - \Psi = 0. \tag{4.6}$$

The straight roll solution of (4.6) is

$$\Psi = e^{\boldsymbol{k} \cdot \boldsymbol{X}/\varepsilon}, \ k = 1; \ \text{or} \ \Theta = \boldsymbol{k} \cdot \boldsymbol{X} \tag{4.7}$$

The regularized phase grain boundary (PGB) solution is simply the sum of two exponentials

$$\Psi = e^{\boldsymbol{k}_+ \cdot \boldsymbol{X}/\varepsilon} + e^{\boldsymbol{k}_- \cdot \boldsymbol{X}/\varepsilon} \tag{4.8}$$

where $\boldsymbol{k}_\pm = (\cos\varphi, \pm\sin\varphi)$. The corresponding phase has gradient

$$\nabla\Theta = \boldsymbol{k} = \left(\cos\varphi, \ \sin\varphi \tanh\left(\frac{Y}{\varepsilon}\sin\varphi\right)\right). \tag{4.9}$$

The free energy of the roll solution is zero. For the exactly self-dual regularized PGB solution (its Gaussian curvature is zero), the self-duality allows us to express the free energy as twice one of its terms:

$$E^\varepsilon = \frac{2}{\varepsilon}\int_\Omega (1 - k^2)^2 d\boldsymbol{X} = \frac{2}{\varepsilon}\int_\Omega \sin^4\varphi \cdot \text{sech}^4\left(\frac{Y}{\varepsilon}\sin\varphi\right) dX\, dY$$

which, if φ is constant in X, gives

$$\lim_{\varepsilon \to 0} E^\varepsilon = \frac{8L}{3}\sin^3\varphi, \tag{4.10}$$

This coincides with the general result stated in Theorem 3.1.

We now use these self-dual solutions to explore how one might uncover configurations which would minimize (3.1) on Ω within the class $\mathcal{A}^\varepsilon$ of functions Θ satisfying boundary conditions $\Theta|_{\partial\Omega} = \alpha(s)$, independent of ε, and $\beta^\varepsilon(s) = \partial_\nu \Theta^\varepsilon_{SD}|_{\partial\Omega}$ where Θ^ε_{SD} is the solution of (4.4) corresponding to the boundary values $\alpha(s)$.

In [9] we show that as long as $\alpha(s)$ has a Lipschitz constant less than 1, $\beta^\varepsilon(s)$ is uniformly convergent and therefore satisfies our admissibility conditions for the class $\mathcal{A}^\varepsilon$. In the next section we will observe that within such an admissible class of boundary conditions, the self-dual solutions limit to viscosity solutions of the unregularized phase equation which, in our present model, is just the eikonal equation. For this reason we refer to these boundary conditions as *Dirichlet viscosity boundary conditions.*

5   Weak Solutions

For admissible Dirichlet viscosity boundary conditions, described at the end of the previous section, we consider the sequence

$$\Theta^\varepsilon_{SD} = \varepsilon \ln \Psi^\varepsilon,$$

where Ψ satisfies the Helmholtz equation (4.6) with boundary value $e^{\alpha(s)/\varepsilon}$. It can be shown that $\lim_{\varepsilon \to 0} \Theta^\varepsilon_{SD}$ is the unique continuous viscosity solution v to

$$|\nabla v|^2 - 1 = 0 \quad \text{on} \quad \Omega, \quad v\big|_{\partial\Omega} = \alpha(s). \tag{5.1}$$

By a steepest descent calculation [1] we give a prescription for constructing the gradient of this viscosity solution which is the appropriate object to consider when making comparisons between the self-dual solutions and minimizing sequences of the RCN energy E^ε. This prescription is

$$\nabla v(\boldsymbol{X}) = -\frac{\boldsymbol{X} - \boldsymbol{X}(\bar{s})}{|\boldsymbol{X} - \boldsymbol{X}(\bar{s})|},$$

where $\boldsymbol{X}(s)$ is the arclength parameterization of $\partial\Omega$ and $\bar{s}$ is the point on the boundary where the expression $|\boldsymbol{X} - \boldsymbol{X}(s)| - \Theta(\boldsymbol{X}(s))$ attains a minimum. Except on a set of measure 0 in Ω, this minimum point is unique and simple. The variational equations for $\bar{s}$ are

$$\left(\frac{\boldsymbol{X} - \boldsymbol{X}(\bar{s})}{|\boldsymbol{X} - \boldsymbol{X}(\bar{s})|} - \nabla v(\boldsymbol{X}(\bar{s})) \right) \cdot \boldsymbol{X}'(\bar{s}) = 0.$$

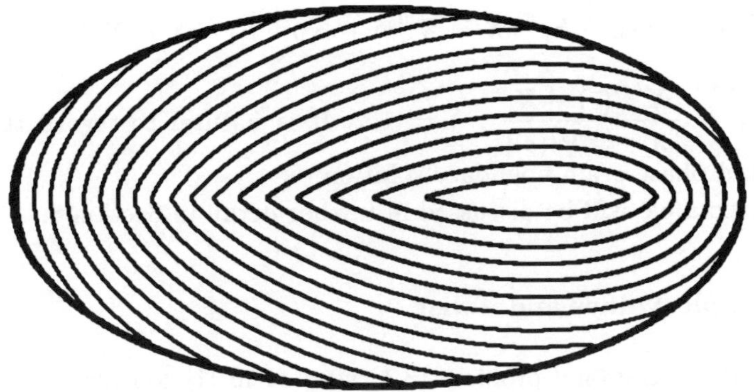

Fig. 3   Eikonal solution with non-constant boundary conditions, and horizontal defect locus

Figure 3 illustrates the application of this construction for the boundary data given by restricting on Ω the linear function mx, $x = X/\varepsilon$ for $m = 0.5$. The wavevector $\boldsymbol{k} = \nabla v = \dfrac{\boldsymbol{X}(\bar{s}) - \boldsymbol{X}}{|\boldsymbol{X}(\bar{s}) - \boldsymbol{X}|}$ has magnitude one everywhere. The defect locus is a straight line segment along the x-axis and hence is within the class to which part 3 of Theorem 3.1 applies.

Another example, and one which is pertinent for the remainder of this paper is to consider again an elliptical domain but with constant boundary data. In fact we can take $\alpha \equiv 0$. This corresponds to the boundary of the domain being a roll. In a physical setting, let us suppose that the boundary is uniformly heated so that $\theta = 0$ and the temperature $F(\theta)$ is maximal there. Then the construction is the obvious one of locating the next temperature maximum a distance of 2π in, along the boundary normal. This construction will fail as the normals meet on a line segment along the major axis (see figure 1). Regularization results in a PGB along that segment; it is centered about the origin between two points which are the curvature centers of the ellipse and which we refer to as terminal disclinations. Associated with each point on this open segment there are two distinct points on the boundary which have the same distance to the given point on the segment. Therefore for $\boldsymbol{X}$ on (or within a wavelength distance of) the line segment joining the two terminal disclinations, from the steepest descent argument, $\Psi(\boldsymbol{X})$ is well-approximated as the sum of two exponentials

$$\Psi(\boldsymbol{X}) \approx \exp \frac{1}{\varepsilon} \, \boldsymbol{k}_1 \cdot (\boldsymbol{X} - \boldsymbol{X}_S^{(1)}) + \exp \frac{1}{\varepsilon} \, \boldsymbol{k}_2 \cdot (\boldsymbol{X} - \boldsymbol{X}_S^{(2)}), \qquad (5.2)$$

where $k_j = \dfrac{\boldsymbol{X}^{(j)}(\bar{s}) - \boldsymbol{X}}{|\boldsymbol{X} - \boldsymbol{X}^{(j)}(\bar{s})|}$, $j = 1, 2$. It is natural to regard this as a modulated phase grain boundary solution.

This example will be discussed in further detail in Section 8.

6   Multiple Values and Twist

Because the variational problem (3.1) studied in [1] considers only variations over fields which are conservative, it cannot realize certain types of
defects, such as disclinations, which are commonly seen in pattern-forming
systems far from threshold. The reason is topological. Modulation equations such as CN or RCN are derived under the assumption that the field
is *locally* gradient. Away from defects, striped patterns locally look like the
level curves of a function such as $\cos(\theta)$. The phase θ itself need not be
well defined (for instance, $\theta \to -\theta$ leaves the local pattern unaltered). The
existence of these ambiguities is intimately related to the types of defects
that can occur. For instance, wavevector fields (f, g) with $f - ig = z^{1/2}$ or
$z^{-1/2}$ where $z = x + iy$ give examples of harmonic vector fields for which
there is not a *single valued* potential θ. Traversing a closed circuit coun-
terclockwise about the origin of these fields one would find that the wave
vector returns to itself but rotated by $\mp\pi$, in contrast to the conservative
fields z and z^{-1} where traversing the closed circuit would yield a rotation
of $\pm 2\pi$. The defects of the former fields which are locally but not globally
gradient are referred to as disclinations. The best way to think about this
is to regard these fields as *director fields*, that is to say unoriented vector
fields.

To include such fields, one needs to consider spaces consisting of double-
valued vector functions on Ω, the double-valuedness corresponding to the
orientational ambiguity of director fields. One way to do this concretely
would be to consider single-valued functions on domains which are Riemann
surfaces rather than simply connected planar domains. In [11], we carried
out such numerical and analytical studies on a double cover of a disc Ω
with vectorial boundary data whose winding number is half integral. In
this way, we were able to obtain stationary solutions of the RCN equation

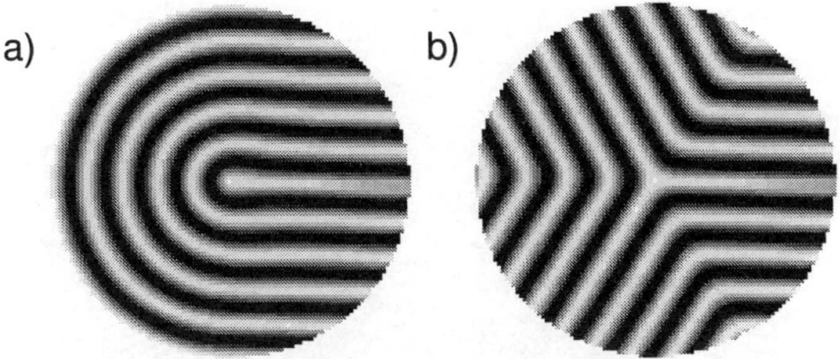

Fig. 4   a) Convex disclination b) Concave disclination numerically calculated as solutions of the RCN equation.

corresponding to *concave* (V) and *convex* (X) disclinations, which are the canonical point defects of two dimensional patterns (see Figure 4). These are ubiquitous in natural patterns, necessarily involve director fields $\pm k$ and, as we shall shortly see, play a central role in finding the true minimizers of (3.1) in the convection context.

One important consequence of the double-valuedness of the phase is that the class of functions that was considered in the preceding sections is too small. Minimizing vector field configurations may, and often will be unstable to director field perturbations.

An examination of some computed solutions to the Swift-Hohenberg equation, or the Boussinesq equations or of experiments of Rayleigh-Bénard convection shows that these systems seem to produce solutions which have disclinations. In addition, while it is possible to describe many of these solutions in terms of a well defined, continuous single valued θ (at least away from the point defects), these descriptions involve requiring sharp "ridges" where $k = \nabla\theta$ flips direction by 180 degrees. The energetic penalty for such a switch would be very high if we consider only vector fields, but, in fact there is no such cost, because there is no difference between $\pm k$. The "correct" energy sees no cost associated with these "ridges". The solution whose energy is minimal is one which is not a single valued function θ with ridges, but instead lives on some Riemann surface that covers the physical domain with degree 2. An illustration of this kind of "solution" is given in Figure 5. The figure shows the graph of the 2-valued phase $\pm\theta$ for the "stadium solution" whose level curves are plotted on the midplane. The director field, $\pm k$, becomes a single valued vector field on this Riemann

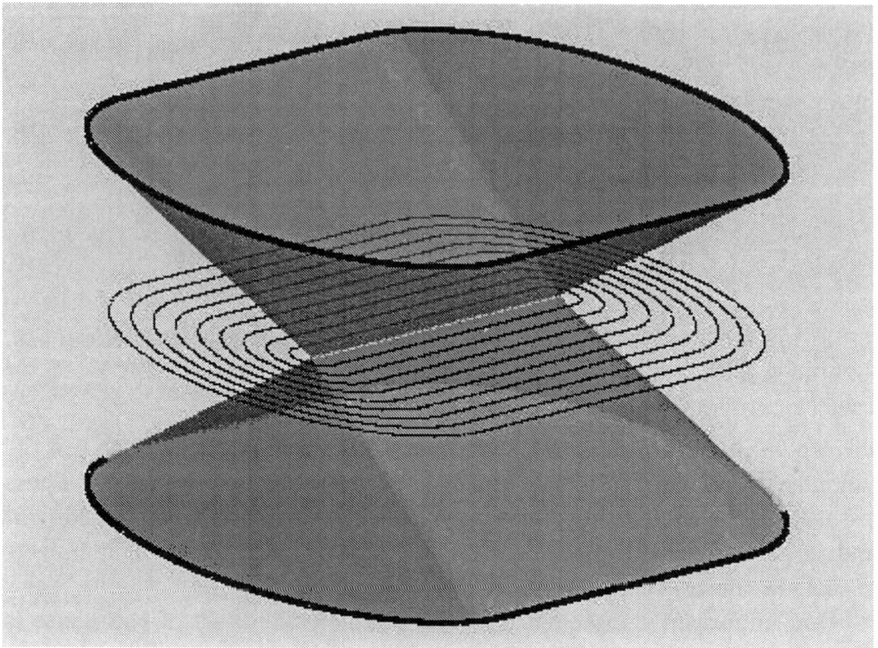

Fig. 5   The solution of the phase diffusion equation over a stadium, is properly defined on a double cover.

surface.

On the other hand, there is remarkably good agreement between the eikonal solutions that we have found in the earlier sections and the computed and experimental solutions on ellipses, except in a neighborhood of a portion of the major axis of the ellipse near the center. Where the eikonal solution predicts that we should have a phase grain boundary where rolls bend (fairly sharply), the computed and experimental results show a string of pairs of V and X disclinations. These disclinations have a spacing that is large in the center, and becomes closer as we move further out. At a certain point, the string of V X pairs is replaced by a phase grain boundary where the rolls behave as in the eikonal solution.

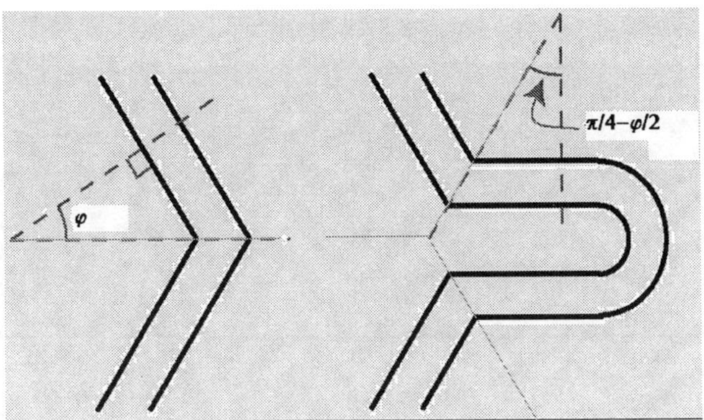

Fig. 6   a) A single phase grain boundary may have higher energy than b) a configuration of three phase grain boundaries with a pair of disclinations (phase grain boundaries are shown as thin solid gray lines).

7   Predicting the critical angle

Let us now examine the stability of a phase grain boundary in a context where the wave vector is a director field and not a vector field. When the roll angle is too sharp, the configuration shown in Figure 1 is unstable to perturbations in which (V, X) disclination pairs are created. On the lines joining these, the sign of k reverses but such a reversal adds no cost to the free energy. In order to understand this instability, let us return to the pure phase grain boundary shown in Figure 6a. In (4.10) we calculated the asymptotic energy cost per unit length of this grain boundary to be $\frac{8}{3}\sin^3(\varphi)$. Now, imagine that from the origin on, we change the configuration to that shown in Figure 6b. We have introduced two new phase grain boundaries, each locally longer than the horizontal PGB by a factor of $1/\cos\left(\frac{\pi}{4} + \frac{\varphi}{2}\right)$, but the new angle $\frac{\pi}{4} - \frac{\varphi}{2}$ between the k vector and the new phase grain boundary may be smaller. The free energy per unit length along the pair of phase grain boundaries shown in Figure 6b is

$$\frac{2}{\cos\left(\frac{\pi}{4} + \frac{\varphi}{2}\right)}\sin^3\left(\frac{\pi}{4} - \frac{\varphi}{2}\right) \cdot \frac{8}{3} = \frac{8}{3}(1 - \sin\varphi). \tag{7.1}$$

But for large φ, $\sin^3\varphi > 1 - \sin\varphi$; equality occurs at $\varphi_c \simeq 43°$. Therefore a phase grain boundary configuration whose k angle φ is too large is expected to destabilize into the configuration shown in Figure 6b. The vector field solution has destabilized to a director field solution in which a concave-

Fig. 7   Solutions of Swift-Hohenberg with boundary conditions forcing the rolls to form at a specified angle, as labeled across the top.

convex disclination pair protuberance has formed. The contribution from the circular end about the convex disclination will be proportional to $\varepsilon \ln \frac{1}{\varepsilon}$ and therefore much smaller than the grain boundary contribution.

In Figure 7, we show the results of a simple numerical experiment solving the Swift-Hohenberg equation with $R = 1$, in which we slowly increase the angle φ in a phase grain boundary. The roll angles are fixed on the lateral

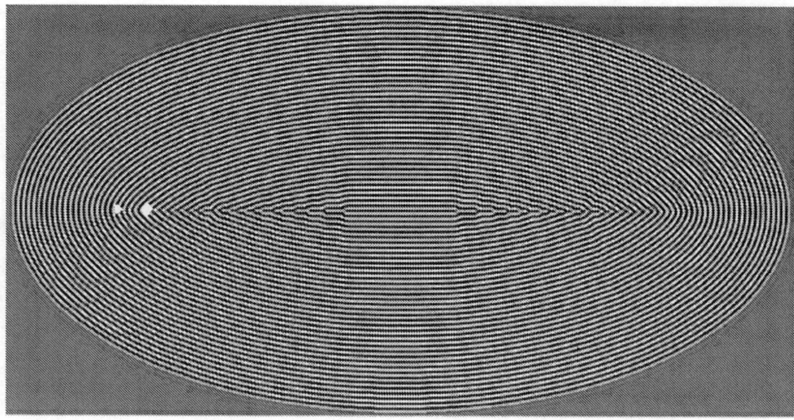

Fig. 8 Rolls from numerical simulation of Swift-Hohenberg for an ellipse with major to minor axis ratio of 1.9. The white triangle marks the center of curvature, the white diamond marks the point where the angle of the phase grain boundary in the eikonal solution surpasses the critical angle, as discussed in the text.

boundary and ramps on the stress parameter are applied at the end parts of the domain. The first appearance of a protuberance indeed appears at an angle $\varphi \approx 45^o$.

8 Swift-Hohenberg numerics

Patterns of very large aspect ratio in an elliptical geometry have been obtained from the numerical integration of the Swift-Hohenberg equation (1.1). We use a pseudo-spectral method in a square periodic domain of width 300π with a maximum resolution of 1024^2 grid points. In order to mimic an elliptical boundary condition, the stress parameter R is ramped from its value 0.5 inside an ellipse contour whose major axis ends at a distance of 5 grid points from the sides of the computational domain, to -0.5 outside this contour. The transition takes place within 1 grid point. Small oscillations are seen outside the ellipse, whose influence is negligible on the dynamics inside the ellipse. The initial conditions are taken to be the cosine of the phase obtained by solving the eikonal equation (5.1) with a zero phase boundary condition on the ellipse boundary. The arbitrary constant that multiplies the solution is chosen to impose k=1. This initial condition is displayed in Figure 1. In Figure 8 we display the roll pattern

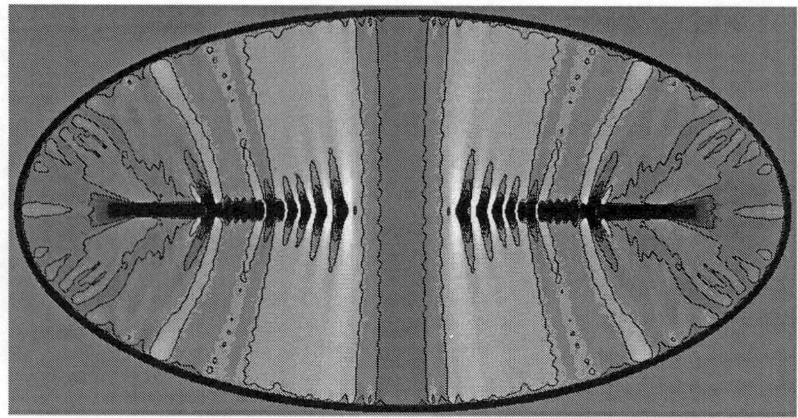

Fig. 9   Local wavenumber magnitude corresponding to figure 8, lighter grays correspond to higher values of k. The contours shown are $k = .995$, $k = 1$ and $k = 1.005$.

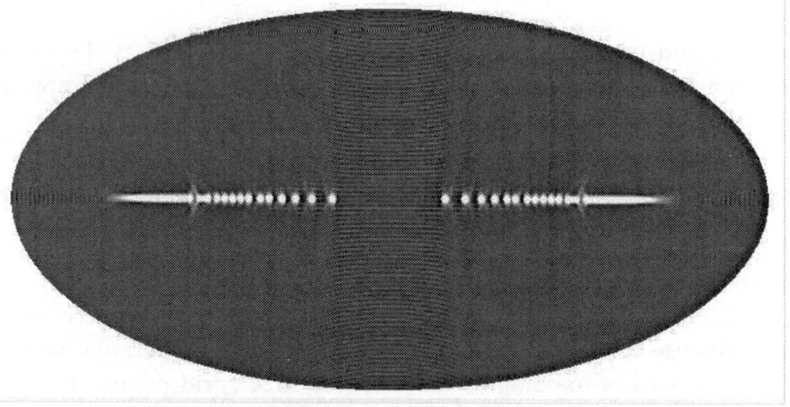

Fig. 10   Smoothed local energy density corresponding to figure 8; lighter grays correspond to higher values.

for an ellipse where the ratio of major to minor axes equals 1.9, after a time $t = 156000$. At this time, the pattern seems almost stationary, although it is difficult to be absolutely certain that the energy would not continue to decrease. A conspicuous feature of the solution is the straight roll domain in the central part followed by a series of dislocations, all located on the major axis, symmetrically with respect to the origin, which help to match the inner rolls to the outer part of the grain boundary predicted by the theory. This solution differs from that of the weak solution because for the latter, the angle made by the rolls in the central part of the grain boundary is too sharp. In that case, as explained in the previous section, a transition occurs that permits the decrease of the energy by forming pairs of concave and convex disclinations (or dislocations). A diamond shaped mark is inserted on the major axis at the location where theory predicts that the roll angle on the PGB reaches the critical one ($43°$). It is easy to find this position analytically. For an ellipse of major and minor axes a and b respectively (the eccentricity is then given by $e = \sqrt{(a^2 - b^2)/a^2}$), the angle of the PGB at a position X along the major axis (the origin being at the ellipse center) is given by

$$\sin \alpha = \sqrt{\left(1 - \frac{a^2 X^2}{(a^2 - b^2)^2}\right) \Big/ \left(1 - \frac{X^2}{a^2 - b^2}\right)} \qquad (8.1)$$

If φ_c denotes the critical angle, the position X_c along the major axis of the ellipse where this angle is reached reads

$$X_c = \frac{e^2 a}{\sqrt{1 + (1 - e^2)\tan^2 \varphi_c}}. \qquad (8.2)$$

As seen on Figure 8, the observed position is a good match to the predicted one.

It should be remarked that the terminal disclination which should appear on the major axis at the center of curvature ae^2, is hardly visible on the SH simulations. This is because PGBs here are regularized shocks. Their position is marked with a triangle symbol.

The local wavenumber, as shown in Figure 9 reaches a value very close to one inside the ellipse, except in the neighborhood of the dislocations. Near the PGB's we see lower values of k. Above and below the line of dislocations, k is greater than one. (The grayscale in the figure is chosen so that values of k above 1.015 are white, and those below 0.985 are black). We calculate the local wavenumber by comparing the magnitude of the gradient of the computed solution of Swift-Hohenberg, to that of computed solutions of Swift-Hohenberg with specified wavenumbers over a range near $k = 1$. The gradient's magnitude squared is first smoothed by convolution with a bump function with width slightly larger than 2π to remove roll scale features, then this value is compared to the values computed in the

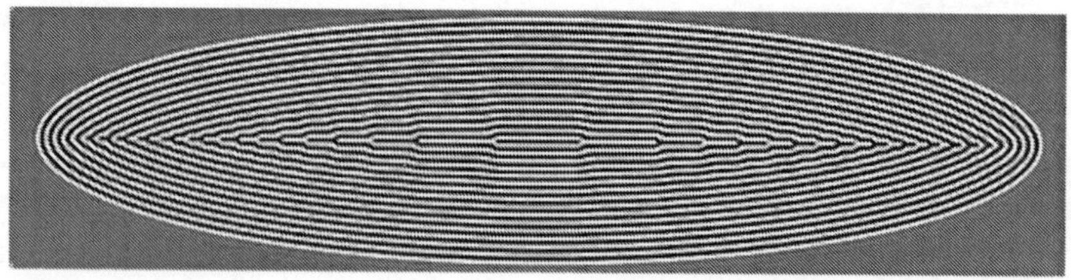

Fig. 11   Rolls from numerical simulation of Swift-Hohenberg with an ellipse whose major axis is 4 times longer than its minor axis.

same way for straight parallel roll solutions of the same equation. This approach is motivated by the ansatz that the solution of SH can be written as $F(\theta(x, y))$, where F is a periodic function. The result of calculating the squared magnitude of the gradient in this case produces a factor of k^2 times a term whose average is independent of k.

Figure 10 shows the result of calculating the local energy density for Swift-Hohenberg, and then smoothing the result by convolving with a bump function to remove features at the scale λ. The result is not sensitive to the details of the choice of the "bump" function. This energy plot allows a precise localization of the dislocations and shows their gradual packing as the end of the PGB is reached.

Another simulation has been performed in the case where the ratio $a/b = 4$. The resulting roll pattern is shown in Figure 11. Here again, the PGB is replaced by a series of dislocations. It appears that the distance between dislocations depends on the original angle between the rolls at the PGB: tightly spaced dislocations reproduce weak roll bending, while widely spaced ones replace sharply bent rolls.

9   Comparison to other numerics and experiments

Similar simulations were performed using the original Oberbeck-Boussinesq equations with top and bottom no-slip boundary conditions ([4]). In Figure 12 we exhibit the roll pattern for such a simulation with a Rayleigh number equal to 2000, a Prandtl number $P = 100$ and a ratio of major to minor axes equal to 1.9, as in the first SH simulation [4]. Sidewalls are heated in order to maintain rolls parallel to the elliptical boundary. A remarkable agreement is found with the SH simulations, indicating that the latter model reproduces the principal features of the solution.

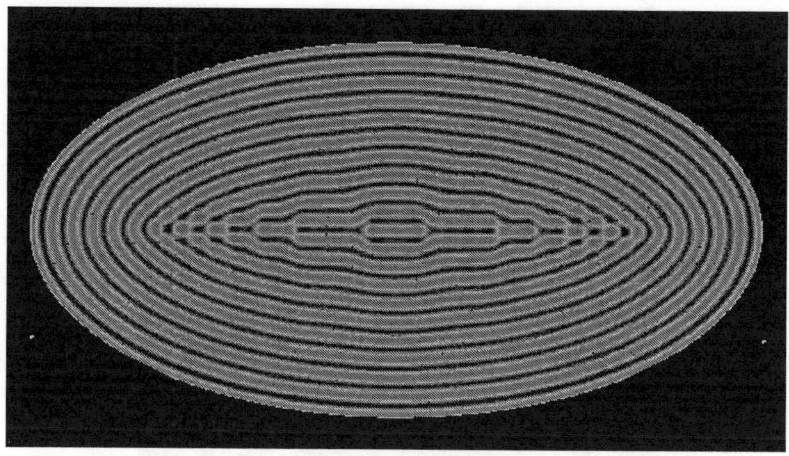

Fig. 12   Roll pattern from numerical simulation of Oberbeck-Boussinesq equations courtesy of Mark Paul ([4])

Experiments have also been performed [5] in ethanol at a Prandtl number of 14.7. Although the inverse aspect ratio is not as small as in the previous simulations, the same features are again observed, showing the robustness of the arguments based on the phase diffusion equation. In particular one clearly sees the protuberances at the end of the inner rolls. Dislocations do not form because of the large aspect ratio. A SH simulation has been performed with the same number of rolls and the same eccentricity (ratio $a/b = 1.5$). A comparison is shown in Figure 13. In addition, it is found in the experiments [5] that the size of the inner straight roll decreases as the stress parameter is increased. This particular result is not found with the SH equation. This behavior is likely to be related to the decrease of the selected wavenumber in Oberbeck-Boussinesq as the Rayleigh number increases. The only direction where this adjustment can take place without roll annihilation is along the major axis. It is achieved by reducing the length of the inner roll, at the expense of a deviation from the selected wavenumber at other places. This can happen as long as the local wavenumber does not fall outside the Busse balloon. When it does, a pair of dislocations form [5], breaking the up-down symmetry. This hypothesis has been tested with the SH equation in the following manner: using as initial conditions the results of a simulation where the selected

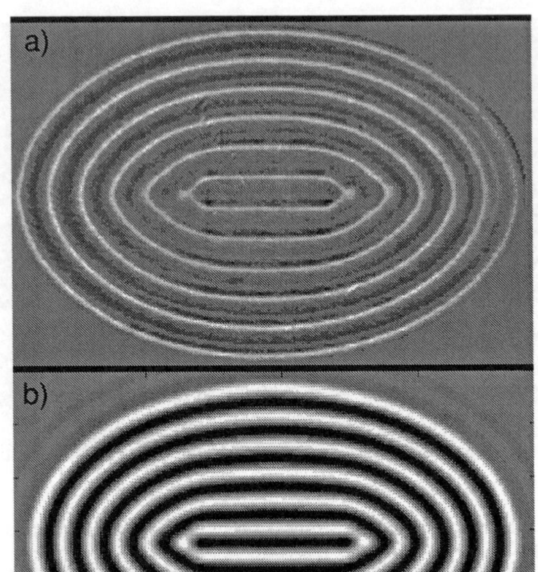

Fig. 13   a) Image of convection rolls from an experiment courtesy of Meevasana and Ahlers [5] with ethanol b) Numerical simulation of Swift-Hohenberg with similar geometry

wavenumber is exactly one, we integrated the SH with a linear operator modified in such a way as to select $k = 0.95$. As predicted the inner rolls shrank as depicted in Figure 14

10   Conclusion

In this paper we have presented strong numerical and experimental evidence that the correct variational model for pattern formation far from threshold is a modification of the regularized Cross-Newell phase diffusion equation that permits variations over a class of vector fields which are only constrained to be *locally* gradient. Moreover, we have been able to analytically study a local comparison between the energy cost for systems which do or do not count the cost of an orientational reversal. Based on this we make

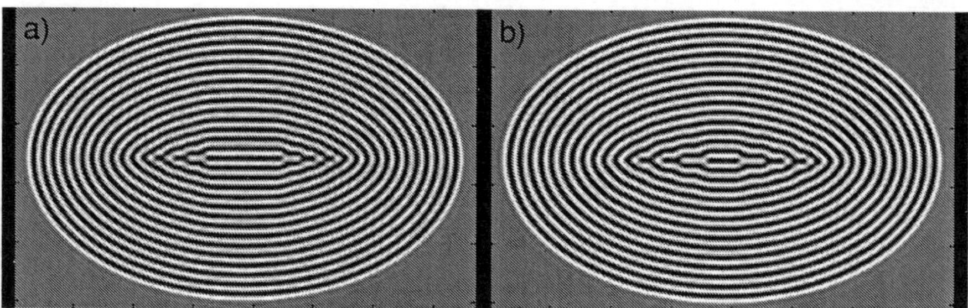

Fig. 14   a) Result of integrating Swift-Hohenberg using $k_B = 1$, this was used as the initial condition for b) which was Swift-Hohenberg for $k_B = 0.95$

a prediction of the transition angle at which a PGB will destabilize into a chain of concave-convex disclination pairs or dislocations. This prediction is seen to be well supported in numerical simulations and experiments. In other words the energy should be modified to discount the bending energy cost associated with sign reversal of the local wavevector. Such a model is currently under investigation and will be reported on in future work.

It is worth noting that there is an explicit formula for a string of dislocations whose character is quite similar to the dislocations seen in figure 7 and figures 8 and 11. For example, a string of dislocations located along the x-axis, at positions $\pm x_n$ is

$$
\theta = y - \text{sign}(y)\epsilon \log \left(1 + \sum_{n=1}^{N} \left(\frac{e^{n\pi} - e^{(n+1)\pi}}{2} \right) \text{erf}\left(\frac{x - x_n}{\sqrt{2\epsilon|y|}} \right) \right. \tag{10.1}
$$
$$
\left. - \sum_{n=1}^{N} \left(\frac{e^{n\pi} - e^{(n+1)\pi}}{2} \right) \text{erf}\left(\frac{x + x_n}{\sqrt{2\epsilon|y|}} \right) \right).
$$

This is an exact solution of the phase equation with a "weak bending" assumption [11]. In fact, if a multi-dislocation with spacing seen in figure 8 is patched to the outer portion of the eikonal solution, this initial condition settles to a solution of the same form with the position of the dislocations essentially unchanged. (The only real changes are along the interface where the solutions are patched together.)

Another natural extension of the RCN problem is to study minimizers of the free energy,

$$
\mathcal{F} = \int_\Omega \left(|\nabla k|^2 + \frac{1}{\varepsilon^2}(k^2 - 1)^2 + \frac{1}{\mu^2}(\nabla \times k)^2 \right) dX, \tag{10.2}
$$

in the limits $\varepsilon, \mu \to 0$. We note that the case where ε is set to infinity and only limits in μ are considered corresponds to a natural model [13] for micro-magnetic systems whose asymptotic minimizers are identical to those described in [1]. Such a study may also have implications for models of defects in superconductors [15] and nematic liquid crystals [16].

Acknowledgments

The authors wish to thank Joceline Lega for many helpful discussions and in particular for generating figure 3. The authors also wish to thank Guenter Ahlers and Mark Paul for helpful discussion and for generosity in sharing their results. NME would like to acknowledge support from NSF grant DMS-0073087 and ACN would like to acknowledge support from NSF Grant DMS-0072803.

References

N. M. Ercolani, R. Indik, A. C. Newell, and T. Passot, The Geometry of the Phase Diffusion Equation, *J. Nonlinear Sci.*, 10:223—274 (2000).

M. C. Cross and P. C. Hohenberg, Pattern Formation outside of Equilibrium, *Reviews of Modern Physics*, 65: 851—1112 (1993).

Y. Pomeau, Caustics of nonlinear waves and related questions, *Europhys. Lett.*, 11:713—718 (1990)

Mark Paul, *private communication*

W. Meevasana and G. Ahlers, *private communication*

F.H. Busse, On the stability of two-dimensional convection in a layer heated from below, *J. Math. Phys.* 46:149—150 (1967).

M. C. Cross and A. C. Newell. Convection patterns in large aspect ratio systems. *Physica D*, 10:299—328, 1984.

L. Ambrosio and C. DeLellis and C. Mantegazza, Line Energies for Gradient Vector Fields in the Plane, *Calc. Var.* 9:327-355 (1999).

N.M. Ercolani and M. Taylor, The Dirichlet-to-Neumann Map, Viscosity Solutions to Eikonal Equations, and the Self-dual Equations of Pattern Formation, *preprint*, 2002.

A. DeSimone and R.V. Kohn and S. Muller and F. Otto, A Compactness Result in the Gradient Theory of Phase Transitions, *Proc. Roy. Soc. Edinburgh Sect. A*, 131:833-844 (2001).

A. C. Newell, T. Passot, C. Bowman, N. M. Ercolani, and R. Indik, Defects are weak and self-dual solutions of the Cross-Newell phase diffusion equation for natural patterns, *Physica D*, 97:185—205, (1996).

T. Passot and A. C. Newell, Towards a universal theory of patterns. *Physica D*, 74:301-352, 1994.

W. Jin and R. Kohn, Singular perturbation and the energy of folds, *J. Nonlinear Sci.*, 10:355—390, (2000).

L. C. Evans, *Partial Differential Equations* John Wiley, New York, 1964.

A. Jaffe and C. Taubes, *Vortices and Monopoles* Birkhauser, Boston, 1980.

F. H. Lin, Nonlinear theory of defects in nematic liquid crystals; phase transition and flow phenomena, *Commun. Pure Appl. Math.*, 42:789-914, (1989).